CONCEPTS OF
CONTEMPORARY
ASTRONOMY

CONCEPTS OF CONTEMPORARY ASTRONOMY

second edition

Paul W. Hodge

Professor of Astronomy
University of Washington

McGraw-Hill Book Company
New York   St. Louis   San Francisco   Auckland   Bogotá   Düsseldorf
Johannesburg   London   Madrid   Mexico   Montreal   New Delhi
Panama   Paris   São Paulo   Singapore   Sydney   Tokyo   Toronto

CONCEPTS OF CONTEMPORARY ASTRONOMY

Copyright © 1979, 1974 by McGraw-Hill, Inc. All rights reserved.
Printed in the United States of America. No part of this publication
may be reproduced, stored in a retrieval system, or transmitted,
in any form or by any means, electronic, mechanical, photocopying,
recording, or otherwise, without the prior written permission of the
publisher.

1 2 3 4 5 6 7 8 9 0   D O D O   7 8 3 2 1 0 9

This book was set in Optima by Black Dot, Inc.
The editors were C. Robert Zappa, Michael Gardner, and James S. Amar;
the designer was Charles A. Carson;
the production supervisor was Leroy A. Young.
New drawings were done by J & R Services, Inc.
R. R. Donnelley & Sons Company was printer and binder.

Library of Congress Cataloging in Publication Data

Hodge, Paul W
   Concepts of contemporary astronomy.

   Includes bibliographies and index.
   1. Astronomy.   I.   Title.
QB45.H68 1979   520   78-16990
ISBN 0-07-029147-0

CONTENTS

PREFACE

In the preface to the first edition of this text, I felt obliged to apologize for adding to the already large number of good books on elementary astronomy. My excuse for going ahead was the fact that I felt that there was still a need for a book that was issue-oriented, that turned its discussions of astronomical facts and concepts around the large-scale and basic questions that underlie and motivate astronomical research. Since then many more good books have appeared, but there is still need for one that does what I attempted originally in this book.

The second edition retains that orientation. It is built around three vital astronomical questions: How did the solar system form? How do stars evolve? and What is the nature of the universe at large? In rewriting, I have attempted to tie each chapter and each subject even more closely to these basic questions, and have tried to show, even more pointedly than before, their interrelationships.

Naturally, the second edition has been updated to cover many of the exciting new astronomical developments since publication of the first edition. I have also rethought much of the pedagogy and attempted to relate each topic more closely to the student's personal interests and experiences. I have not bowed to my personal scientific conservatism to the point of leaving out the more sensational modern topics in astronomy, such as the questions of extraterrestrial beings, of multiple universes, or even of astrology, as I know from classroom experience that such matters are of interest to many and can form an avenue of approach for some who might otherwise be lost along the way.

One further change has been a reorganization of some of the material. Much of the background material given early in the first edition has been condensed and moved to the appendixes so that instructors can assign it or not more freely, according to their individual teaching plans. The solar system material has been rearranged to emphasize more clearly the similarities and differences between the planets and their relationships to interplanetary material. A large amount of tabular data has been placed in a series of appendixes, which also include a new glossary and sources of bibliographic and audio-visual materials. The index has been expanded, and the star charts have been amplified in order to make them more useful to students who are using small telescopes or binoculars. End-of-chapter questions and problems have been divided into easy and hard ones; those with an asterisk are more difficult and time-consuming than the others.

Many people, including both instructors and students, have helped to make a fairly large number of improvements in the material. Two particularly perceptive editors at McGraw-Hill were also instrumental in forming the second edition into its present shape. Several institutions and individuals have generously provided new data and illustrative materials for this edition. I am pleased that convention provides me with this opportunity to thank them all.

Paul W. Hodge

CONCEPTS OF CONTEMPORARY ASTRONOMY

CONCEPTS OF THE SOLAR SYSTEM

1

FITTING IN

The purpose of this chapter is to introduce you to the universe and to show how human beings fit into the general astronomical scheme of things.

**1-0
GOALS**

Black holes, quasars, moon rocks, pulsars, the big bang, alien civilizations— these words indicate some of the excitement of modern astronomy. They conjure up visions of strange lands, unexplored frontiers, and remote, almost unbelievable events. They are part of the realm of astronomy. To study them is to discover the universe and to understand them is to gain a cosmic view.

**1-1
THE REALM OF
ASTRONOMY**

The universe of astronomy is vast. The world's largest telescopes have seen only its nearest parts clearly. The depths beyond are still largely unexplored, and only through detection of strange optical light and weak radio signals can we even guess at what lies out there near the limits. Among the most distant things are the *quasars*, brilliant, remote, tiny, and baffling, apparently superexplosions in the centers of galaxies which are themselves too distant for us to see. Dashing away from us almost at the speed of light, these explosions are so far from the earth that it takes billions of years for the light to reach us.

A little nearer at hand, the largest objects in the universe are the *clusters of galaxies*, groups of up to 10,000 bright spots of distant brilliance, held together

Fig. 1-1. From observatories like this one, astronomers' views of space extend all the way from the planets to the deepest reaches of the universe. (*Kitt Peak National Observatory.*)

in the depths of space by some still unknown history. Some clusters are so large that it takes 100 million years just for light to cross from one side to the other.

The lights that make up such giant assemblages are themselves immense objects by any familiar standards. They are the *galaxies*; some are quiet, old, and spheroidal to perfection, while others are flung out in spirals and bursting with activity. The variety in galaxy shapes is amazing, and we have only in these last 30 years or so seen how our own Galaxy fits in with the rest. We need not be ashamed of it; it is big, one of the biggest in our local neighborhood of space, and well-populated, containing some 400 billion stars. Spread out like a giant pinwheel, its spiral arms are dotted with exotic objects: huge clouds of dust, complex interstellar molecules, hot hydrogen gas glowing brightly in ultraviolet light, and cool hydrogen emitting faint radio noise.

The *stars* of our Galaxy range in age from old, cool stars, even older than our sun, to bright, hot, newborn stars just taking shape. We find dead stars, collapsed stars, exploded stars, star fragments—every evidence that the stars

Fig. 1-3. Our sun is just one of hundreds of billions of stars in a galaxy like this one (which is known as NGC 4736). Billions of galaxies are seen in deep space, and extend as far as telescopes can see. (*Kitt Peak National Observatory.*)

are not unchanging, that they evolve and eventually die. We can predict when death will come to the sun and we can see nearby examples of sunlike stars that have died, their feeble remaining glow testimony to the future frozen fate of our own earth.

Elsewhere, in all their splendor, we find brilliant glowing *gas clouds* lit by the bright new stars in them and colored by the rainbow of spectral lines that divulge their makeup to us. In such young environments radio signals are sent out by the glowing gas, which is some 10,000 degrees hot, and we can thus see them even when hidden from us optically by thick screens of dust.

Our *sun*, which is almost 5 billion years old, is not even halfway through its lifetime as a star. Steady in its life-giving outflow of energy, it burns from a central thermonuclear furnace set at a temperature of 10 million degrees. A close look with solar telescopes and satellites shows us that the sun is boiling at its surface and covered with the shapes of huge bubbles from hot gas beneath. Here and there are fierce magnetic storms, the sun spots, and occasionally giant flaring outbursts of energy that fling particles away from the sun in streams and out into space. Many of these hit the earth, affecting our weather subtly, and perhaps altering our climate.

The earth and the eight other *planets* are basically cool bodies. They are heated mostly by the sun's rays, but they are also warmed inside from other

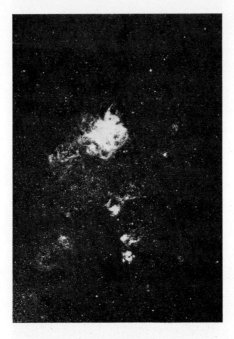

Fig. 1-4. Stars are born from giant clouds of gas and dust. This remarkable complex is one of the brightest clouds of gas known. It lies in one of our nearest neighbor galaxies, the Large Magellanic Cloud. (*Cerro Tololo Interamerican Observatory.*)

sources—the leftover heat of formation, gravitational shrinking, and radioactivity. Space exploration of the planets has greatly increased our understanding of their puzzling differences. We know better why our earth is unique, why there are no people on the other planets, and how earth's development paved the way for our gradual evolution. We understand the thick pall of clouds that keeps the surface of Venus forever hidden and deathly hot. We can figure out why Mars has huge volcanic peaks, far higher than any on earth, even though it

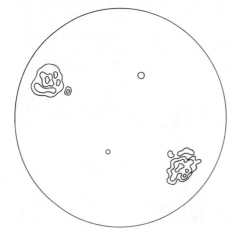

Fig. 1-5. Our sun as seen from space in ultraviolet light with the research satellite OSO 8. Two bright spots show areas where solar activity and heat are at a maximum.

is a much smaller planet. We understand giant Jupiter's strange makeup, which is mostly liquid hydrogen with a cloud-laden atmosphere of hydrogen gas, helium, ammonium, and methane. The history of the solar system and our understanding of how it must have formed is one of the most intriguing issues of contemporary astronomy.

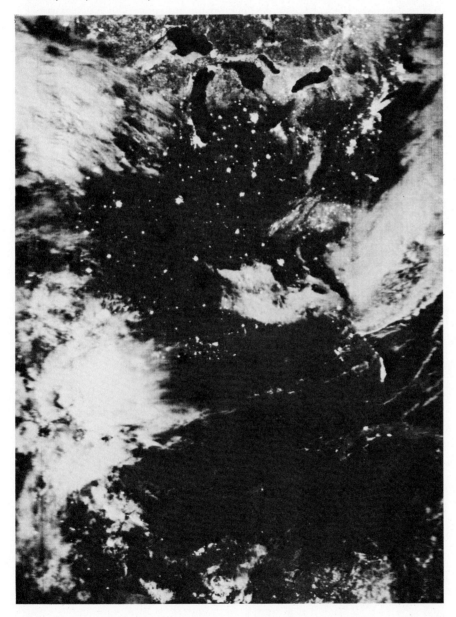

Fig. 1-6. From space the earth at night is a constellation of cities. This view shows the eastern United States from New York to Florida. The Great Lakes are visible (*NASA.*)

1-2 HOW DO WE FIT IN?

We live in a mind-boggling sea of space that is inhabited by strange and wonderful things. How do we, as intelligent beings, fit into it?

We are part of the story, perhaps one of its most interesting and remarkable chapters. As members of the human race, we are apparently the only highly intelligent beings in our corner of space. We have looked around, peered out from our earthly platform, and discovered the universe. We have mapped it and deciphered much of its history. Maybe we are the first to do so. Maybe the universe, since it began 15 billion years ago, has slept in ignorance of itself these aeons until suddenly in a few years mankind's curiosity and collective intelligence has opened its cosmic eyes and the universe has seen itself for the first time. Maybe we are the brains of this outfit.

There is simply no evidence yet that life exists elsewhere, on strange planets orbiting distant stars. Many astronomers who study the early history of the earth and sun believe that earthlike planets might be common in our galaxy, but they remain undetected. Intriguing stories about flying saucers and ancient astronauts fire our imaginations, but when we look hard at such stories, the evidence is unconvincing. We continue to search; our radio telescopes listen for distant messages and our optical telescopes test for the gravitational effects of planets around nearby stars, but success is not yet ours. We are like

Fig. 1-7. Geologists as well as other scientists are involved in many aspects of astronomical research today. Geologist Harrison Schmidt, shown here next to a huge lunar boulder, explored the moon on the Apollo 17 mission. (*NASA.*)

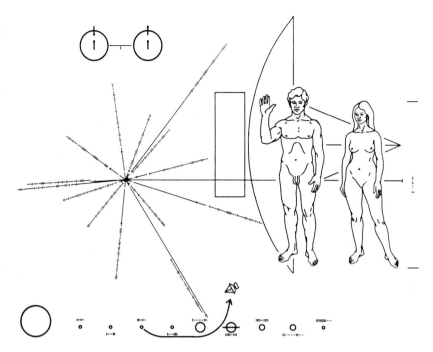

Fig. 1-8. Human beings' first feeble reach beyond the solar system was the Pioneer 10 spacecraft, which carried this coded message into interstellar space. If it ever should be picked up by distant explorers of the future, they will gain some idea of our shape and size (behind the human figures is the spacecraft sketched in for scale), the nature of our planetary system (bottom figures), and the location of the sun (the diagram at the left pinpoints the sun's position with respect to several pulsars). (*NASA.*)

Robinson Crusoe, stranded on a cosmic island, not knowing whether or not we are alone until we can see the footprints in the sand.

In the meantime we are reaching out. Our telescopes are searching for clues to the many riddles of distant space. Did our sun form because of the violent explosion of a nearby star? Astronomers have found evidence that many of the atoms that make up the earth (and you and me) were formed in other stars that died billions of years ago and spewed their insides outward into interstellar space, where they were captured later by the forming sun and planets. Our cosmic roots lie in these ancient supernovae, the likes of which we can currently see exploding in other galaxies, where more atoms are being built up and more star formation is being triggered. Perhaps billions of years from now some of these atoms will end up in beings like us, who will peer at our Galaxy with their telescopes and wonder whether we ever existed.

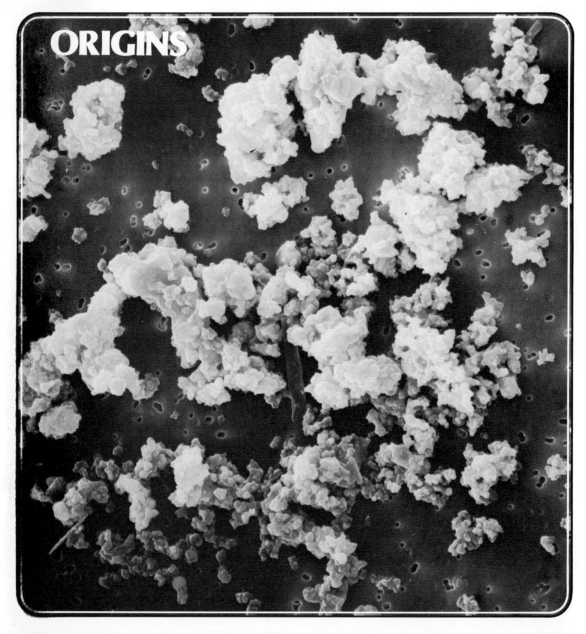

ORIGINS

This chapter introduces the planets and their neighbors by recounting our progress in understanding one of the most intriguing of astronomical puzzles: How did the solar system come about? What circumstances led to the formation of the planets, and how common are these circumstances in the universe? Are we here because of a freak event, or are planets like ours formed so routinely that they surround many stars?

**2-0
GOALS**

One of the principal arguments for sending rockets to the moon and spacecraft to the planets has been that such exploration is necessary in order to understand how the earth and its counterparts in space came about. From ground-based studies we had learned quite a bit about the various bodies in the solar system, but we still knew too little to answer this basic question. Now, after two decades of space flight, six manned visits to the moon, and automated approaches to, or landings on, four other planets, we know a great deal more. Now we are close to an understanding of how the solar system formed originally and of why each planet is so different from every other planet.

The origin and development of the solar system has been a subject of speculation for centuries. Each new fact seemed to add to the complexity of the problem. It is clear that the solar system did not originate in a simple manner, in spite of the fact that many of the theories which have attempted to explain it are framed in simple terms. If a theory of the origin of the solar system is to be truly complete, it must explain all the facts. This is still extremely difficult, not only

**2-1
IMPACT OF THE
SPACE
PROGRAM**

Fig. 2-1. The Skylab Space Station was designed to carry out many experiments from Earth orbit, including solar astronomy, measurement of interplanetary dust, planetary and stellar physics, and terrestrial resource analysis. (*NASA.*)

because all the known facts amount to such a large and bewildering sum of data, but because many vital facts are not yet known.

This part of the book deals with the known properties of the solar system and of the bodies that comprise it. Where possible, the relevance of this information to our understanding of the way in which the solar system came about will be emphasized. We are a long way from a universally acceptable understanding of the remote events that brought about the creation of our sun and its planets, but a basic pattern is emerging.

2-2 PECULIAR REGULARITIES IN THE SOLAR SYSTEM

In examining the various theories of the solar system's origin, it is necessary to keep in mind that there are certain gross properties of this system that require immediate explanation in any theory. These are the strange and conspicuous regularities that occur among the locations and motions of the planets and the satellites. Precisely because of their nonrandom nature, it is clear that such regularities are important clues about the conditions present at the time of the system's formation.

The first important fact is that the orbits of all the planets are in nearly exactly the same plane. The solar system is extremely flat. This important fact is probably the result of the original form of the preplanetary material, which may have also been very nearly flat.

The second important fact is that the planetary orbits are very nearly circular. Although in each case the ellipse that forms the orbit departs slightly from perfect circularity, the orbits are very much more nearly circular than might have been expected on a random basis.

The third important fact is that the orbits of the planets are nearly in the same plane as the rotation of the sun itself. The poles of the sun, therefore, nearly coincide with a line perpendicular to the plane of the planetary orbits. This fact suggests that the rotation of the sun is somehow connected generically with the orbital motions of the planets.

The fourth important fact is that the planets themselves rotate in the same direction as they revolve around the sun (with two exceptions which deviate from this rule in only minor ways). This fact seems to argue that the orbital motions and the rotations of the planets are connected in some way by a common direction of motion of the preplanetary material.

The fifth interesting fact is that the distances of the planets from the sun are not just arbitrarily arranged, but in fact show a pattern which can be expressed in terms of a simple mathematical law. This law is not a law of physics in any

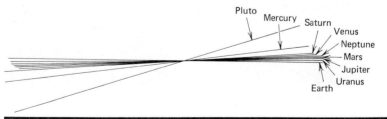

Fig. 2-2. The orbital planes of the planets are all nearly lined up, making the solar system a very flat phenomenon.

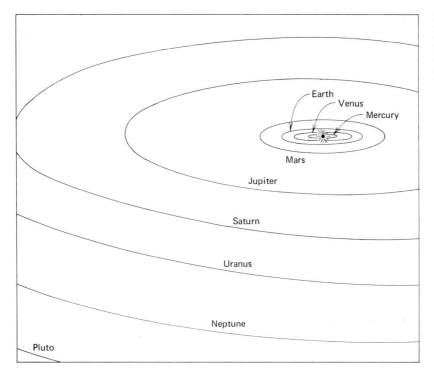

Fig. 2-3. From above and afar, the planetary orbits would look like this. The inner planets are much closer to each other than the outer ones.

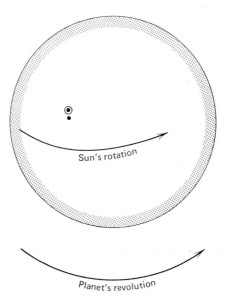

Fig. 2-4. The direction of the sun's rotation is the same as the direction of the planet's motions.

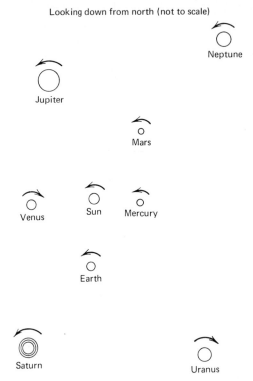

Fig. 2-5. Almost all the planets rotate in the same direction as their revolution in their orbits. Uranus and Venus are exceptions.

Looking down from north (not to scale)

Neptune

Jupiter

Mars

Venus   Sun   Mercury

Earth

Saturn   Uranus

sense, but must instead result from the primordial conditions in the pre-solar system material. The formula that describes the planetary distances is called *Bode's law*, and has the following form: A sequence of numbers is written beginning with 0, then 0.3, and from then on each number is doubled (0.6, 1.2, 2.4, etc.). To each of these numbers a constant value of 0.4 is added. The result is a list of the distances of the planets from the sun in terms of the earth's distance, taken as 1. Except for Neptune, the agreement between the values for the distances calculated from Bode's law and the actual distances of the planets is surprisingly good. It is so good, in fact, that it led in the nineteenth century to the immediate acceptance of the first asteroid, which filled what was otherwise a void in the law between the orbits of Mars and Jupiter.

Another important solar system fact that must be explained in terms of a proper theory is the way in which each satellite system, such as those of Jupiter (fourteen moons) and Saturn (ten moons), is nearly identical in its arrangement

Fig. 2-6. Jupiter's satellite system is like a miniature solar system. The inner bright satellites all revolve in the same direction and in the same plane as Jupiter's rotation.

Jupiter   V   Io   Europa   Ganymede   Callisto

(Nine others ➤)

to the system of planets. Some of them also follow Bode's law in their distances from the planets that they orbit. Their orbits are normally nearly coplanar, circular, and in the plane of the planet's rotation. We don't know the direction of rotation of the satellites in general, but the direction of revolution in their orbits is usually the same as the direction of the planet's rotation, as well as the revolutionary direction of the planetary orbits around the sun. Therefore, in terms of rotational directions, the planets and the satellites all show a strong tendency toward uniformity.

Only a few orbits and rotational directions run counter to the general trend. These are called *retrograde motions.* In most cases, the retrograde motion is clearly the result of events that have occurred more recently than the formation of the planets. For example, the outer satellites of Jupiter, about one-half of which have retrograde motions, are most likely captured asteroids. In the case of such capture, either direct or retrograde orbits can occur with nearly equal probability.

A final important and all-encompassing property of the solar system is the fact that the planets and satellites contain almost all the rotational motion (what physicists call *angular momentum*) of the solar system. The sun, which contains almost all the mass of the solar system, includes only a very small fraction of the total angular momentum.

These few simple relationships and regularities among the solar system bodies can be explained at least in general terms by all the major theories of the system's origin. The degree of success is in some cases highly debatable, but it is generally conceded that several of the ideas suggested by theorists can satisfy most requirements. For example, von Weizacker, who in 1944 suggested that turbulent eddies formed in the contracting envelope of the solar nebula, was able by this means to explain the fact that the planets showed coplanar orbits, that they rotated in the same direction as their orbital revolutions occurred, and that they showed the pattern of distance from the sun that had been given by Bode's law.

2-3 MEASURING PROPERTIES OF THE PLANETS

So that you may understand how some of the basic features of the solar system are determined, a description of the way we measure the distances, orbits, sizes, and rotation rates of the planets is given here. All these properties of the planets provide us with clues about the manner in which the planets were formed billions of years ago.

Distances

The distribution of the planets is an interesting feature of the solar system and has important implications for its origin. But the distances between the planets are of vital importance to astronomers for another reason as well. Interplanetary distances form the foundation of the distance scale used for the whole universe. Before astronomers can measure distances anywhere else in space, they must

Fig. 2-7. Radar ranging to the moon establishes the lunar distance.

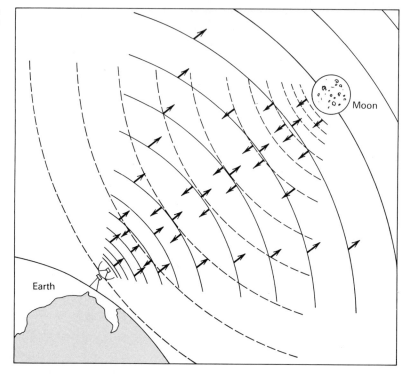

first have established the scale of the solar system, and particularly the size of the earth's orbit, firmly.

In the past, measuring distances in the solar system was a very difficult task, but recently a new and simpler method has emerged. A few years ago, radar methods proved to be so much more accurate than the older classical methods (involving optical triangulation) that they became the prime source of information about distances. To measure a distance by radar, a radar pulse is sent to some nearby object, such as the moon, from which a signal is reflected back to the earth to reach the radar-detecting instrument. And then, since the velocity of the radar pulse is known (it is the velocity of light in outer space), and since the amount of time that lapses between the sending and receiving of signals can be measured very accurately, the distance to the object can be calculated. This was first used for measuring the distance to the moon in the early 1960s, with the result that the moon at mean distance was determined to be 384,402 kilometers away (with an accuracy of ±1.5 kilometers). More recently, radar has been used to measure the distance to the planets Venus and Mars. Because of improved techniques, the accuracy of those measurements is even greater; for instance, the distance to Venus is known to an accuracy of 0.0001 percent. The most accurate distance measured so far is that to the lunar surface, where a reflector was set up by the Apollo astronauts for this purpose. With the giant telescope at McDonald Observatory in Texas, astronomers used optical radar (lasers) to measure the distance to the reflector on the moon to within a centimeter. This is an accuracy of about 0.000000003 percent.

Everything in the solar system is in motion; the moon, the planets, their satellites, the sun, the comets, and all other interplanetary bodies are moving in some measurable direction with some measurable velocity. A careful look shows that these motions are not random but systematic, and they have certain similar properties. These properties were first recognized correctly by the astronomer Kepler (1574–1630), whose years. of painstaking calculations resulted in three statements about the ways in which all planetary motions are similar. These are now called *Kepler's laws of planetary motion*, and are as follows:

Planetary Motions

1. The planet's path (called its *orbit*) is in each case an ellipse with the sun at one of the foci of the ellipse. (An *ellipse* has two foci that are similar to the center of a circle; for a circle, the distance from the center to any part of the circle is the same, while for an ellipse, the distance from one focus to any part of the ellipse and then to the other focus is always the same.)

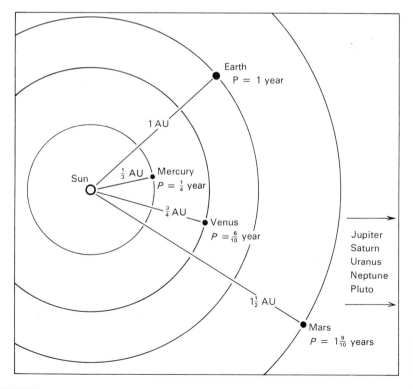

Fig. 2-8. The relationship between periods and orbital sizes for the inner planets.

Fig. 2-9. Circles and ellipses of various eccen-
tricities.

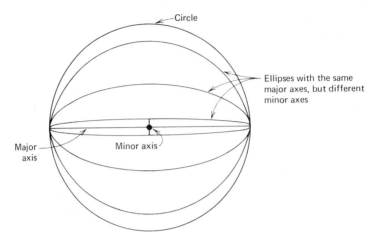

2. The velocity of a planet depends on where it is in its orbit; the planet moves in such a way that a line from it to the sun sweeps out equal areas at equal times. When it is close to the sun, it moves faster than it does when farther from the sun.
3. The general motion of a planet also depends upon its distance from the sun; the time it takes for the planet to make a complete revolution around the sun (called its *period*) is greater if its distance from the sun is greater, according to the formula

$$P^2 = a^3$$

where P is the planet's period in years and a is its semimajor axis (half the distance across the major axis of the ellipse) in units of the earth's orbit.

Newton's Law of Gravitation   Although Kepler's laws describe the motions of the planets accurately, they are not real physical laws because they do not have general application to nature.

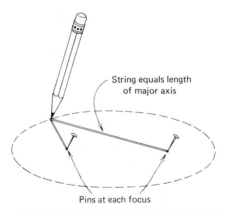

Fig. 2-10. Constructing
an ellipse.

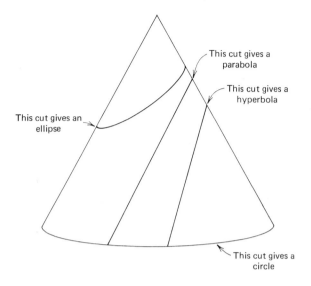

This cut gives a
parabola

This cut gives a
hyperbola

This cut gives an
ellipse

This cut gives a
circle

Fig. 2-11. Conic sections
can be produced by slic-
ing a cone at different
angles.

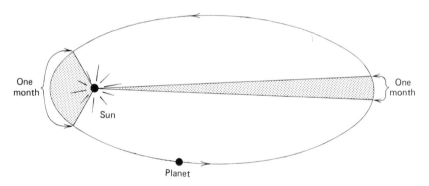

One
month

One
month

Sun

Planet

Fig. 2-12. A line connec-
ting a planet to the sun
will sweep out equal
areas in equal time.

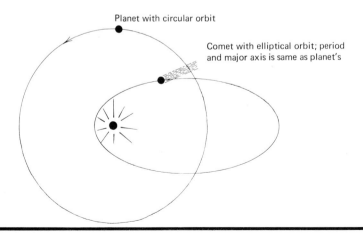

Planet with circular orbit

Comet with elliptical orbit; period
and major axis is same as planet's

Fig. 2-13. The semimajor
axis of a comet and a
planet with the same
period.

Fig. 2-14. The principle of inertia: a body at rest tends to remain at rest, a body in motion tends to remain in motion.

A body at rest tends to remain at rest

A body in motion tends to remain in motion

Nevertheless, they provide a vital clue to the problem of finding such a general law. They take on particular importance when they are compared with physicists' observations of objects in motion on the earth's surface. An effect first noticed by Galileo is that bodies possess what is called *inertia*, which is the ability either to stay at rest or to stay in motion unless a force alters the situation. A parked car, for instance, remains in its place unless acted upon by some force, such as its engine. On the other hand, a moving automobile, even after its engine has been turned off, continues to move in the same direction. Eventually it comes to a stop only because of *friction* (a force). This concept of inertia is an important one, because it leads to a generalization about nature that connects the observations of moving bodies on the earth's surface to the observations of astronomy that are embodied in Kepler's laws.

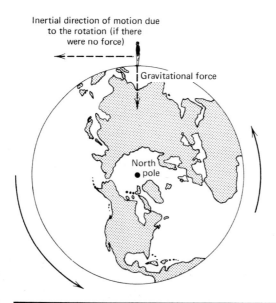

Fig. 2-15. Inertia and rotation.

These generalizations were first stated explicitly by Sir Isaac Newton, and they are known as *Newton's laws of motion.* They are:

1. Unless acted upon by force, any body at rest remains at rest, and any body in motion continues this motion in a straight line and with a uniform velocity.

2. Whether a body is at rest or in motion, any change in motion is proportional to the force acting upon the body and is in the direction of this action.

3. For every action, there is an equal and opposite reaction.

We have already pointed out a familiar example of the first law of motion. The second law of motion is less familiar and a little less easy to verify. It really implies a definition of the term *force,* and leads directly to a definition of the term *acceleration.* The acceleration of a body is the rate of change of motion for the body. If the driver of an automobile that is at rest or in uniform motion places a foot on the accelerator, the car moves faster and faster. The amount of acceleration involved depends upon, is in fact exactly proportional to, the amount of force applied by the motor. This proportionality between the force and the resulting acceleration may be expressed as follows:

$$\text{Force} = \text{constant} \times \text{acceleration}$$

The value of the constant of proportionality varies with different objects. A very light car, for instance, requires a smaller amount of force for a particular acceleration than does a very heavy one. Experiments on the earth have shown, in fact, that this constant is directly proportional to the weight of the

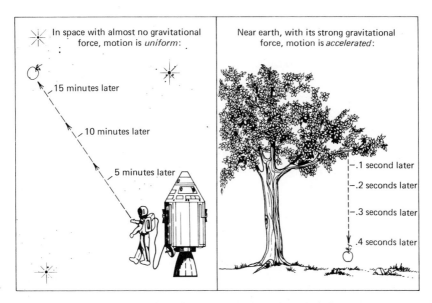

Fig. 2-16. Force results in acceleration.

object. Newton's laws, however, were hypothesized to hold true in any part of space, whether on the surface of the earth or in outer space. In outer space between stars, an object has no weight; yet a force on it will produce an acceleration just as on the surface of the earth. The constant is defined, therefore, as the *mass* of the body; the mass of any object is a measure of the amount of force needed to produce a certain acceleration. We define units of mass, force, and acceleration such that the mass and the weight of any object are equal at sea level on the earth. Thus, the automobile of our example will have both a mass and a weight of, say 1,000 pounds, but if placed in interstellar space it would have a weight of 0, though its mass would still be 1,000 pounds.

Kepler's first law regarding the planets stated that they did not move in straight lines, and if we accept Newton's laws of motion, then we conclude that the planets must be acted upon continuously by a force. From Newton's second law, we conclude that this force must be in the same direction toward which the planets are being pulled, away from their "initial" straight-line direction. As the diagram in Fig. 2-17 shows, this indicates that the force must be coming from the direction of the sun. Furthermore, the second law says that the force must be proportional in size to the amount of the change of motion that is experienced by the planet. Therefore, a planet that experiences very great changes of motion, namely the planet with a small, highly curved orbit near the sun, must be experiencing a stronger force than a very distant planet that is moving more nearly in a straight line. This indicates immediately that the force, which apparently has its source at the sun, decreases in intensity as it gets farther away from the sun. Newton found that the amount of decrease is considerable, and by working mathematically with Kepler's laws, he was able to determine a formula for it that applied exactly for all the planets. He also found that this formula fit data for the moon and other satellites in their orbits; scientists have since found that it explains the motions of the stars and galaxies as well. It is "universal," and therefore a truly general law of nature. Because the force between bodies is called gravitation, Newton's formula is called the

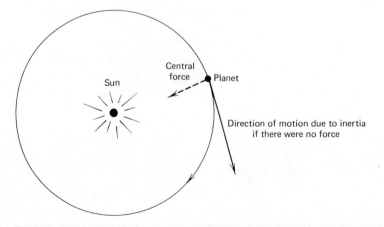

Fig. 2-17. The path of a planet indicates the presence of a central force.

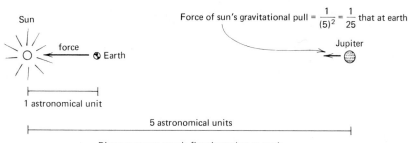

Sun

force ← ◉ Earth

1 astronomical unit

5 astronomical units

Distance to sun equals five times that at earth

Force of sun's gravitational pull = $\frac{1}{(5)^2} = \frac{1}{25}$ that at earth

Jupiter

Fig. 2-18. The force is greater the closer the planet is to the sun.

Universal Law of Gravitation. It states that the force between *a* and *b* equals a universal constant *G* times the mass of *a* times the mass of *b*, divided by the square of the distance between them *(r)*:

$$F = G\ \frac{m_a m_b}{r^2}$$

The force was found by Newton to depend upon the *square* of the distance, and to decrease for greater distances. Thus the planet Venus, which is about three-quarters of the earth's distance from the sun, experiences a force due to the sun's gravity of some $(4/3)^2 = {}^{16}/_9$, or almost two times larger than the earth's. This stronger force causes Venus' motion to be more curved than the earth's (a greater departure from a straight line) and to be faster.

In order to remain in orbits, the planets must have originally had some motion with respect to the sun. If any of them had originally been stationary, the force of gravity would have pulled them directly into the sun, to their destruction. On the other hand, if when formed they had been moving at a speed very much greater than they now have, they could have moved right on past the sun and away, without being captured into an orbit. The planet's

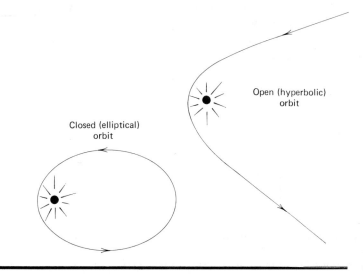

Open (hyperbolic) orbit

Closed (elliptical) orbit

Fig. 2-19. Open and closed orbits.

permanent orbits around the sun result from their having had a certain amount of motion with respect to the sun at the beginning.

The masses of the planets and the sun enter into Newton's laws, just as the distances do. However, the masses of all of the planets are so much smaller than the mass of the sun, that differences in orbits due to the different planetary masses are minor. The earth, for example, appears to revolve around the sun, but actually both objects revolve around a point between them, called their *center of gravity*, which is located according to the ratio of the masses of the two bodies. For the earth and the sun, the center of gravity is only about 400 kilometers from the center of the sun, which is 700,000 kilometers in radius. The center of gravity between the earth and the moon is likewise imbedded below the surface of the earth, but with many double stars, where the masses of the two components are nearly equal, the revolution occurs around a center of gravity that lies midway between them.

Newton's universal law of gravitation is found to apply to almost the entire range of astronomical areas, and its discovery and use have been essential to progress in understanding the universe. Only for very extreme cases of high velocities and violent conditions, found in some abnormal objects in space, must the law be modified by Einstein's law of general relativity, which provides a more accurate and more universal way of interpreting nature under such conditions.

Planetary Sizes   If the distance of a planet is known, its size can be determined by observations. In the past, the size was measured visually at the telescope by using a *micrometer*, a calibrated eyepiece that allows the astronomer to determine angular distances very accurately. Once the angular size is measured, the astronomer need only know the true distance to the planet at the time of measurement in order to calculate its physical size. Fig. 2-20 illustrates that this is done by means of simple geometry; the angular size and distance uniquely determine the shape of the triangle. The astronomer normally uses trigonome-

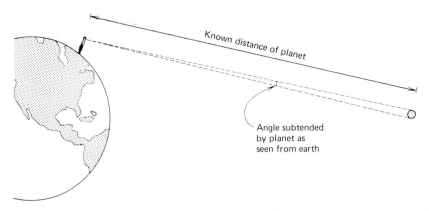

Known distance of planet

Angle subtended
by planet as
seen from earth

Fig. 2-20. Measuring
planetary sizes.

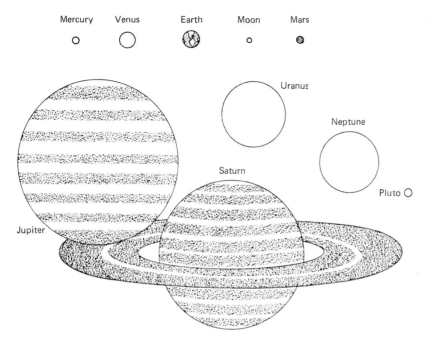

Mercury   Venus   Earth   Moon   Mars

Jupiter

Saturn

Uranus

Neptune

Pluto

Fig. 2-21. The relative sizes of the planets.

try to calculate the size of the triangle represented by the planet's physical size, but this could also be done geometrically by drawing triangles.

In the last 10 years the sizes of the nearer planets—Venus, Mars, and Mercury—have been determined with far greater accuracy than with the above method by using radar. From large numbers of different radar-ranging measurements, taken at different angles and at different positions in the orbits of the planets and of the earth, it is possible to solve the question of the sizes of the planets to within a few kilometers. Jupiter is by far the largest planet, more than ten times as big as the earth, while Mercury is the smallest, one-third the diameter of the earth, and not much larger than a satellite. In fact, Titan, the largest of Saturn's satellites, is nearly the same size as Mercury.

All the planets rotate around an axis, like the earth, but each with a different period of rotation. The rotation periods can be measured by three different methods, depending upon the proximity of the planet and its surface features. If the planet is close enough so that positions of permanent features on its surface can be measured accurately, then the rotation period is found by measuring the motion of these features for a period of time. For Mars, for example, for which the rotation period is found to be 24½ hours (almost the same as the rotation period of the Earth), visual observations of its surface features established the rotation period with high accuracy long ago. Similarly, Jupiter, although cloud

Planetary Rotation

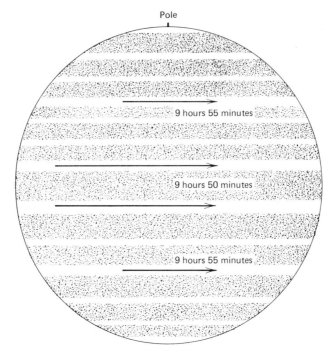

covered, has certain distinguishable features on its surface (see Chap. 3) that can be seen consistently through several rotations. In the case of Jupiter, two somewhat remarkable facts emerge about its rotation period. First, it turns out to be a very short period (only 9 hours and 50 minutes), which means that the surface is moving with a very high velocity considering the planet's large size. Second, it is found that the rotation period is different at different latitudes on Jupiter's surface. Features close to the poles take nearly 10 hours to rotate, some 10 minutes more than the rotation period at the equator. Saturn is very similar to Jupiter in both these respects, with an equatorial rotation period of 10 hours and 14 minutes, and somewhat longer periods at latitudes toward the poles. Saturn's rotation period is measured by means of small faint cloud features, which are seen only occasionally on its surface.

The same method has also been used to determine the rotation period of the two inner planets, Mercury and Venus, but with very poor results. In the case of Mercury, the problem is the extreme difficulty in making out the surface features of this planet, not only because of its small size, but also because its close proximity to the sun makes it a difficult object to observe. Mercury is very seldom high above the horizon during the nighttime, when the terrestrial atmosphere is relatively quiescent. Because of very poor visibility when telescopes look through large thicknesses of the earth's atmosphere, as is the case when an object is viewed close to the horizon, observing conditions are

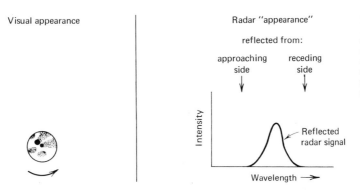

Visual appearance

Radar "appearance"

reflected from:

approaching side

receding side

Intensity

Reflected radar signal

Wavelength →

Fig. 2-23. The rotation rate of the planet Mercury, which had been determined from visual data, was based on very scanty and difficult observations. Radar data are much more reliable.

very seldom favorable for Mercury. For many decades, the rotation period of Mercury was thought to be 88 days (equal to its orbital period), indicating that this planet always keeps the same face pointed toward the sun, with the opposite face in eternal darkness. However, when radar measurements were made of the surface, this period was found to have been in error. With Venus, the problem is that the surface itself had never been seen because of continual cloud cover, and only very faint, hazy cloud patches are visible. Radar measurements have shown that the various tentative visual determinations of the rotation period of Venus were also all in error.

The Doppler Effect

A second method for measuring the rotation period of planets involves the use of the effect that the rotation has on the radar signals returned from its surface. If one side of the planet is moving toward us and the other side moving away from us with a given velocity due to its rotation, then the signal reflected from these two sides will be slightly altered in wavelength by this motion. This is called the *Doppler effect*. It is an exceedingly useful phenomenon in astronomy because it allows the astronomers to detect the motions of astronomical objects with high accuracy. The Doppler effect is perhaps more familiar to most people in the case of sound; it causes a high-pitched sound for a source of noise coming toward the listener, and a lower-pitched sound for a source going away. A train's horn heard coming toward the station always sounds higher than the same horn after the train passes. Similarly, a diving airplane makes a high whine on its descent toward the listener, and a lower "whoosh" after it has gone by. This change in pitch for sound is a result of the crowding together of the sound waves as received by the listener when the source of sound approaches, and of the opposite stretching of the wavelengths as the source retreats.

A similar effect is found with light. The wavelengths of light given off or reflected by an object moving toward the observer appear to be shorter than wavelengths at rest or from a receding source. Thus, radar signals from a rotating planet show some waves with too short a wavelength (reflected from

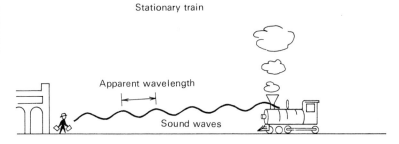

Fig. 2-24. The Doppler effect produces a change in the wavelengths of waves from the source moving toward or away from an observer.

Stationary train

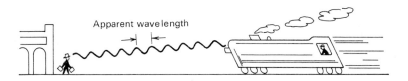

Approaching train

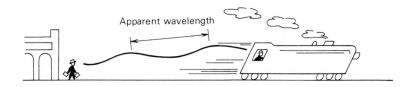

Departing train

the approaching side of the planet) and other wavelengths with too long a wavelength (reflected from the receding side of the planet). The amount of this signal distortion allows the radar astronomer to determine the rotation periods accurately. For Mercury, radar results in the 1960s showed that the previous measurements, which had stood for a long time, were completely in error. The rotation period is not 88 days, but 59 days—approximately ²/₃ of the orbital period. For Venus, radar demonstrated that the period is 243 days, not so very different from its orbital period (225 days). This indicates a very remarkable situation on the planet. The rotation period is in the opposite direction to the period of revolution around the sun (it is *retrograde*), and the motion of the sun in the sky as seen by a hypothetical observer on the surface of Venus is exceedingly slow, taking the equivalent of 58 Earth days from sunrise to sunset.

The two outer giant planets, Uranus and Neptune, are too distant for surface details to be distinguished; therefore, the Doppler shift is used to find their

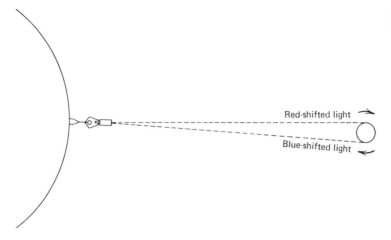

Fig. 2-25. Doppler measurements of a rotating planet can measure its period of rotation.

Red-shifted light

Blue-shifted light

rotation periods, but by means of optical rather than radar signals. A spectrograph is placed at the telescope so that it views the planet across its equator, and the Doppler shift of the reflected sunlight at different places on the disk of the planet is measured. For these two planets, the rotation periods are short—-about 22 hours for Uranus, and 23 hours for Neptune. (These values are quite uncertain and are currently being checked at several observatories.)

Finally, for the planet Pluto, which is so distant that its disk is exceedingly difficult to make out even with the largest telescopes and the best observing conditions, the rotation period has been measured by means of photoelectric measurements of its brightness. These have shown that in its brightness there is a periodicity involving a small change in luminosity, with a regular recurrence of the pattern over a 6-day period. This suggests strongly that 6 days is the rotation period for this planet. In 1978 a small satellite of Pluto was discovered, and it apparently has a 6-day period of revolution, being synchronous with the planet's rotation.

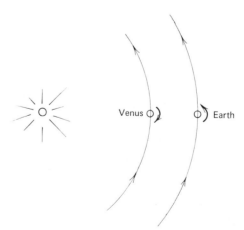

Venus   Earth

Fig. 2-26. The Earth's rotation is direct, while the direction of rotation of Venus is retrograde.

**2-4
SOME
THEORIES
OF THE ORIGIN**

It is appropriate now to introduce some theories of *cosmogony* that deal with the origin of the solar system. Since the first modern cosmogony was put forward by Descartes in 1644, there have been many hundreds of theories advanced to explain the facts. Approximately twenty different versions have been developed in sufficient detail and with sufficient comparison with observations so that we can consider them separate theories, and not just hypotheses. Some of the ideas that make up these chief theories are summarized in the following paragraphs.

Descartes' idea was that the solar system formed into bodies with primarily nearly circular orbits because of the vortex (whirlpoollike) motions in the pre-solar system object. He explained the planets in terms of primary vortices and the systems of satellites around the planets as secondary vortices. In some respects, although very crude in detail and in physical insight compared to present concepts, Descartes' theory is a forerunner of some of the most modern and popular ideas of this century.

In the eighteenth century, the scientist Buffon suggested that the planets could be explained as the by-products of a collision between the sun and a giant comet. In his view, such a collision would lead to a system of bodies that would all be revolving around the sun in a nearly common plane and would be rotating in the same direction as their orbital motion. Since those are two important features of the observed solar system, Buffon's idea gained considerable support and has been the pattern after which several other theories involving collisions have been based.

The philosopher Emmanuel Kant suggested in 1755 that the solar system might be explained naturally as the result of the condensation of a huge

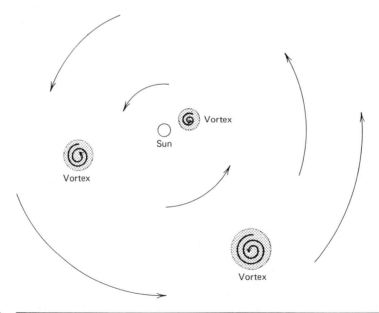

Fig. 2-27. Descartes' vortices were involved in one of the earliest theories of planet formation.

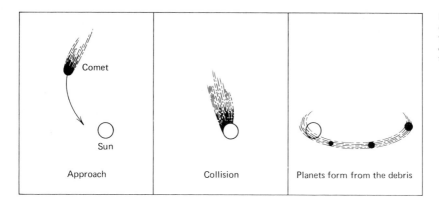

presolar gas cloud. He thought of this cloud as condensing from interstellar gas and rotating in the same direction that the planets now rotate. Kant pointed out that condensations in the nebula might lead to the formation of solid bodies that would then not only share in revolving around the sun with the gaseous nebula, but would also have rotation of their own in the same direction.

Laplace in 1796 developed Kant's idea further, suggesting that a rotating nebula that preceded the solar system gradually contracted, throwing off rings of gas each time the velocity of rotation reached too large a value for stability. Each ring of gas was then suggested to have condensed to form a planet and satellite system.

At approximately the turn of the twentieth century, several ideas were put forward that developed further and in more detail the concept of the solar system as a product of the collision between the sun and another body. Bickerton, Chamberlin, Moulton, Arrhenius, Jeffreys and Jeans all had versions of theories built on this general model. Rather than dealing with a comet colliding with the sun, as Buffon had suggested, these theories involved the

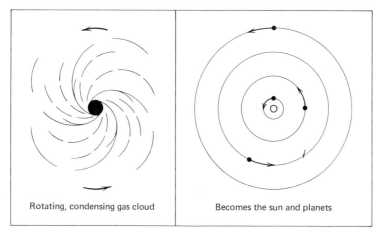

Fig. 2-29. Laplace's hypothesis about the solar system origin is very much like most modern theories.

collision or near-encounter of a star with the sun. The theories differ in many details, such as whether the planets were formed from material pulled out tidally from the sun by the other star, or whether the collision was a head-on collision—leaving just one star and a long, drawn-out filament as the result. In some cases, it was suggested that the star need only have come fairly close to the sun to have caused sufficient disturbance to form the planets, satellites, and other small bodies. In other cases, a grazing collision was suggested as having pulled material physically out of the sun to form the planets. For all these theories, it was found by the astronomers that the exact description of what might have happened under the proposed conditions presented an extremely difficult problem in physics. In many cases, it was possible only to outline the probable features of the event with very little quantitative understanding or development.

Other theories include the idea (which Russell put forward in 1935) that previous to the formation of the planets, the sun was a double-star system in which one star exploded or disrupted, forming a long filament out of which the planets condensed. Lyttleton suggested a year later than the sun had previously been a triple-star system, and that the other two bodies collided, coalescing into an unstable object that departed from the solar system and left behind a filament of material that could then condense into the present planets. Even more recently, Hoyle suggested in 1944 that the sun may have been a double star, and that the other star exploded as a *supernova*. This explosion involves the collapse of the star and subsequent hurling of much of its matter into interstellar space in a violent manner. The supernova was thought of as having expelled most of its matter out beyond the solar system, leaving only enough to form the planets.

There is a general tendency to prefer an explanation of the solar system that does not require an unusual and improbable event, such as a star-sun encounter. For that reason, the nebular hypothesis is the most attractive. In science, it is always preferable to find a probable natural explanation of an event in preference to an ad hoc, unexpected, or improbable event. It should, of course, be kept in mind that the unexpected does occur occasionally, and that there is no guarantee that an unobserved, unexpected event has not occurred. Especially in the case of the solar system, the phenomenon is possibly quite unique; therefore, its cause may also be a unique event. If it is

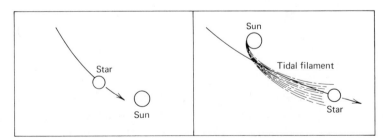

Fig. 2-30. A star-sun collision as a means of forming the planets has also been suggested, but is not favored presently.

eventually found that planetary systems are exceedingly rare, and that they in fact cannot be formed except by an exceedingly rare event such as a stellar collision, then we will be forced to conclude that our system formed in this manner, and that it is no matter of chance that we ourselves are in this unique system. On the other hand, if it is found that planetary systems are common in the universe, as is quite possible, then an ad hoc hypothesis with regard to its formation is much less attractive.

The many remarkable facts about the planets that were gleaned as a result of direct spacecraft exploration have recently been combined with prior knowledge to build up an increasingly clear picture of the probable events that led to the planets' formation. Age-dating of meteorites and lunar rocks has shown that the planets formed 4.6 billion years ago. By studying certain very primitive types of meteorites, we can even date some of the complicated events that preceded planetary formation—including the creation of some of the elements themselves.

**2-5
CURRENT IDEAS**

Our star, the sun, probably originated the same way as other stars in our Galaxy, as a result of the condensation of gas and dust in an unusually dense area. We see many places in nearby space where this process is going on now. From the abundance of certain kinds of materials, for instance atoms of aluminum that have a certain mass, we can see that shortly before the sun formed there probably was an exploding star nearby (called a *supernova*) that contaminated the gas and dust with the products of its upheaval. Perhaps this high-powered blast of material triggered the collapse of the gas and dust cloud that eventually formed the sun and planets. Something must have, for we calculate that without some kind of nudge, the cloud would have simply remained a cloud, distended and uncondensed.

On the basis of the angular momentum (which can be thought of as the amount of *rotational* motion; a more precise definition is given in the glossary, Appendix A) in the solar system, some astronomers argue that about twice the sun's mass must have condensed into the primitive solar system cloud, but then, as the sun went through its forming stages (described in Chap. 10) about half this material was hurled from the system. Near the sun, where temperatures in the cloud were high, only the heavier materials could remain; therefore the inner planets, like the Earth, are made up of heavy, rocky materials. Out past Mars, where the heat from the forming sun was less intense, larger planets like Jupiter could form from the light elements, mostly hydrogen.

Evidence so far suggests that the way in which the planets actually formed was fairly gradual. Pieces of dust collided in that early cloud and stuck together. Gradually these little collisions led to bigger and bigger chunks. Eventually the bodies were large enough to have some gravitational influence on nearby smaller objects, and then they began to collect material from all

Fig. 2-31. The early solar system material may have included spherical droplets like this particle, melted by solar outbursts. In this case, the melting was due to collision with the earth's atmosphere; this tiny meteoritic sphere was retrieved from deep sea sediments. (*University of Washington Interplanetary Dust Laboratory.*)

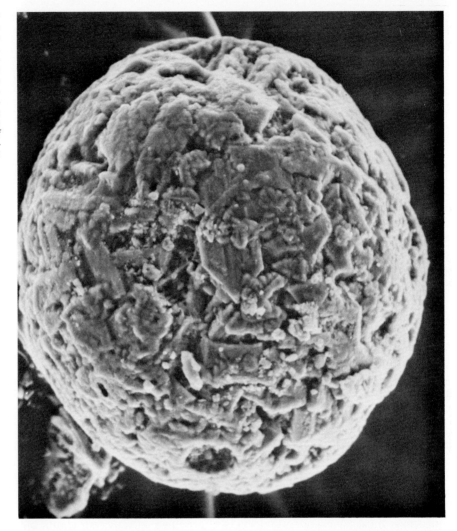

around them by means of gravity. The inner planets formed mostly from rocky chunks, but the outer planets also contained large amounts of icy material and gravitationally collected gas.

Gradually the larger ones from these hundreds of bodies swallowed up more and more of the smaller ones, until only a few very large ones, which became the present planets, were left. How they achieved their present structures, how their various kinds of atmospheres formed, and how their surfaces were molded from succeeding events are questions that will be addressed in the following chapters.

The above is a rather sketchy summary of current ideas. Much present work is still speculative and uncertain, even though many highly detailed mathemat-

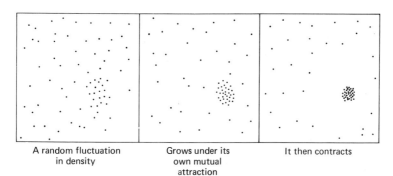

A random fluctuation   Grows under its   It then contracts
in density   own mutual
                           attraction

Fig. 2-32. The growth of a gravitational instability.

ical calculations have gone into it. We have nevertheless come a long way toward understanding our origins, largely because of the great progress made recently in exploring the planets and the other bodies of our solar system.

QUESTIONS

1. Cite four regularities in the motions or positions of the planets that help to unravel the mystery of how they formed.
2. From what we know so far about the origin of the solar system, how unique do you think our system is and why?
3. Why does Jupiter have so much more hydrogen and helium than Earth does?
4. How would you use triangulation to find the distance from earth to the moon?
5. How long does it take for a radar pulse to travel to the moon when the moon is at its mean distance?
6. During the Apollo program, what was the soonest that Mission Control in Houston could expect an answer to a question addressed to astronauts on the moon?
7. If a planet is eventually found beyond Pluto, and its period is 350 years, what is the size of its orbit?
8.* How much would you weigh on Mars? On Jupiter?
9.* How far from earth's center of gravity is the center of gravity of the earth-you system?
10.* Occasionally, Mercury is seen passing across the sun in silhouette. Why does this not happen more frequently? Calculate the ratio of Mercury's apparent size to that of the sun when this happens. The event is called a "transit of Mercury."

EXPERIMENTS AND OBSERVATIONS

1. On a large sheet of paper, draw a scale model of the planetary orbits from Mercury to Mars (use circles to approximate the elliptical orbits). Mark off 30-day intervals on each orbit, and then determine (by sighting from the Earth) what the motions of each of the other planets will appear to be. Under

what conditions will Venus look largest? When will Mercury be observed most easily? When does Mars appear to describe a loop in the sky, as seen from the Earth? What is the nearest planet to the Earth at closest approach?

2. Having landed on a planet in a distant star system, you proceed to study the other planets of the system. The nearest one, found to be only 1,000,000 kilometers away by your radar, subtends an angle of 10°. How big is it in kilometers? Use a scale drawing on graph paper and a protractor to find the answer.

3. From the star map for the current month, find which planets are visible at this time (monthly star maps are printed in popular astronomy magazines and many newspapers and nature magazines). Plot the position of a planet as you see it among the stars. Keep a record of its movements over as long a time as possible, plotting it every available clear night. Will you be able to tell whether it has moved in one night? In ten nights? In one month? Does its brightness change noticeably from night to night? From month to month?

THE PLANETS

3-0
GOALS

In this chapter the nine planets of our solar system are compared, and from the comparison you should gain a more compelling appreciation for our earth. After you have read the chapter, ask yourself: If we should make earth uninhabitable through ecological carelessness or nuclear war, what other planet, if any, can we turn to for a home?

Another goal for this chapter, as for all chapters in Part One of this book, is an understanding of how the different properties of the different planets came about. How did the circumstances of their origin and evolution lead to their present condition?

3-1
TEMPERATURES

One of the most vital and important features of a planet is its surface temperature. Measuring a planet's temperature is a relatively simple thing to do. All planets emit radiation that is partly sunlight reflected directly from their surfaces, and partly radiation emitted by the surface material because of its warmth. The radiation from a planet is measured over many wavelengths, and then the reflected solar light is subtracted, leaving only the radiation emitted by the planet itself. From experiments in the laboratory and from theoretical studies of the properties of radiating bodies, scientists have long known that the properties of any radiation that is emitted by an object are dependent upon the object's temperature. With perfect radiators, objects which absorb all the light falling on them, the radiation emitted is called *blackbody radiation*. The term derives from the fact that such a perfect absorber (which is also a perfect radiator) has a dull black surface. Although the planets are not quite perfect absorbers or radiators, they do absorb much of the energy incident on them (sunlight), and therefore the properties of blackbody radiation are similar to those found for the planets.

Blackbody
Temperatures

The temperature of a blackbody can be determined by measuring the amount of radiation emitted by the body at all wavelengths and then comparing the result with calculated "blackbody curves." These curves were first derived in detail by the German physicist Max Planck and are often called Planck curves.

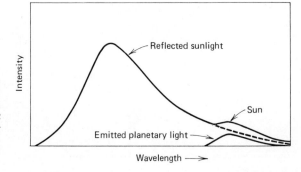

Fig. 3-1. The spectrum of a planet is made up of sunlight reflected from the surface of the planet plus light at longer wavelengths that is emitted because of the planet's own heat.

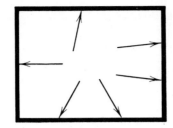

Black walls
absorb all
light incident
on them

Fig. 3-2. A blackbody is a
perfect absorber and
perfect emitter of radia-
tion.

Even a relatively small section of the radiation curve for a planet will give
information about its temperature, because each Planck curve has a characteristic shape (see Fig. 3-3).

There are two means of comparing portions of a planet's radiation with the
blackbody curve, depending upon the properties of the Planck curve. First, it is
possible to gauge the temperature of a body by determining the wavelength at
which most of the radiation is emitted. This maximum point in the radiation
curve is very sensitive to the temperature; in fact, it is inversely proportional to
it. The effect of this proportionality, which is referred to as *Wien's law*, is
noticeable even in fairly familiar circumstances. If an object is heated up
gradually, it first turns a deep, dull red and then changes to bright red, orange,
yellow, and finally becomes "white hot." These colors are an indication of the
wavelengths of maximum radiation emitted by the object at different temperatures; it first begins radiating mostly in the deep red, long-wavelength region
and then, when it becomes exceedingly hot, its maximum radiation moves
toward the blue end of the spectrum (shorter wavelengths). It is not possible to

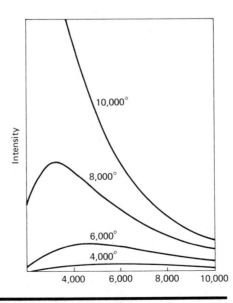

Fig. 3-3. Blackbodies of
different temperatures
emit radiation with dif-
ferent intensity and
wavelength distributions.

Fig. 3-4. Wien's law relates the temperature of a blackbody to the peak wavelength of its emitted radiation.

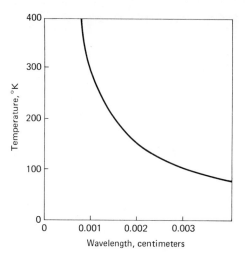

see radiation from very cool objects because the maximum radiation from them is at too long a wavelength—detectable by means of infrared detectors or radio wavelengths, but not by the eye. Every object in nature radiates energy roughly according to the Planck curves—even this book, for which the radiation is most intense at a wavelength of about 10 microns (100,000 Angstroms), if the book is approximately at room temperature.

The planets are all relatively cool bodies compared to stars, and their radiation maxima all lie in the infrared and radio region of the spectrum, outside the visible range. This fact makes it a relatively simple job to subtract the sun's radiation, which is reflected directly from the planetary surface, because the sun's temperature is approximately 6,000° and most of its radiation is at much shorter wavelengths than the planet's own radiation. There is always some solar radiation, however, at the long-wavelength regions, and it must still be subtracted. Note in the Planck curves of Fig. 3-3 that the hotter objects radiate more light at all wavelengths than cooler objects, even at the points of maximum radiation for the cooler bodies.

The total amount of radiation emitted by a blackbody is also related to the temperature, as the German physicist Stefan showed many years ago. The German theoretician Boltzmann proved theoretically that the Stefan relationship, which stated that the total amount of energy is proportional to the fourth power of the temperature, is exactly what would be expected from the Planck curves for a blackbody. The Stefan-Boltzmann law, as it is now commonly called, provides another means of measuring the temperature of an object, namely by determining the total amount of energy that is radiated at all wavelengths.

Temperatures measured in these ways are given for the planets in Secs. 3-4 to 3-8. Mercury is very hot because of its proximity to the sun. Its temperature is high enough so that many substances, such as lead, would melt under the midday sun.

For Venus, the quoted temperature is determined at very long wavelengths by means of radio and refers to the surface of the planet. At visual wavelengths, astronomers have been able to determine the temperature at the top of the visible cloud layer of Venus, which is much cooler than the surface beneath. The high temperature on the surface of Venus is believed to be due to a probable *greenhouse* effect. In a terrestrial greenhouse, the glass is transparent to sunlight, which has its maximum energy at wavelengths of about 0.5 microns. The glass is opaque to the longer wavelengths emitted by the plants and furnishings of the greenhouse, which have maximum energy at wavelengths of 10 microns or so. Therefore, the energy from the sun is allowed into the greenhouse, where it is absorbed, but the reemitted energy does not escape. The energy is trapped, and the greenhouse is made warm. (In actual practice, this effect does not contribute greatly to the warmth of greenhouses on Earth, which are warmed primarily by inhibited convection.) In the case of Venus, the greenhouse effect is believed to act similarly so that some solar energy manages to reach the surface through the cloud layer, but less of the planetary energy that is reradiated escapes.

For Mars, the temperatures are much more moderate, ranging from values that would be quite comfortable for humans to values that correspond to the very low temperatures sometimes encountered in the polar regions of the Earth. The measurements show that for several hours around noontime on the planet, the temperature remains above freezing at the equator. However, because of the thin atmosphere on Mars, the temperature of the air is much lower at various heights above the surface, and is probably seldom above freezing at ankle height.

The more distant planets are exceedingly cold, and their clouds consist of ice crystals of various substances, such as ammonia and methane. The temperatures are lowest for the more distant planets because of the fact that the amount of solar energy received by an object is inversely proportional to the square of its distance. As Fig. 3-5 shows, this is a result of the fact that light moves into an increasingly larger volume as it moves out from the sun, so that any surface placed at a distance from the sun will receive an amount of the light that depends on its distance. Consider the light emitted at a particular instant by the sun. After that light has traveled away from the sun, to the distance, say, of the earth's orbit (1 astronomical unit), it has an area of $4\pi r^2 = 4\pi 1^2 = 4\pi$ astronomical units². If this same burst of light is followed out to a distance of 2 astronomical units from the sun, the area of the surface of that volume will be $4\pi r^2 = 4\pi 2^2 = 16\pi$ astronomical units², four times as great as at earth's distance. Therefore, an object the size of earth would receive only one-fourth as much light at that distance compared to what Earth receives. The planet Pluto, with an average distance of nearly 40 astronomical units, receives,

Fig. 3-5. The inverse square law for light explains why the nearer planets to the sun are much warmer than the more distant ones.

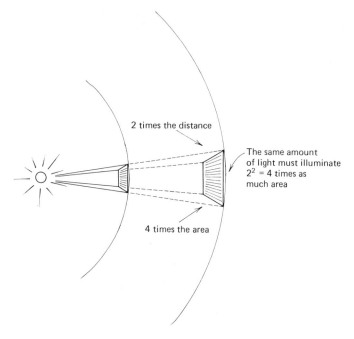

2 times the distance

The same amount of light must illuminate $2^2 = 4$ times as much area

4 times the area

therefore, only $(^1/_{40})^2 = ^1/_{1,600}$ as much light from the sun as does earth (per unit area). A square inch on the surface of Pluto would have only $^1/_{1,600}$ as much energy to heat it up during a given amount of daylight as compared to a square inch on Earth, and the temperature of that surface would be expected to be proportionately that much lower.

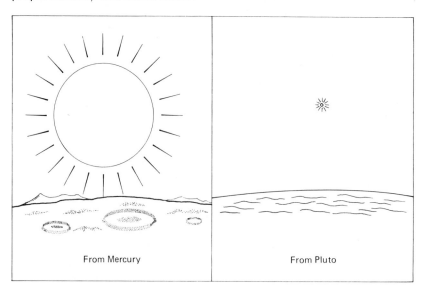

From Mercury

From Pluto

Fig. 3-6. The appearance of the sun as seen by an observer on the planet Pluto and on the planet Mercury.

The exact measures of planetary temperatures indicate that there are small deviations from the expected values. These deviations can be attributed to the optical effects of the atmospheres and the materials on the surface of the planets, which prevent the planets from acting exactly like blackbodies, and to internal heat. The measured temperatures can be determined to within 5 or 10°, even for the distant planets Uranus and Neptune.

Matter can exist in one of three forms—solid, liquid, or gaseous. When the planets formed, they probably were produced out of a mixture of matter in these three forms. Most of it was probably in the solid form for the inner planets, near which the gas had been pushed out and away by radiation from the sun, and in the gaseous form for the outer planets, where the gases could remain because of less radiation pressure from the sun. If we examine the extents of the present atmospheres of the planets, we find that the four inner planets have relatively small atmospheres or none at all and the outer planets (with the possible exception of Pluto) are still largely gaseous, with very extensive atmospheres. This differentiates the planets and tells us a great deal about their mode of origin and the conditions in the solar system when they formed.

3-2 PLANETARY ATMOSPHERES

The three primary characteristics that differentiate one atmosphere from another are the surface pressure of the atmosphere (the amount of pressure that the atmosphere above exerts on a unit area of the surface), its surface temperature, and its chemical composition.

Surface Pressure

The surface pressure is determined by the total amount of gas in the atmosphere of the planet, by its radius, and by its mass. The pressure is due to the gravitational pull of the planet on the atmosphere. Because gravity depends in part upon the distance between the gravitating body (the planet) and the object attracted (the atmosphere), the radius of the planet enters into the determination of the surface pressure; the radius gives the distance between the bulk of the planet's mass and the atmosphere. The mass of the planet is important because the gravitational attraction is proportional to the mass. The total amount of gas in the atmosphere is important because it determines the total atmosphere's mass above a unit area of the surface.

The pressure at the surface of the earth because of the atmosphere is called a pressure of "1 atmosphere," a standard unit which we will use in this book to describe other planets' atmospheres. If the earth's mass were made twice as big as it is and its radius kept the same, the atmospheric pressure would be twice as great. On the other hand, if the total amount of air in the terrestrial atmosphere were half of what it is, the atmospheric pressure would be approximately half the real value. If the earth's radius were changed to just half its present size, but the total mass of the earth were kept the same, then the atmospheric pressure

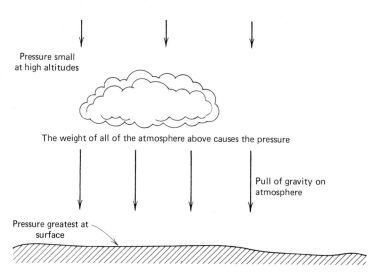

Fig. 3-7. The surface pressure of an atmosphere depends upon how much gas there is above the surface and upon the amount of pull of the planet's gravity.

Pressure small at high altitudes

The weight of all of the atmosphere above causes the pressure

Pull of gravity on atmosphere

Pressure greatest at surface

would be about four times its present value, because of the decreased distance between the atmosphere and the bulk of the earth gravitating beneath it.

Temperature The temperature of a planetary atmosphere depends upon the sources of energy and on the planet's ability to absorb that energy. There are two primary sources of energy; first is the sun, which is a very much more important source for the nearer planets than for the more distant; and second is the planet itself, which reemits energy that has been absorbed by the surface from the sun. Potentially each of these sources of energy can put energy into the atmosphere to increase its temperature, but the energy will not remain in the atmosphere unless the atmosphere itself absorbs it. In the earth's atmosphere, for example,

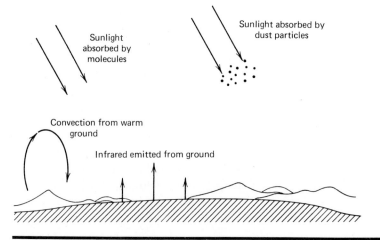

Sunlight absorbed by molecules

Sunlight absorbed by dust particles

Convection from warm ground

Infrared emitted from ground

Fig. 3-8. Sources of atmospheric heating.

most of the light from the sun passes through unabsorbed and reaches the surface of the earth. Similarly, some of the longer (because of its lower temperature) wavelength radiation that the earth reemits passes back out through the atmosphere unabsorbed and is lost into space. Much of it, however, is absorbed by the molecules of air (oxygen, nitrogen, and water vapor) and the dust in the air. This energy serves to heat up the atmosphere by means of what is called *radiative heating*. It is similar to the kind of heating that one experiences when standing close to a roaring beach fire, when the air around one is not particularly warm, but nevertheless one is heated by the radiation emitted from the fire. In this example, the method of heating is more efficient than that found in the atmosphere, because a person absorbs radiation more completely than the transparent air can. It is also true that some of the warmth of the atmosphere is imparted by convection—the movement of the air from one part of the atmosphere to another, in particular, from the warm surface to colder, higher regions.

The physical property that causes an atmosphere to become heated is its opacity to radiation, and the degree of this depends upon the chemical composition of the atmosphere. Each different molecule has different absorbing properties, and some molecules can absorb energy much more efficiently than others. If a molecule absorbs energy from a particular wavelength, then it can absorb that solar energy as it falls onto the planet. The atmosphere in that case is not perfectly transparent, since some of the light would be absorbed by the atmosphere before it reaches the surface.

    Most of the absorptions in the terrestrial atmosphere occur at longer and shorter wavelengths than visible light; therefore, our atmosphere is fairly

Physical Properties

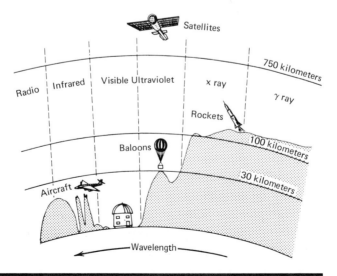

Fig. 3-9. The transparency of the earth's atmosphere.

transparent to sunlight. Clearly, it is no coincidence that our planet's atmosphere is transparent to visible light, as the eyes of the animals on the earth must have developed sensitivity to the particular wavelengths that they commonly see. If our eyes were sensitive only to far-ultraviolet light, for example, we would find them useless, as this kind of light is not transmitted through the atmosphere. It is almost all absorbed high in the upper atmosphere, and the earth would appear to be virtually continuously dark to a person who was only able to see ultraviolet light. The amount of absorption by an atmosphere depends upon the density of the gas doing the absorbing, so that at the top of the atmosphere where the density is very thin, little energy is absorbed, while at the bottom near the surface where the density is its greatest due to the pressure caused by the weight of all the atmosphere above it, the absorbing properties are greatest. There the temperature is high, because of the great amounts of energy absorbed.

Chemical Composition

The chemical composition of a planetary atmosphere also determines its gross properties. We now have some understanding of the origin and evolution of planetary atmospheres, and these ideas can explain the very great differences in chemical composition for the different planets' atmospheres. The composition of an atmosphere helps to determine both its pressure and its temperature because of the different properties of different atoms and molecules. An atmosphere made up mostly of heavy molecules, like oxygen (O_2) and nitrogen (N_2), as in the case of the earth, would have a greater surface pressure than a similarly sized atmosphere made up of the light molecule hydrogen (H_2). As mentioned above, completely different absorbing properties of different molecules also affect the atmosphere in determining the temperature and its variation with height.

Evolution

The chemical compositions of the planetary atmospheres have told scientists a great deal about the evolution of conditions on the planets. Three of the four inner planets, Venus, Earth, and Mars, have very thin atmospheres, made up primarily of oxygen or oxygen compounds (O_2, CO_2, H_2O) and some nitrogen (N_2). The outer planets, on the other hand, consist mostly of hydrogen and hydrogen compounds (H_2, NH_3, CH_4) and helium. It is now realized that the atmospheres of the outer planets are nearly original, "primordial" atmospheres, and have changed relatively little since planetary formation. For the inner planets, on the other hand, the present atmospheres are secondary, since they were formed later in their planetary histories. The most abundant elements in the universe, in the sun, and probably therefore in the pre-solar system material, are hydrogen and helium. But only the outer planets are massive enough and cold enough to hold an atmosphere made up of such light elements, which are not held by gravity as effectively as heavy elements

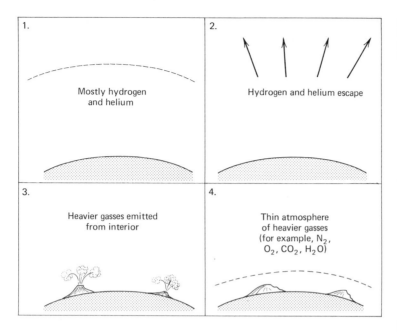

1. Mostly hydrogen and helium

2. Hydrogen and helium escape

3. Heavier gasses emitted from interior

4. Thin atmosphere of heavier gasses (for example, N_2, O_2, CO_2, H_2O)

Fig. 3-10. A schematic history of the evolution of the earth's atmosphere.

Fig. 3-11. Gases from the interior of the earth are continuously being released in the atmosphere, as at Halemaumau in Hawaii.

because of their smaller mass. Thus for the inner planets, any primordial hydrogen-helium atmosphere would be lost—the light gases would have escaped into space. The inner planets, Venus, Earth, and Mars have subse-

quently generated their own atmospheres of heavier elements, mostly carbon dioxide (CO_2) in the cases of Venus and Mars, and mostly O_2 and N_2 for Earth. These gases are produced by outgassing from the interiors of theses planets through volcanic activity. In the earth's case, the evolution of life forms led to the production of large amounts of O_2. Volcanologists measure great quantities of various gases as they are being emitted by terrestrial volcanic eruptions, which occur frequently enough on earth to explain the entire atmosphere. Mercury is too small in mass even to keep such an atmosphere for long, and by analogy with the moon, its small size may also mean a smaller degree of volcanic activity. Neither the moon nor Mercury has been found to have an appreciable atmosphere.

3-3 PLANETARY INTERIORS

One of the questions that is relevant to the problem of the solar system's origin is the question of the chemical and physical nature of the planetary interiors. Ideally, we would like to know the detailed structure, temperatures, densities, pressures, and chemical constituents of the planetary interiors, since these will have a clear bearing on the means by which the planets originated and were put together by the forces that were acting on the preplanetary material.

Unfortunately, it is impossible to obtain direct information about the interiors of planets by sampling or exploration. It has not even been possible for scientists to probe significantly into our own earth's interior. The temperatures and pressures become so great that the problem has remained insurmountable. Instead, it has been necessary to obtain what information we can get more passively. In the same way that a person can get some knowledge of the contents of a parcel without unwrapping it (by weighing it, measuring its size, shaking it, determining its shape, measuring its temperature, etc.), we must learn from indirect clues about the nature of the planetary interiors. There are several kinds of observational data relevant to the interiors of the planets: planetary masses, their densities, their shapes, their rates of rotation, their surface temperatures, and the chemistry of related objects, such as meteorites and the moon.

Masses of Planets

The total mass of a planet has obvious relevance to the planet's nature and must be known before any theoretical model of a planet's structure can be derived. Masses of planets can be measured in either of two ways, both of which are highly accurate. One method is to determine the mass by observing the periods and semimajor axes of the orbits of any planetary satellites. Then, simply from Newton's law of gravitation, it is possible to derive the mass of the planet immediately according to the following equation:

$$M = \frac{4\pi^2\, a^3}{GP^2}$$

where *M* is the mass of the planet (in most cases the satellite's mass is negligibly small in comparison with the planet's mass), *G* is the constant of gravitation, *a* is the semimajor axis of the orbit, and *P* is the period of the satellite's orbit. The mass of the planet determines the characteristics of a satellite's orbit completely. For a satellite with a certain size of orbit, the period will be short if the planet is very massive, and long if the planet has a low mass.

Consider, for example, three satellites of three of the major planets. The innermost satellite of Jupiter, called V, has a mean distance from its planet of approximately 180,000 kilometers. Mimas, one of the satellites of Saturn, has a very nearly equal mean distance from its planet (185,000 kilometers). Similarly, Ariel, one of the satellites of the planet Uranus, has a nearly equal distance from its planet (190,000 kilometers). In spite of the fact that all three of the satellites have orbits that are nearly exactly the same in size, their periods are very different. The period for Jupiter V is 12 hours, the period for Mimas is 22 hours, and the period for Ariel is 60 hours. This is due to differences in the masses of the planets. It is clear that Jupiter is considerably more massive than Saturn, and that Saturn in turn is more massive than Uranus.

Another example is a comparison between earth's moon and the satellite Dione of Saturn. Both satellites have a mean distance from their planet of about 380,000 kilometers. The period of the moon's revolution around the earth is 27 days, but the period of Dione's revolution around Saturn is only 2.7 days. Clearly Saturn is much more massive than earth. Appendix F shows the relative masses of the planets and confirms the conclusion reached in these examples.

How are masses determined when planets have no satellites? The best method is to place an artificial satellite around the planet. We have already measured very accurately the mass and mass distribution of the moon by placing a spacecraft in orbit around it. Something like this has also been done in the cases of the planets Venus and Mercury, which have no known natural satellites.

But before spacecraft were available, our knowledge of the masses of these

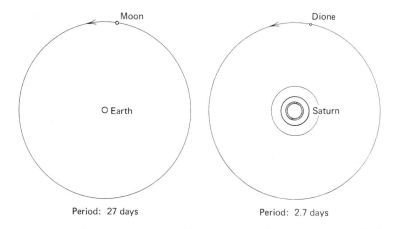

Period: 27 days   Period: 2.7 days

Fig. 3-12. The orbits of the satellite Dione and of the Moon.

planets came from a long-term study of the effects that these planets have on the orbits of other nearby planets. These effects are called *perturbations* on the orbits, and they can be determined by extremely precise measurements of the positions of the planets in their orbits over long periods of time—over centuries where possible. Because the sun's mass is so enormous compared to that of the planets, it dominates the motions of the planets in their orbits almost completely and the effect of planetary perturbations on each other is extremely small. Nevertheless, perturbations are large enough so that accurate measures of the planets' masses can be obtained from them.

For example, four measures of the mass of the planet Venus were made possible in recent years by the perturbations that Venus had caused on the orbits of four different kinds of objects. The motions of Earth were found to be affected slightly by the relative positions of Earth and Venus. This led to a determination of the mass of Venus that is accurate to three figures, with a value equaling approximately 82 percent of earth's mass. Similarly, the planet Mercury's orbit is measurably affected by Venus, and this also leads to a figure of about that value for the mass of Venus. Thirdly, an even more accurate figure for the mass of Venus comes about as the result of a study of the asteroid Eros, which occasionally passes closer to the planet Venus than does either Earth or Mercury. Because of the proximity of Eros to the planet, and because of its much smaller mass, Venus produced a more marked effect on its orbit than on either Earth or Mercury; therefore, a more precise determination of the mass of Venus resulted from the study. All these results agreed with the measures for the

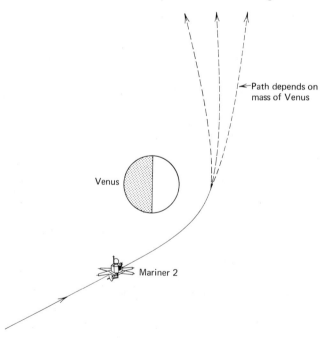

Fig. 3-13. The flyby spacecraft Mariner 2 provided an accurate measure of the mass of the planet Venus.

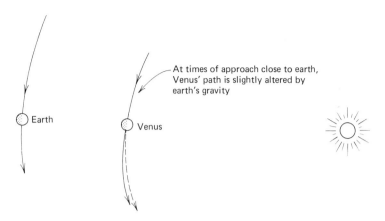

At times of approach close to earth, Venus' path is slightly altered by earth's gravity

Earth

Venus

Fig. 3-14. The perturbations caused by the earth on the orbit of the planet Venus are a means of measuring its mass.

planet Venus that had been determined by space probes, which came even closer to the planet and showed an even greater effect.

Densities of Planets

The second important observational datum relevant to planetary interiors is their mean density. This is obtained simply by dividing the measured mass by the volume, which can in turn be calculated from the measured dimensions of the planet. When the densities of the planets are calculated (see Appendix F), it is found that they are by no means the same from planet to planet; therefore, all the planets are not similar in their interiors. Mercury, Venus, Mars, and Earth do seem to have roughly the same densities, but the outer planets are very much less dense. The four major planets all have small densities, ranging from 2.3 grams per cubic centimeter for Neptune, down to 0.7 grams per cubic centimeter for Saturn. (Saturn's mean density is even less than that of water.) A comparison with the known densities of various substances at normal temperatures and pressures shows that the earthlike planets have densities that are not so very different from that of ordinary rocks; the density of rock is larger under the high pressures expected at the centers of planets, where rocks would be most compressed. Based on the evidence produced by meteorites (see Chap. 5) and on models of earth's magnetic field, it is believed that the interior of earth contains a core made up primarily of iron, which also can make its density higher than the mean density of normal rocks. On the other hand, the lower density of Mars suggests that it may contain primarily only rocklike substances, with no metallic iron core.

The major planets have such low densities that rock, iron, and other heavy materials can make up no more than a very small percentage, if any, of their compositions. They must be made up mostly of much lighter elements. It is presently believed that they are made up primarily of the light elements hydrogen and helium, which are found in their interiors in liquid and solid form.

Planet Shapes   The third kind of evidence regarding the nature of planetary interiors is their measured oblateness. As Fig. 3-15 shows, the planet Jupiter is very conspicuously oblate; that is, its diameter across the equator is very obviously larger than its diameter from pole to pole. It therefore has a flattened look that is the result of its rapid rotation about its axis.

For planets like Mars, which are so nearly spherical that it is difficult to measure oblateness simply by looking at the disk of the planet, it is still possible to determine oblateness by measuring its effect on the orbits of the planet's satellites. For example, in the case of Mars, the two satellites Phobos and Deimos move in orbits that are changing slowly because of the slightly oblate distribution of mass within Mars. The oblateness of Mars causes a rotation of the orbital planes of the two satellites. Thus, by observing the satellites over an extended period of time it is possible to detect even a very small oblateness— one too small to detect by direct measurement. In fact, the oblateness of Mars as determined in this way turns out to be only 0.5 percent. That is, the polar diameter of Mars is approximately 99.5 percent of the length of its equatorial diameter.

The nonsphericity of the earth was detected long ago by studying the motion of its natural satellite, the moon. More recently, with the large numbers of artificial satellites that are much nearer the earth (and therefore much more affected by the irregularities in the gravitational pull of the earth that are caused by its nonsphericity), we have an even better idea of the earth's shape. It is measurably oblate, with the difference between the length of its polar diameter and the length of its equatorial diameter measuring approximately 0.3 percent, which amounts to approximately 40 kilometers (27 miles). However, the earth is not perfectly oblate, but instead has an irregular shape, with bulges and depressions in various parts on the surface. For example, there is a depression off the Atlantic coast and another off the Pacific coast of southern North America—both of which are lower by about 150 feet than would be the case if the earth were a perfect oblate spheroid. An even lower point, 300 feet below the idealized surface, lies just south of India and occupies most of the Indian

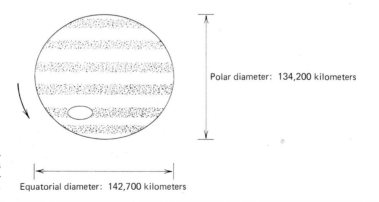

Polar diameter: 134,200 kilometers

Fig. 3-15. The polar diameter of Jupiter is smaller than the equatorial diameter.

Equatorial diameter: 142,700 kilometers

ocean. High points exist in Western South America, the East Indies, and Western Europe.

In the case of the earth, the nonsphericity can also be measured by means that have not been available for other bodies in the solar system. One of these is the direct measurement by gravity meters of the earth's gravity pull at different positions on its surface. Another consists of measurements of the distance along the surface that correspond to a specific number of degrees of latitude (e.g., for a flattened earth, the number of kilometers corresponding to 1° of latitude near the equator will be larger than the number near the poles). Both these methods give results that agree with the measurements made by satellites.

Calculations show that the oblateness of a body depends primarily upon three things: its rotation rate, mass concentration, and radius. The faster a body rotates, the greater its oblateness. This is because the material near the equator of a rapidly rotating planet experiences an outward force that to some extent counteracts the inward force of gravity. This outward *centrifugal* force is a result of the inertia of the material, which, according to Newton's laws of motion, tends to continue in a striaght-line direction. Any part of the equatorial region of a rotating planet feels, therefore, a relative tug in an outward direction. This outward tug counteracts somewhat the inward pull of gravity, causing the net force downward on a portion of material to be smaller. Near the poles, where the velocity of rotation is very small, the centrifugal force is small. The difference in this force between the different parts of the planet therefore leads to an equatorial "bulge"—to oblateness.

A second factor that plays an important part in the establishment of a planet's oblateness is the concentration of mass in its interior. If much of the matter of the planet is distributed in a dense core at the center of the planet, and the outer parts of the planet are therefore of very little total mass and low density, then the oblateness is much smaller than would otherwise be the case. This is because the amount of material and mass is small out near the surface where the velocity is great, and the material and mass are therefore of little consequence. On the other hand, in the case of a planet that has its mass distributed evenly throughout, so that the density of material varies only slightly or not at all from the center outward, the outer parts will have considerable mass, and therefore considerable inertia. The centrifugal force will then be sufficient to distort the planet's shape considerably. Such a planet will have a larger equatorial bulge than a planet with most of its mass concentrated in its

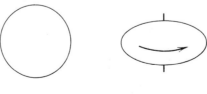

Stationary   Rapidly rotating

Fig. 3-16. The relationship between oblateness of a planet and the degree of rotation.

Fig. 3-17. The relationship between the oblate-
ness of a planet and the
degree of concentra-
tion of mass toward its
center.

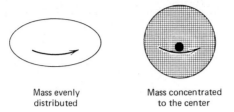

Mass evenly
distributed

Mass concentrated
to the center

center. Calculations have shown that in the case of a completely homogeneous planet, with its density the same everywhere, the ellipticity is two and one-half times greater than in the case of a similar planet (with the same size, mass, and rotation rate) that has all its mass near the center.

Another important item for consideration is the radius of the planet. A planet's rate of rotation on its axis is not a measure of the actual space velocity of material on the planet, but rather of the amount of time this material takes to make a circuit around the axis of the planet. We must also know the planet's radius before it is possible to know how fast the material is moving in space. For example, the period of rotation for the earth is 24 hours, but we must know the size of the earth before we can establish how fast we on the surface of the earth are moving as a result of this rotation. Because the radius of the earth is approximately 6,000 kilometers, the circumference is $2\pi r$, which is about 40,000 kilometers. Therefore, the surface near the equator of the earth has a velocity of about 1,600 kilometers (a little over 1,000 miles) per hour. A supersonic jet airplane can fly westward near the equator and keep the sun above the horizon indefinitely, while an ordinary jet transport, such as a 707 flying only 600 miles per hour, can follow the sun only by flying westward on a path near the poles.

Considerations of the oblateness, radii, surface gravities, and rotation rates of some of the planets have given us information on the distribution of matter in their centers. It has been found, for example, that Mars has its material distributed almost uniformly from the center outward. There is no great concentration of mass near the center of Mars; it is much more uniform than Earth in the distribution of its material. On the other hand, the major planets

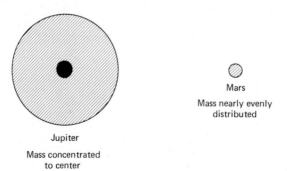

Mars

Mass nearly evenly
distributed

Jupiter

Mass concentrated
to center

Fig. 3-18. The derived
concentrations of mass in
Mars and in Jupiter.

such as Jupiter and Saturn, for which good measures of oblateness are available, have very highly concentrated mass distributions. A large percentage of their masses is limited to their very central regions. The density of material in the outer parts of these planets is very much smaller than at the center.

A complete model of a planetary interior, based on observations and on the physical laws that apply, can come only when the detailed behavior of the material making up that planet is known for the conditions that might exist there. This behavior is called the *equation of state* of the material. It can be established by very complicated calculations or by laboratory experimentation, or both. For the interiors of planets, where interior pressures must be extremely large (up to almost a million times the atmospheric pressure at the surface of the earth), experiments are very difficult because there is not sufficient experimental evidence for all the materials and pressures that might be applicable. In such cases, theoretical calculations can be used. The result can be a rather complete picture of the planetary interiors that depends only upon our guess as to what material exists there.

Evidence about the material in the planetary interiors comes from their mean densities and from several other considerations, including the compositions of their atmospheres, their surface temperatures, their oblatenesses, and meteorite compositions. If we assume that the material is in equilibrium, i.e., that it is not unstable and contracting or exploding, then we can use what are called *the equations of hydrostatic equilibrum* to show how the gravitational downward pull on material is balanced by the upward pressure of the material beneath it. Only for the outer crust of the earthlike planets, where the mechanical strength of the material—the solidified rock—is greater than the pressure, will these equations not apply. When the equation of state of the material is combined with the equations of hydrostatic equilibrium, it is possible to construct a model of the interior of any planet. The model will indicate the temperature, the pressure, and the density of the material at all positions from its center to its outer surface.

The nature of the innermost planet, Mercury, was somewhat mysterious for many years. It is an extremely difficult planet to observe from Earth because it is always close to the sun in the sky. Much of what little we used to know about its surface was the result of almost impossible observations made in the old-fashioned way (astronomers peering through telescopes), usually in the daytime. Radio and radar observations suggested a similarity to the moon, but just how much similarity was not clear until a spacecraft, Mariner 10, visited Mercury in 1974.

3-4 MERCURY

The automated cameras on Mariner 10 sent back the first good views of Mercury's surface, and initially it was difficult to see any difference at all between Mercury and the moon. The main visible features were craters that ranged in size from mere meters to hundreds of kilometers across. Some areas

Fig. 3-19. The planet
Mercury photographed
by the Mariner 10 space-
craft. In most respects,
Mercury looks quite sim-
ilar to our moon.
(NASA.)

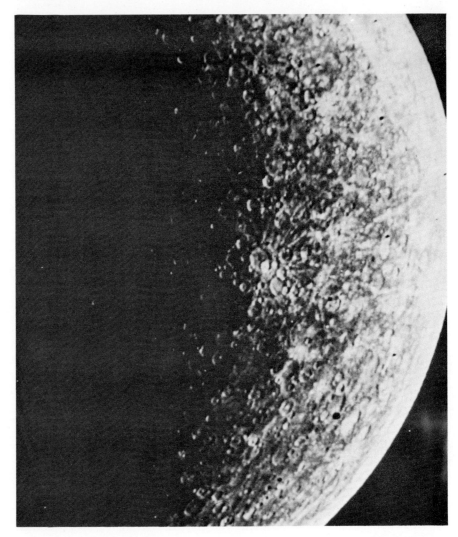

had smooth dark fields of lava, which were relatively free of large craters, and a few small volcanic cones were visible. The large craters were all clearly formed by the impact of giant meteorites, as were the lunar craters, and surrounding them was a blanket of small secondary craters, which had been made by the material ejected from the larger craters as they formed explosively.

Subsequent study of the Mariner 10 photographs has shown us three interesting differences between Mercury and the moon. First, we see that on Mercury the secondary craters around giant craters are all restricted to a much closer area surrounding their parents. This can be understood as a result of higher gravity on Mercury, where the ejected material was pulled back to the

Fig. 3-20. Craters in this Mariner 10 close-up of Mercury range up to over 100 km in diameter. Apparently they are all impact craters. (*NASA*.)

surface sooner than similar material in the lunar case. Second, there are some long, curving, scalloped cliffs on Mercury, unlike anything on the moon, and they probably result from the early, slow shrinking of the planet because of its large iron core. And finally, there is an area of strange, chaotic, hilly country on Mercury that is very unlike the moon's surface. This peculiar terrain lies antipodal (just opposite on the planet) from a giant crater, some 1,400 kilometers across, called Caloris Basin. It is likely that the huge impact that caused Caloris Basin produced "earth" quakes that spread out along the surface of Mercury from that point, and then all met together again at the opposite point, causing huge quakes that wrinkled up the surface into the strange pattern that we now see.

While the outside of Mercury is like the outside of the moon, its inside is much more like that of Earth. Its large density and a surprisingly strong magnetic field (detected by Mariner 10) indicate that it has a large iron core, estimated to extend at least 1,800 kilometers out from the center, thus making up about 75 percent of the planet's radius. When Mercury was formed, its closeness to the sun apparently meant that only the heavier elements could be held onto, and the lighter elements were pushed out by intense solar radiation. The result is a small, sun-scorched planet of high density, with a wealth of iron and other heavy elements.

Except for a negligible amount of helium temporarily captured from the

Fig. 3-21. Mercury's surface shows evidence of
ancient volcanism, which
probably ceased about 3
billion years ago. Long
shallow valleys may be
frozen lava channels.
(*NASA.*)

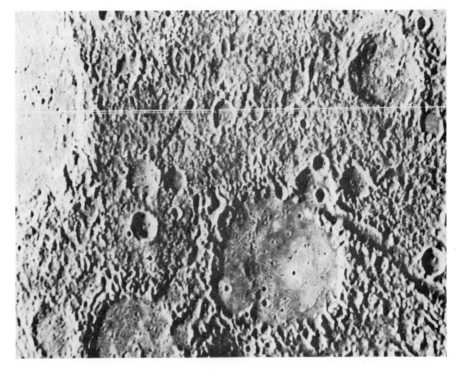

Fig. 3-21. Mercury's surface shows evidence of ancient volcanism, which probably ceased about 3 billion years ago. Long shallow valleys may be frozen lava channels. (*NASA.*)

stream of gas ejected by the sun, Mercury has no atmosphere; all of it was lost near the beginning of Mercury's history. The surface is extremely hot, with noontime temperatures averaging 425°C (765°F). The dry, airless surface bakes in this noonday sun for almost two earth-weeks at a time. Mercury is not a hospitable planet.

**3-5
VENUS**

Continuously cloud-covered, Venus has revealed few of its secrets to terrestrial telescopes. Since Galileo first pointed his tiny telescope at Venus, we have known that it shows phases like the moon. But beyond the progression of the phases, which are caused by the planet's circling of the sun inside earth's orbit, little else has been seen. Very short wavelength (ultraviolet) light showed some structure (Fig. 3-22), but it was fleeting and irregular, clearly just cloud forms unrelated to the surface.

Some idea of just how high the visible clouds of Venus must be above the surface came as soon as we could compare its visually measured radius with that measured by radar. Since the radio waves used in radar penetrate through the clouds to the surface, the measured difference of 40 kilometers (25 miles) must be the depth of the atmosphere from the cloud tops to the hard surface. The Venus atmosphere must be much thicker, then, than Earth's, where the cloud tops are seldom higher than 10 kilometers above the surface.

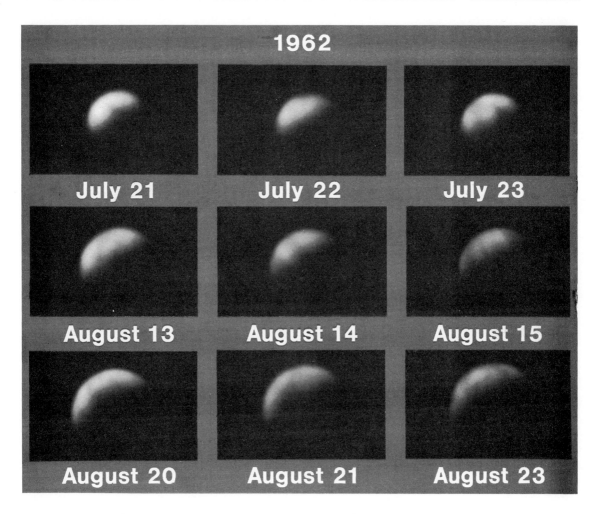

1962

July 21   July 22   July 23

August 13   August 14   August 15

August 20   August 21   August 23

The thickness of the atmosphere has been thoroughly confirmed by the many space probes that have been sent to Venus. Four Soviet spacecraft have landed safely on the surface by parachuting through the thick and cloudy atmosphere. These spacecraft radioed back the properties of the atmosphere: temperatures near the surface of about 475°C; pressures about ninety times the earth's surface pressure; and a composition very different from the earth's, including about 95 percent carbon dioxide (CO_2), a little nitrogen (N_2), a very small amount (0.01 percent) of water vapor (H_2O), and maybe a tiny bit of oxygen (O_2). By contrast, Earth's atmosphere is mostly N_2 (about 80 percent) and O_2 (about 20 percent).

Close-ups of the clouds are included in the thousands of pictures taken by Mariner 10 on its way past Venus to Mercury. These close-ups show large-scale patterns similar to earth's and suggest high speeds of up to 100 meters per

Fig. 3-22. Photographs of the planet Venus at six phases, showing some structure in the cloud layers. (*Lick Observatory.*)

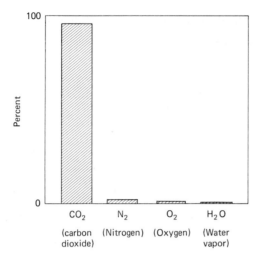

Fig. 3-23. The composition of the atmosphere of Venus.

second among the upper clouds. Spots and mottlings indicate that, as in earth's atmosphere, convection currents are common, especially where the noontime sun is heating the high atmosphere.

Using large telescopes and special devices designed to detect the absorption of the different chemical elements (Appendix D), spectroscopy from earth has shown the presence of various other gases in small proportions. Poisonous carbon monoxide (CO) is present, and various acids, including hydrochloric (HCl) and hydrofluoric (HF), show up in the spectra. Sulfuric acid is thought to be one of the substances most likely to make up the clouds. It may be that below the visible cloud tops there are zones in the atmosphere where steamy acid rains fall from the clouds, only to evaporate when they reach hotter regions below. Acid rains, poisonous gases, crushing pressures, and broiling temperatures—the atmosphere of Venus will not be a comfortable one to explore.

The radar signals show that Venus always presents the same side to us on earth at its closest approach. Whenever the two planets are nearest, with Venus passing between us and the sun, Venus' rotation brings the same face toward us. This remarkable fact is certainly not a mere coincidence; it is speculated that Earth, over the billions of years of the solar system's existence, has gained control of Venus' rotation gravitationally by tidal action. It is apparently similar to the case of the moon, where our (earth's) gravitational pull, through tidal action, has slowed the moon's rotation so that now it always keeps the same face toward us.

The surface of Venus was first seen in 1975, when two Soviet spacecraft, Venera 9 and 10, landed and each sent back a single picture before the spacecraft died in Venus' harsh environment. Each showed rocky ground. In Venera 9's case, the surroundings were strewn with sharp-edged boulders, while in Venera 10's picture, taken at a landing site some 2,000 kilometers

Fig. 3-24. The clouds of Venus, here photographed close-up by the Mariner 10 spacecraft. Global circulation patterns are evident. (*NASA.*)

away, the rocks were thin and flat and seemed to overlay a darker layer of rough material, maybe lava. The composition of the surface material was found to be roughly similar to that of basalt, a volcanic rock common on the surface of both earth and the moon.

Radar images of Venus show a great deal of large-scale detail, including several large circular forms that are probably impact craters like those on the moon. Unlike the moon's, though, the Venus craters are shallow and rounded, probably because of the greater weathering caused by its atmosphere.

We have barely begun our exploration of Venus, but already we know that it is not a place where we might envision setting up colonies in the near future. As

Fig. 3-25. The structure
of the atmosphere of
Venus. The clouds may
be partly sulfuric acid
and partly sulfur par-
ticles.

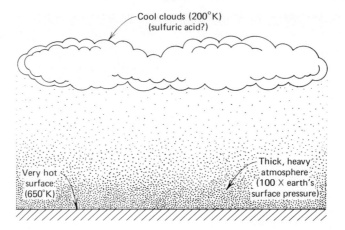

Cool clouds (200°K)
(sulfuric acid?)

Very hot
surface
(650°K)

Thick, heavy
atmosphere
(100 × earth's
surface pressure)

an abode for humans it is singularly unattractive. Furthermore, our studies of it result in the chilling realization that earth could have been every bit as forbidding if its temperature had been just a little bit higher. Then the oceans would not have formed, the carbon dioxide would not have been trapped in the rocks, and oxygen would never have been supplied by the development of life. Earth could have had an oppressive, hot, poisonous atmosphere very much like that of Venus. We should remember this fact when we think about the pollution of our own atmosphere. If we increase the amount of sunlight-absorbing dust in the atmosphere, it could raise earth's temperature. With Venus as an example, we must be sure not to let the increase go too far.

**3-6
MARS**

Mars has long intrigued both astronomers and the general public. Its fairly clear atmosphere and its frequent close approaches to earth have made it seem the most earthlike planet, though it is actually considerably smaller than earth. Its

Fig. 3-26. A reconstruc-
tion of the view of the
surface of Venus as seen
by the Venera 9 lander,
based on the photograph
published by the USSR.
Large rocks were seen to
be scattered about the
surface.

Fig. 3-27. A photograph of Mars taken by the spacecraft Mariner 7 just before its flyby. The bright circle is the huge volcano Nix Olympica, and the dark feature that has a somewhat linear appearance is one of the largest of the objects previously called canals, when discovered with earth-based telescopes. (*NASA.*)

atmosphere has no oxygen; it is mostly CO_2 and N_2, with a small and variable amount of water vapor, but though there is no free oxygen, at least there are no acids or poisonous gases. At the surface there is only 0.02 percent of the atmospheric pressure that we have at the earth's surface, so the atmosphere is more like what we find high above Mt. Everest. The temperatures tend to be

Canals of Mars

Fig. 3-28. The canals of Mars as such do not exist, but are now explained as largely illusory features that were caused by the human eye's tendency to connect into lines objects that are really random patches. Compare this diagram, based on earlier drawings of Mars, with Fig. 3-27.

even more bitterly cold than on Mt. Everest, ranging from about −125°C at the poles in winter to just a little above freezing at the equator at noon. These facts, gathered slowly at first by telescopic studies, and more rapidly by recent spacecraft, tell us that Mars is a somewhat worse place for life than Earth's worst places. The first successful Mars landers, Vikings 1 and 2, failed, in fact, to find even the most primitive forms of life there, though the last word on the topic was at least slightly equivocal.

It is the remarkable surface of Mars that has intrigued humanity for centuries and that still fills us with curiosity. Surely, this curiosity about the land forms on Mars will eventually make it the first planet to be visited by humans. It was the first to be mapped in detail, first by telescopes and now by orbiting spacecraft. Most of the names on the map of Mars were given in the nineteenth century and had roots in classical history and mythology. Thus we find "Elysium," "Utopia," and "Chryse."

Dark markings on the surface, as well as changes in the surface features, have been seen for over a hundred years. These led to a great deal of speculation about the nature of the Martian surface and the possibilities for life there. Among the most famous Martian surface features were the so called

Fig. 3-29. The ice of the Martian polar caps is shown here, spread over an area that contains many craters. (*NASA.*)

"canals," long thin lines joined together at "oases," that seemed to many early astronomers to resemble feats of engineering—intimating the presence of intelligent life at work on the planet. More recently, our more detailed survey of the planet by space probes has shown that canals as such do not exist. There is no chance that Mars harbors intelligent, highly developed life forms like Earth's; but the results do show the remarkable and surprising variety of geological forces at work on the surface of the planet.

The most conspicuous feature of the surface of Mars is its red color, which has been recognized since the time of the ancients, and no doubt gave rise to the choice of the name Mars. The color is similar to that of red desert sand on earth, such as that in central Australia. The lack of very much water in the atmosphere and on the surface suggests that the surface is, to a large extent, desertlike. Frequent immense dust storms indicate that the material on the surface is a fine material—sand or dust.

The most conspicuous Martian surface features visible through a small telescope are normally the polar caps, which are a brilliant white and stand out from the rest of the disk. It was recognized long ago that these polar caps shrink and grow with the seasons, very much like the terrestrial snow cover. On the winter hemisphere of Mars, the white cap can cover the ground completely as far as Martian latitude 60°, with an extent of up to a million square kilometers. When spring arrives, the polar cap shrinks slowly in size until it disappears almost entirely in Martian summer. Then in autumn, the polar areas are often

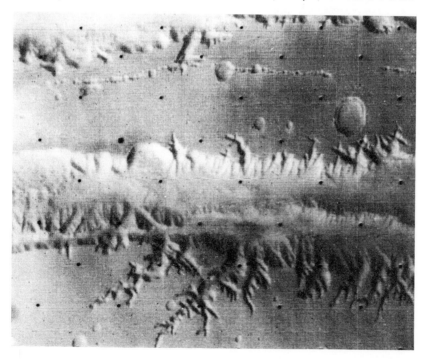

Fig. 3-30. A view of the immense Martian canyon known as the Vallis Marineris. Note the line of craters that seems to parallel it. These are probably sink holes caused by caving in of material over a smaller crack. (*NASA.*)

covered with a thin haze of clouds. This condition often prevails until midwinter, when clearing shows the cap at its maximum size again. Studies of the polar caps by space probes, especially Vikings 1 and 2, which orbited Mars and mapped its surface in very great detail in 1976 and 1977, show that the polar caps are covered with large shallow craters of apparent meteoritic origin. These areas are mostly fairly old geologically. Shadows due to winds show up in the "snow," which has been determined to be both water ice and dry ice (frozen carbon dioxide).

The rotation poles of Mars are tilted almost to the same degree as the terrestrial poles, and so the Martian surface experiences seasons very much like the terrestrial ones. Because the Martian day is only slightly longer than Earth's day, Mars' surface also experiences similar alternations between light and dark. Because Mars' year is very nearly two terrestrial years, each season lasts about twice as many days as the seasons on Earth. From Earth, astronomers have watched the changing of the Martian seasons, with the most conspicuous change being in the size of the polar caps. Alternations in the relative brightness and darkness of the gross surface features have also been observed. One of the most intriguing questions about the planet is the reason for these seasonal changes—whether it is due to changing wind patterns affecting the distribution of wind-blown dust or to some other cause.

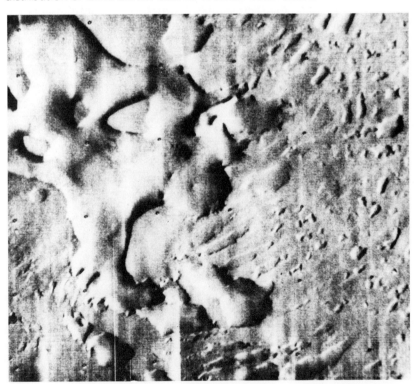

Fig. 3-31. A canyon area of the surface of Mars, showing erosional features looking something like the Badlands of South Dakota, but probably produced by wind erosion. (*NASA.*)

The Mariner and Viking space probes showed that the surface of Mars has an exceedingly varied nature when seen close up. Four principal kinds of geological features have been recognized on Mars. These are the huge and ancient plains covered with meteorite craters, the chaotic jumbled hilly areas, the areas of deep and intricate canyons, and the areas marked with immense volcanic mountains. The first three Mariner probes, Mariners 4, 6, and 7, were flybys which passed close enough to Mars to return photographs of certain portions of its surface, and by chance, these recorded principally only the first of the above kinds of features. Therefore, it was thought possible that Mars might be similar to the moon in primarily being covered with meteorite craters and in having very little recent geological history of any other sort. The age of these cratered plains is estimated to be on the order of 3 billion years, judging from the density of craters and comparison with the moon. The meteorite craters range in size from those barely visible on the photographs (a few kilometers across) to objects more than 200 kilometers in diameter, and they show the same broad flat floors and slightly raised rims that are characteristic of the craters on the moon.

Mariner 9, which reached Mars in 1971, showed that these very old crater plains are by no means the only kinds of features on the planet. Although they

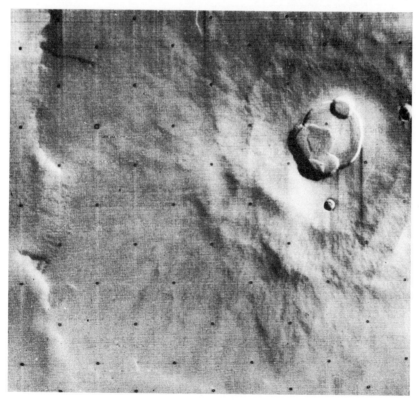

Fig. 3-32. A portion of the huge volcano Nix Olympica. The deep caldera or crater is at the right and a scarp, marking the edge of the sloping crater walls, is at the left. (*NASA.*)

Fig. 3-33. An area of the surface of Mars covered by lava flows, and showing some lava channels. (*NASA.*)

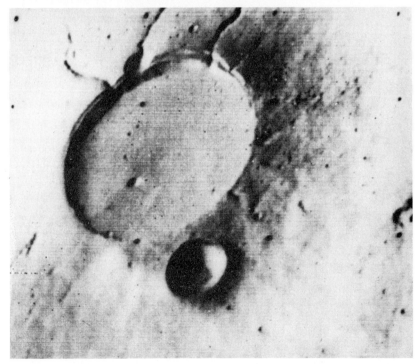

Fig. 3-34. The Martian volcano with the largest central caldera, or crater, known as Arsia Silva. (*NASA.*)

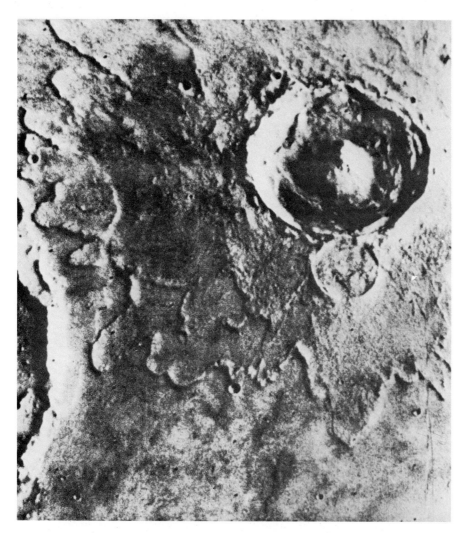

Fig. 3-35. A Viking spacecraft orbiter photograph of a Martian impact crater, showing the large splash marks around it, and suggesting a mudflow that was probably caused by the melting of subsurface ice by the impact. (*NASA.*)

do dominate certain areas of the planet, there are areas with considerable evidence of more recent geological activity. These are areas that have no visible craters and, instead, are covered with what appear to be low rolling hills without obvious pattern. This lack of craters means that they must be young, because meteorites of large size bombard the planet continuously and craters would be seen unless recent geologic activity had erased them.

Among the most surprising Martian surface features that were revealed by Mariner 9 are the canyons, one of which is much deeper and longer than the Grand Canyon of Arizona. Surprise about these canyons is primarily due to the fact that study of the Martian atmosphere many years ago indicated that there is almost no water on the surface or in the air. Therefore there is no agent that

Fig. 3-36. Complex Martian canyons arranged in an intricate network that is explained by crustal movement. These canyons are near the Martian equator in Noctis Lacus. (*NASA.*)

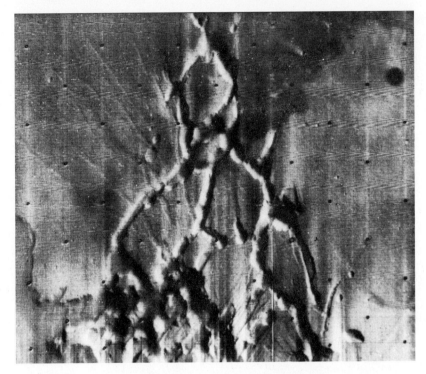

could produce a canyon like most of those on earth. This problem remains a mystery for some of the canyons, which may have to be explained by the hypothesis that at one time, fairly recently, there was more water than we now observe. The largest canyon system, however, was clearly formed mostly by faulting (the slipping and separating of rocks). The cracks of the surface opened up to form a canyon, and under the pressure of gravity the rocks and materials at the surface slipped down into this crack to produce a remarkable system of tributaries. This canyon, in the Coprates area near the Martian equator, is roughly similar in origin to the Great Rift valley in Africa.

Other highly surprising features seen on Mars' surface are the immense volcanic craters. These greatly exceed, both in area and height, the volcanoes with which we are familiar on earth. Four of the largest volcanoes are clustered in one portion of the surface, and one of these, Nix Olympica, is an object that has been observed with great interest by terrestrial astronomers for many years because of its variable appearance through the telescope. Mariner 9 photographs showed that its variations are due to the coming and going of a cloud cap, and possible frost at its summit. It may also be that in the past there has been volcanic activity that may have been detected from the earth. Nix Olympica is more than 500 kilometers across and rises 25,000 meters (75,000

Fig. 3-37. The first picture taken by the Viking 2 lander, showing the barren, rock-strewn surface of the plains named (long ago) Utopia (See also Fig. 7-7). (*NASA.*)

feet) above its base. By comparison, a typical volcano on Earth is some 10 kilometers across and on the order of 3,000 meters high above its base. The largest volcanoes on Earth are in the Hawaiian chain, where the twin peaks of Mauna Kea and Mauna Loa occupy an area with a diameter (counting portions of these mountains under the ocean surface) of 150 kilometers and rising some 10,000 meters above their subocean base. The Martian volcanoes show many of the features that are familiar to terrestrial volcanologists, including a central caldera (crater) and lava channels and lava tubes along the flanks of the mountains. The "south spot" volcano, also called Arsia Silva, has the largest caldera on Mars, over 100 kilometers across from rim to rim.

3-7 JUPITER

The planet Jupiter does not exhibit a solid surface because of its perpetual cloud cover. In fact, a solid surface of the sort with which we are familiar may not exist, even under the clouds. Markings on the face of Jupiter are, nevertheless, very easy to see and have been studied in considerable detail. The most conspicuous markings are the cloud belts which extend like lines around the planet, all parallel to its equator. They are thin and somewhat variable in shape, and the fact that their rotation periods depend on their latitudes (see Chap. 2) suggests that they have been drawn out into lines by the rapidly and differentially rotating planet. The belts change from time to time, although there is a general pattern that has been given names, as shown in Fig. 3-38.

A rather remarkable feature on the visible surface of Jupiter is a large and oval reddish spot called the *Great Red Spot*. It has been observed since 1830 and remains something of a puzzle because of the very peculiar motions that it shows. The Great Red Spot is very large, measuring about 10,000 by 40,000 kilometers. Over the years, its width has remained constant but its length varies from time to time, though it is always almost perfectly elliptical. Its period of rotation is not constant, which indicates that it is not a feature that is permanently tied in some way to a hard surface beneath, but must "float" amongst the clouds. The period range over the last 130 years has been from 9

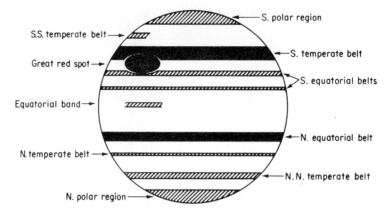

Fig. 3-38. The cloud bands of Jupiter are semipermanent in their latitude and have been given names according to their position on the planet.

hours 55 minutes 32 seconds, to 9 hours 55 minutes 44 seconds. There is a controversy about its nature, but the most likely explanation is that an internal disturbance (or a collision with a comet or other body) set up a stormlike wind pattern that simply has not died out yet.

The major planets' atmospheres are found to be very different from those of the inner planets, with the main constituent being hydrogen and various molecules containing hydrogen. The most conspicuous absorption lines in the reflected solar spectrum from Jupiter, for example, are due to methane (CH_4)

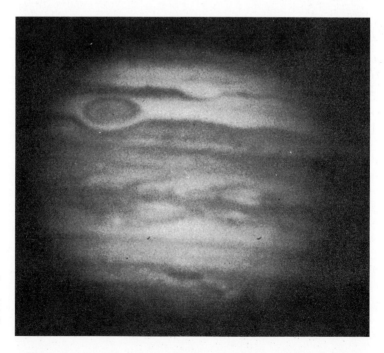

Fig. 3-39. One of the puzzling features of the Jupiter clouds is the Great Red Spot seen here at the upper left. (*Kitt Peak National Observatory.*)

Fig. 3-40. The first flyby of Jupiter, made by the Pioneer 10 spacecraft, revealed finer details in the cloud band structure than had been seen from earth. (*NASA.*)

and ammonia (NH_3). These molecules are extremely rare in the atmospheres of the inner planets. Although they are the most conspicuous features in the spectrum of Jupiter, it is found that they are mere minor constituents of that planet's atmosphere. A somewhat less conspicuous spectral feature, due to molecular hydrogen, shows that hydrogen is approximately a hundred times more abundant than methane in the atmosphere, which in turn is about 20 times more abundant than ammonia.   Fig. 3-42 shows a comparison of the spectra of the different gases detected by spectroscopic means in the atmospheres of Jupiter and the other outer planets. All have a fairly similar mix of elements, with molecular hydrogen being the most abundant detected constituent. What is not shown in the figure is the spectrum of helium, because helium has no spectroscopic features that have been measured. For it, other means must be used to determine abundance, as described in the next section.

The surface pressures that have been determined for the major planets refer to the pressures at the top of the cloud layers. There is no information whatever available from observations regarding the true surface pressure at any hard surface of the planet, if these planets actually have well-defined solid surfaces. In all cases over the area observed the temperature is very low. Jupiter's

Fig. 3-41. The Great Red Spot in a Pioneer close-up photograph. (*NASA.*)

Fig. 3-41. The Great Red Spot in a Pioneer close-up photograph. (*NASA.*)

measures approximately −143°C, and calculations of the variation of temperature indicate that at heights of 20 or 30 kilometers above the cloud surface, the temperature is as low as −180°C.

Although the surface pressures for the planets cannot be known for any positions beneath the cloud layers, it is clear from the total amounts of the various molecules detected above the cloud layers that these planets have very much thicker atmospheres than any of the inner planets. In fact, they may have atmospheres that measure as much as 500 kilometers thick or more—some 50 kilometers thick above the cloud layers and hundreds of kilometers thick in the unseen portions beneath the clouds. Deep under the atmosphere is a hypothetical nongaseous surface that may be a sea of liquid ammonia, possibly liquid

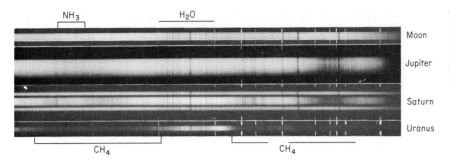

Fig. 3-42. Spectra of Jupiter, Saturn, and Uranus compared to that of the moon. The NH₃ (ammonia) and CH₄ (methane) features are identified. (*Lick Observatory.*)

water, mixed perhaps with water and ammonia ices, all overlaying a global ocean of liquid hydrogen.

One of the more remarkable features of Jupiter's atmosphere is its far outer atmosphere, which apparently includes an *ionosphere*, similar to that of Earth in some respects. Earth's ionosphere consists of a tenuous region of electrically charged particles caused by the effects of high-energy solar radiation falling on the upper air. Radio observations show that Jupiter is also surrounded by *radiation belts* like Earth's *van Allen Belts*, which were discovered by the first U.S. satellite. These belts are considerably beyond Jupiter's atmosphere as normally defined. They consist of magnetically trapped charged particles, undoubtedly from the sun. The fact that these particles are trapped around Jupiter in belts indicates that Jupiter must have a strong magnetic field. Probably related to these radiation belts are the brief radio bursts that occur at certain wavelengths, indicating the presence in the planet's atmosphere of several sources for "noise storms," the exact nature of which is still being unraveled. They can be thought of loosely as resembling thunderstorms on earth, though the conditions are undoubtedly exceedingly different in detail.

Fig. 3-43. Radio bursts from the planet Jupiter, recorded by the University of Texas radio telescope. (*Courtesy of H. J. Smith.*)

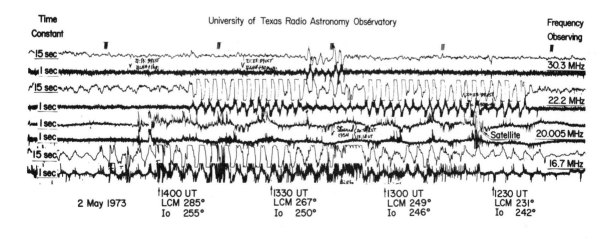

Occultation Experiments

For both Jupiter and Saturn, further information on the atmospheric properties can be obtained by observing occultations of stars or satellites. An *occultation* occurs when the planet passes in front of a distant star or one of its own satellites, as a result of either the planet's or the satellite's motion. By very carefully observing the brightness of the occulted object as the planet's atmosphere begins to occult it, astronomers can measure the amount of absorption of light in the atmosphere at different heights above the surface of the planet. This measurement gives information on the structure of the atmosphere, which is related to its chemical composition and total content. Before spacecraft were available, it was necessary to rely heavily on natural satellites and distant stars.

Pioneer Flyby Results

Two long-range spacecraft, Pioneers 10 and 11, made close flyby visits to Jupiter in 1973 and 1974, and the result was a much clearer picture of its atmosphere, interior, and giant particle-laden magnetic field. The atmosphere was found to be ''boiling'' through convection, with the light zones higher than the dark bands. These zones are rising gas, and the belts are then the descending gas of the convection currents. The planet was found to be radiating away its own heat, which was most likely heat still left over from its initial formation, and it is this internal heat that is carried out through the atmosphere by the convection seen in the clouds.

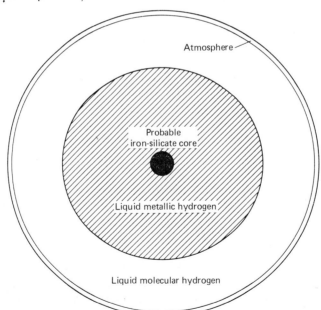

Fig. 3-44. A model of the interior of Jupiter, based on the results from spacecraft measurements of its density and gravitational field shape.

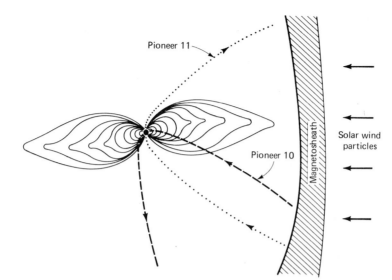

Pioneer 11

Pioneer 10

Magnetosheath

Solar wind
particles

Fig. 3-45. The magnetic field of Jupiter is a huge and complex phenomenon as mapped out by the Pioneer spacecraft.

The Pioneer spacecraft also discovered the surprising fact that Jupiter is almost entirely liquid. It can best be understood as consisting of almost pure hydrogen; a layer of liquid molecular hydrogen probably overlays a core of liquid metallic hydrogen. (The hydrogen is compressed so severely that it acts like a metal). At the very center is a small nucleus of iron-rich rocky material. This nucleus is completely hypothetical, and its size, which is roughly the size of earth, is estimated by assuming that Jupiter has about as much of the heavier elements relative to hydrogen as the sun has. Most likely as it formed from the material of the pre-solar-system cloud, these heavier elements sank to the bottom to form a heavy core. Jupiter was massive enough, unlike the inner planets, to hold onto its hydrogen and that is why it is now so large and so low in density.

The Pioneer spacecraft passed through the immense magnetic field of Jupiter and measured its shape and strength. Fig. 3-45 shows the results. A bow wave, as on a boat, points into the wind that comes from the sun (Chap. 6) and the field extends from the planet to about 100 times Jupiter's radius. Captured in the field are large numbers of charged particles, probably caught from the solar wind. These are so plentiful that they even cause the inner satellites of Jupiter to be radioactive. The field also was found to affect the earth in a surprising way. Pioneer data showed that speeding electrons are spewed out into the solar system ahead of Jupiter's magnetosphere, and that they spiral in even to Earth. Previously, cosmic ray particles detected at the Earth were found to be enhanced every 13 months, and we now know that these are Jupiter's electrons, which have been ejected by its remarkable magnetic field.

**3-8
THE OUTER
PLANETS**

The planets Saturn, Uranus, and Neptune have properties that are somewhat similar to those of Jupiter. Saturn shows belts parallel to its equator that are similar to, but rather fainter than, Jupiter's. The very best photographs of the surface of the other planets are those taken by the balloon-borne and space telescopes, and they indicate a general similarity for all four of the outer giant planets. About Pluto, which is apparently even smaller than Earth, and therefore not one of the giant planets, very little is known.

Discovery

Saturn seen near opposition is a fairly bright object; it has been known since ancient times. Uranus, Neptune, and Pluto, however, are much fainter, invisible to the eye without a telescope (actually, some people can just glimpse Uranus sometimes, if they know exactly where to look). The three trans-Saturnian planets were all discovered in the modern era, one each century, beginning with the eighteenth century. Each discovery was an important event in the history of astronomy and each has a story connected with it.

William Herschel, a young musician in the German baroque tradition in Hanover, moved to England in the mid-1700s to accept a job as a church musician, director, and composer. Musicians then (and to some extent now) move freely from country to country, depending upon the job market and the whims of musical styles. Young Herschel, however, had an avocation that gradually became a consuming interest. After he had read a book on astronomy, he decided that he would like to explore the universe and discover

Fig. 3-46. The planet Saturn, taken when its rings were viewed from above, near their time of maximum openness as viewed from earth. (*Lowell Observatory.*)

Fig. 3-47. The planet Uranus with its brighter satellites. (*Lick Observatory.*)

the size and shape of our system of stars. Since good telescopes at that time were expensive (the famous optician Dollard was just then developing achromatic lenses), Herschel first had to make his own telescopes, which he did in ever-increasing size.

Both in his telescope making and in his research on the stars Herschel collaborated with his sister Caroline Herschel, one of a long succession of famous women astronomers. The two worked on telescopes, making them by day and observing with them by night, and presumably only the English weather prevented the complete neglect of Herschel's job as a musician.

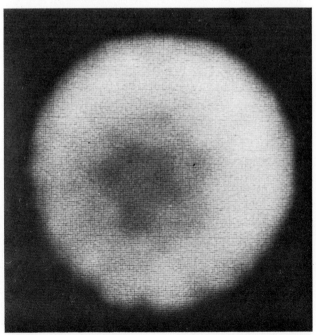

Fig. 3-48. A photograph of Uranus taken with a CCD (charged coupling device), which is an electronic imaging instrument. (*Lunar and Planetary Laboratory, University of Arizona.*)

In 1781, while the Herschels were systematically mapping the stellar universe, they noted a faint, strange-looking object which they first thought might be a comet, because of its motion between the stars. However, Herschel eventually realized that it must instead be a new planet, more distant than the most distant planet then known, Saturn.

King George III happened to have an amateur's interest in astronomy and had been supporting the Herschels' work to some extent, so Herschel named the planet Georgium Sidus. This apparently pleased the King, who had suffered some disappointments about then in connection with the American colonies, and he asked Herschel down to Windsor Castle for a chat. Herschel emerged with an annual salary to support his astronomical work and he soon gave up the secure field of music for the new and chancy field of astronomy, to which he, his sister Caroline, and later his son Sir John Herschel, devoted their lives. Although keeping its royal name for a few years in England, the planet was eventually renamed Uranus, after Urania, the muse of astronomy.

The discovery of the next planet out, Neptune, was made in a very different way. Uranus was observed after its discovery, so that its orbit was accurately known, and in about 1820 astronomers began to have difficulty understanding its motions. Its orbit seemed to be slowly changing in a way that remained a mystery for years. In 1841 John Couch Adams, a student at Cambridge, in England, decided to try to solve the problem. He hypothesized that a more distant planet beyond Uranus was perturbing its orbit by the influence of its own gravitational field. Although proving by 1843 that such a planet could solve the mystery, it took him 2 more years to arrive at a complete analysis, which allowed him to predict the location, the mass, and the orbital shape for the hypothetical planet.

The next steps in this story are complicated, full of human errors, mishaps, heartbreaks and triumph. Adam's professor at Cambridge, Rev. Challis, though director of the best observatory in England for looking at this planet, happened to be a hesitant and indecisive man who chose instead to send his student to the Astronomer Royal, Sir George Airy, at the Greenwich Observatory near

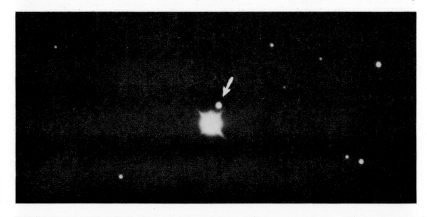

Fig. 3-49. Neptune and its brightest satellite, Triton. (*Lick Observatory.*)

London. When Adams first called, with his predictions under his arm, the Astronomer Royal was out, attending a committee meeting in London that was deciding on the gauge for English railways. When he next called, the Astronomer Royal was at dinner (it is said that he kept unusual dinner hours) and his butler sent him away in what was apparently a properly cool butlerish manner. When finally Adams resorted to sending the computations to Greenwich by post, the Astronomer Royal replied with a letter, transmitting his thanks and asking a simple question that showed that he had thought only carelessly about the problem. Adams, shy and offended, did not answer, and so the Astronomer Royal left the letter in the unanswered file and did nothing for several months.

In the meantime in France, the director of the Paris Observatory, Urbain Leverrier, was working on the same problem, using much the same methods as had been used previously (unknown to him or to almost anybody) by Adams. By June 1846, he arrived at a solution nearly identical to that given by Adams in October 1845. When Airy received and read Leverrier's published paper, he immediately wrote to Prof. Challis at Cambridge, suggesting that a search be undertaken. (No one seems to have said anything either to or about Adams and his prior prediction). Rather than just looking in the direction in the sky that had been predicted by Adams and Leverrier, Challis decided to make an extensive, painstaking, and systematic search over a fairly large section of the sky. He began in August 1846, laboriously mapping the sky, but because he expected to take at least a year for it, he put each night's measures of star positions away in a drawer, to be examined later.

Meanwhile, in France, Leverrier had realized that the predicted large mass for Planet X implied a fairly large disk as seen in a telescope, and he wrote to Johann Galle at the Berlin Observatory in a letter dated September 18, 1846, suggesting that just by looking in the predicted position one might immediately distinguish Planet X from the stars. Galle received the letter on September 23 and discovered the planet that very night. Within a week it had been confirmed, announced to the world, and named Neptune because of its location in the far depths of space.

Later on poor Prof. Challis looked up his nightly records and discovered that he had seen and measured Neptune several times in August, but had not realized it, because he had put away the data to look at later. And poor Airy was forced to explain to an angry nation that blamed him bitterly for losing to France an honor that it was felt should have been England's.

The discovery of Neptune is one of the famous lessons in the history of science. Yet, it is not unambiguously interpretable as a lesson. Discovery, that most exciting and exhilarating of scientific activities, involves a large measure of both good and bad luck (that is, *chance*), as well as judgment. For every John Couch Adams prediction, there are others that have no solid basis and could lead to a wasteful "wild goose chase."

Pluto was discovered in a much more orderly fashion than either of its predecessors. The astronomer Percival Lowell's interests, in the early 1900s, included the search for a trans-Neptunian planet. He and others carried out calculations like those made before by Adams and Leverrier. This time they attempted to explain further slight discrepancies in Uranus' orbit. They could not use Neptune, because it had not yet been observed over a long enough interval for the data to be useful. A new planet was found close to the position predicted by Lowell, but not until 14 years after Lowell's death. Clyde Tombaugh, a 23-year-old astronomer hired to use a new search telescope built in 1929 at the Lowell Observatory, found the planet on the afternoon of February 18, 1930. Its name was suggested by an 11-year-old English girl, who thought it appropriate because Pluto is the god of darkness and the underworld and because the name began with Lowell's initials.

Physical Properties

Because of their vast distance from Earth, the outer planets are difficult to observe and study. Except for Pluto, which has an irregularity in its reflectivity that produces a variable light curve, we do not even know their periods of rotation reliably. For Uranus and Neptune it was shown in 1977 that the rotation periods, measured by optical Doppler shift spectroscopy, are considerably longer than was previously thought, with both lasting about 22 hours.

We now know the sizes of the outer three planets better than might be expected, considering their apparently tiny size as seen through telescopes. By timing occultations of stars by these planets, it has been possible to measure their diameters, and to some extent their shapes. Uranus and Neptune are giants, though smaller than Jupiter and Saturn (Appendix F), while Pluto is small, like the terrestrial planets or like the larger satellites. Because of its small size and its eccentric orbit, which brings Pluto across Neptune's orbit on occasion, some astronomers believe that it might once have been a satellite of Neptune.

The atmospheres of Saturn, Uranus, and Neptune are similar to that of Jupiter in most respects (Fig. 3-42), since they contain mostly hydrogen, helium, and hydrogen compounds, especially methane.

Like Jupiter's, Saturn's upper atmosphere is warmer than can be explained simply by the sunlight it receives. This results from the probable fact that these giant gaseous planets are still slowly contracting after formation; the heating would thus be a result of the release of gravitational energy by the contraction. Pluto, on the other hand, is very cold, probably about 40 K. In 1976 it was discovered that the surface of Pluto is covered with methane frost.

QUESTIONS

1. If Planet x has a Planck curve with its maximum at 10 centimeters wavelength and Planet y with a maximum at 5 centimeters, which is hotter? Both are found to emit the same total amount of energy, after reflected sunlight is subtracted. Which is larger?

2. How much more intense is sunlight on the surface of Mercury than on Earth's surface?

3. What is the surface temperature of Venus in degrees Fahrenheit?

4. Make a table that contrasts the three important properties of an atmosphere as determined for the various planets.

5. Can you think of sources of heating for the earth's atmosphere other than absorption of solar radiation?

6. If human pollution in the atmosphere should in the future create large amounts of dust and gas that absorb radiation, what will be the effect?

7. If this pollution material actually reflects more energy from the solar radiation than it absorbs, what will be the effect?

8. What kinds of measurements would you propose for a spacecraft intending to parachute down into Jupiter's atmosphere?

9. Calculate the mass of Saturn from the comparison given in the text between the orbital periods of Dione and the moon.

10.* From its known density and radius, construct a mathematical model of Mercury, assuming it to be made up of a rock exterior with an iron core. Neglect pressure effects in the interior.

1. By checking in the *American Ephemeris and Nautical Almanac*, or other sources, find when Venus will be near *inferior conjunction* (when it is nearly between us and the sun). Observe it with your college's telescope, or your own small telescope, and attempt to detect how far the points of the crescent extend. You will probably have to observe it at sunset or earlier, or near sunrise. Because of its thick atmosphere, the crescent can sometimes be seen to extend *beyond* 180°. Why?

2. The *phase function* of a planet is a curve that tells how bright the planet is as

OBSERVATIONS AND EXPERIMENTS

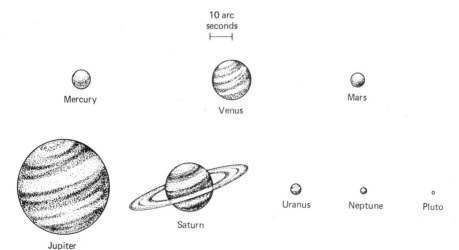

Fig. 3-50. The relative sizes of the planets as seen from earth at their times of closest approach.

seen from various angles with respect to the sun. Its shape depends on the roughness of the reflecting surface. Cover one ball with coarse sandpaper and one with shiny paper of the same color; then compare their phase functions by shining a slide projector beam on them from various angles. Make a "qualitative" comparison of their reflection curves (rough, without numbers), and then do it "quantitatively" with a photographic exposure meter.

3. Using a photographic exposure meter on a telescope, measure the phase function of the moon by observing it for 2 weeks or so. Is your result more like the sandpaper or the shiny paper? If a large enough telescope is available to you, do the same for Venus. (Be sure to measure the brightness of the sky and subtract it if your measures are obtained before it is completely dark.) Compare this result with the experimental curves.

4. Measure the phase function of clouds in the sky on a partly cloudy day. Is your result like the Venus curve?

5. Using a photographic exposure meter, test various materials to find one as bright as Venus and one as dark as Mercury (6 percent reflection of light). Does the *albedo* (percentage of light reflected) of materials depend upon the size of chunks? Try crushing suitable materials to test this. Does the phase function (amount of light reflected at various angles to the direction of the incident light) depend on the chunk size? Measure the light reflected at 0° and 45° from the direction of incident light.

6. Binoculars will show the disk of the planets Jupiter, Saturn, Mars (at its nearest), and Venus. To perceive these disks, which are very small, it is necessary to brace the binoculars on something steady. A small telescope (sturdily mounted) does much better for the planets. A telescope 3 to 12 inches in diameter will show the phases of Venus and Mercury, the polar caps and dark markings on Mars, the belts and the Great Red Spot on Jupiter, the faint bands on Saturn (sometimes), and the planetary disk of Uranus and, perhaps Neptune.

7. With the aid of your college's telescope, devise a means of measuring the polar flattening of Jupiter. You will need to use a micrometer or an eyepiece reticle that is graduated.

8. Measure Jupiter's polar flattening by making use of the photograph in this book.

Fig. 3-51. The relative apparent sizes of Venus, seen from the earth at its various phases.

THE MOON AND OTHER SATELLITES

**4-0
GOALS**

Our moon has a great deal to tell us about the early history of earth and the other planets. Lying close at hand but essentially dead for the last 3 billion years, the moon's surface gives us a record of what happened long ago. The record it provides is earlier than anything we can read in the rocks of our more active and changing planet. This chapter shows how the moon and other satellites can be used to explore our early times.

**4-1
SATELLITES**

In 1610 when Galileo first pointed his telescope toward the bright planet Jupiter, he was surprised to see that the planet was accompanied by four comparatively faint objects that stayed close to the planet and moved across the sky with it. Galileo realized the importance of his discovery because he was able to make an analogy between this family of small companions (clearly revolving around the planet Jupiter) and the solar system, which is made up of planets revolving around the much larger sun. These four objects were recognized as moons of Jupiter, and the similarity between their orbits around Jupiter and the planet's orbits around the sun was one of the arguments that Galileo used to support the then heretical view that earth was not the center of the universe, but rather was one of a family of planets moving around the sun. The four objects he discovered are called *the galilean satellites of Jupiter*, and they are four of the largest moons in the solar system.

Not all the planets have satellites; Mercury and Venus do not have known companions. All the other planets, however, have at least one satellite; Earth has the moon, Mars has two (Phobos and Deimos), Jupiter has fourteen, Saturn has at least ten, Uranus has five, Neptune has two and Pluto has one. The largest satellite in the solar system, Titan, which is one of the satellites of Saturn, is as big as the planet Mercury. The smallest satellites are so small that they are barely detectable, with estimated diameters of only 10 kilometers or so. The four galilean satellites of Jupiter are some 4,000 kilometers in diameter and are bright enough to be glimpsed with binoculars. A small telescope will show them all clearly.

The most thoroughly studied satellite is that of the earth, our moon. Because of the extensive exploration of the moon during the 1960s and early 1970s, we know a great deal about its detailed structure and composition. A total of 750 pounds of lunar material has been returned to the earth for analysis, and many volumes of detailed reports have been written about these lunar samples, with many still to come. The surface of the moon is now mapped thoroughly, with details shown to sizes as small as only a hundred meters or so. The ages of lunar rocks are known for several different places on the lunar surface, and cores extending down into the surface material of the moon have been analyzed to investigate the subterranian material. And finally, the use of high-precision optical radar equipment (called *lasers*) has made it possible to measure the distance of the moon from the earth with an accuracy of an inch or so—a

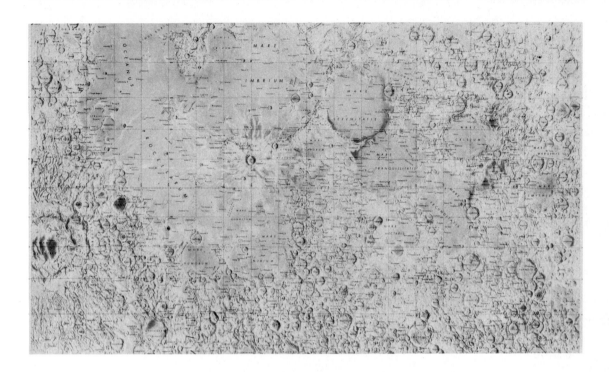

higher degree of accuracy than that known for many distances on the earth itself. The moon is indeed a thoroughly studied object, and many of the important questions that scientists were unable to answer before 1960 have now been answered, sometimes with surprising results.

Fig. 4-1 A map of the near side of the moon, prepared by NASA for the Apollo missions. *(NASA.)*

A view of the moon without a telescope clearly shows that the surface is not uniform. The "man in the moon" is the result of an arrangement of light and dark areas of very large size on the surface.

**4-2
LUNAR LAND
FORMS**

This arrangement was likened by early astronomers to the arrangement of land and sea areas on the earth, and therefore they chose to call the dark areas on the moon *maria*—the Latin word for sea. As telescopes improved in the succeeding centuries, it was clear that the maria did not contain water. There are seven recognized maria, which are distinguished by their dark color, their very large size (averaging some 500 kilometers in diameter), and their relatively smooth and unmarked surfaces. Most of them are roughly circular in outline, and this is evidence that they probably had an ancient origin as immense crater scars, that had been produced by the impact of large bodies on the moon early in the history of the solar system. Modern measurements of the ages of the

Maria

Fig. 4-2 A close-up photograph of the moon,
taken by the Apollo 17
spacecraft. The round
black feature above and
left of center is Mare Cri-
sium, with Mare Fecundi-
tatus below it. *(NASA.)*

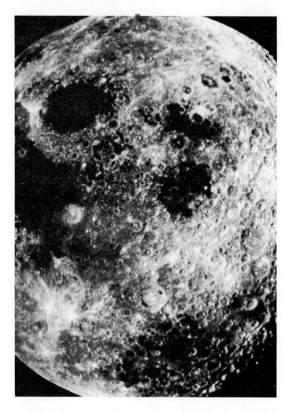

maria and of the nature of the material on the maria floors corroborates this
speculation. The maria apparently formed at a time in the solar system when
there was still a great deal of debris—large chunks of interplanetary material of
the sort that built the planets, the moon, and other bodies of the solar system.
The last few remaining chunks of large size near the earth system formed these
immense circular seas by impacting onto the lunar surface. The subsequent
melting of lunar materials, and the possible volcanic action engendered by
these immense collisions, produced the relatively smooth floors of the maria.
These floors are now marked only by the few small craters that result from
subsequent collisions with smaller bodies.

Craters   Lunar craters are the second most conspicuous kind of feature visible on the
surface through binoculars or a small telescope. They range in size from over a
hundred kilometers down to microscopic sizes detectable only on materials
that are brought back and studied in the laboratory. Almost all the craters result
from collisions of various sizes of interplanetary bodies with the lunar surface.

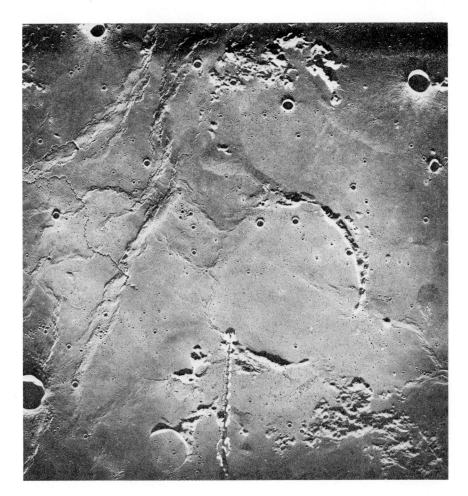

Fig. 4-3 A photograph of lunar craters, taken during the Apollo program from the orbiting support module. *(NASA.)*

They are similar to the meteorite craters on the earth, but the difference is that on the earth the continuous erosion of land forms by rain and wind and other geological processes wipes away the evidence in a relatively short time. On the moon, however, the lack of an atmosphere and the lack of any active geological forces (volcanism or extensive rock movement) mean that the only agency capable of obliterating a crater is another crater that has been produced by a more recent impact. Studies of the impact craters on the moon show that the largest craters are filled with smaller craters, that these smaller craters are in turn filled with even smaller craters, and so on. The number of craters is very large for very small sizes, due to the distribution of sizes of impacting bodies. Studies of interplanetary dust and meteoroids (Chaps. 5 and 6) indicate that there are a few larger bodies and a very large number of smaller bodies in

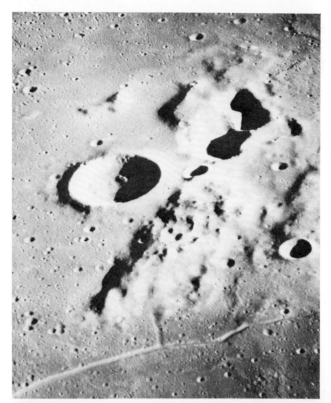

interplanetary space. Therefore, the chances of forming a large crater on the moon at any given time are exceedingly small compared to the chances of forming a small crater.

When a large body impacts onto the lunar surface, it creates a considerable disturbance in that portion of the surface and the surrounding area. Astronomers now know the extent of this disturbance, and they understand most of the mechanics of it through study of terrestrial impact craters and the rocks and soil brought back from the moon. The terrestrial meteorite craters show that the impact of a meteorite body is a very energetic event that can fracture rocks down to several times the depth of the actual crater formed. The material both of the meteorite and of the soil and rocks of the surface may be pulverized to form fine *rock flour* that is scattered about the surface, as well as large amounts of shattered rocks. For a crater that is roughly a kilometer across, material can be thrown out to a distance of several kilometers. For the largest lunar craters, such as the crater Tycho, which is 80 kilometers in diameter, material on photographs can be traced to thousands of kilometers out from the center.

Fig. 4-6 The large crater Copernicus, photographed here by a Lunar Orbiter spacecraft, is seen spread out at the bottom of this photograph. The crater wall has cliffs that are 300 meters (about 1,000 feet) high. The mountains in the center of the crater are also about that high. (NASA.)

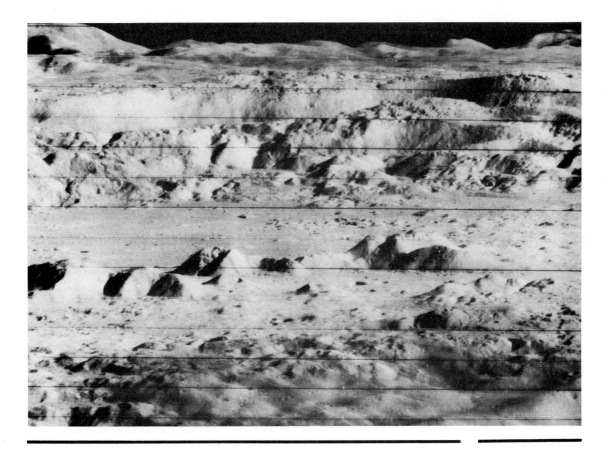

Fig. 4-7 Craters within craters show up in the foreground of this photograph, which also shows Hadley Delta, a mountain near the Hadley Rille. The photograph, which shows many small craters in the middle distance, was taken on the Apollo 15 mission. *(NASA.)*

The craters have raised rims that in some cases rise as high as 3,000 meters above the surroundings for the largest craters. Usually, there is a floor that is deeper than the surroundings, occasionally containing a central mountain peak. The largest craters are very flat (if drawn in profile) and are so large that a viewer standing in the center of one of them would not be able to see the rim even it if were thousands of meters high, because it would be beyond the horizon of the moon.

Fig. 4-8 From the study of craters on the earth that have been formed by meteorite impact, it was learned that craters depend for their structure upon the size of the impacting bodies.

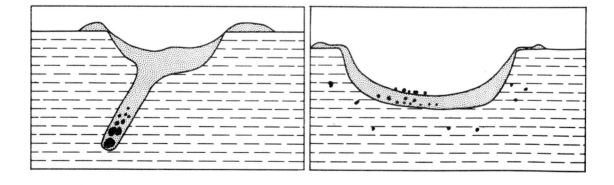

The reason why the lunar craters show effects from a major disruption of the lunar surface can be traced to the fact that an incoming body from interplanetary space gains a great deal of energy upon approaching the moon. If a body moving through space approximately in the same direction and with the same velocity as the moon should gradually come close to the moon, the gravitational pull of the moon on that body would exert a force that would alter its direction and velocity so that it would fall into the moon. The least amount of velocity that a body can have would be that given to it by this gravitational pull (not counting any velocity different from the moon that it might have had prior to the encounter). This minimum velocity of collision is found to be about 3 kilometers per second (~7,000 miles per hour). If an interplanetary body roughly the size of a house were to fall onto the surface of the moon with this velocity, the collision would have the same amount of energy as 10 thousand *tons* of TNT, and it would cause a considerable disturbance.

Both theoretical calculations and experiments with large explosives on the earth give information on the nature of the disturbance that is created by such a collision. For small objects, from microscopic size up to the size of a large boulder, most of the energy of the collision goes into digging out the crater; the size of the crater is roughly the same as the size of the incoming body, at least no greater than two or three times its diameter. For much larger bodies, the collision is a much more explosive event, and the crater can be considerably larger than the colliding body. The energy not only digs out a hole for the crater, but produces a shock wave that is propagated through the rock of the entire region beneath and to each side of the crater, causing a jumble of broken rocks underground. A rim is formed by material thrown out from the crater area, and debris from the explosion is scattered over a large area.

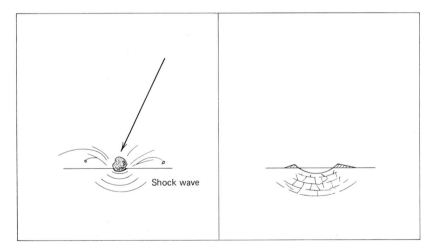

Fig. 4-9 The formation of crater by impact.

Lunar Mountains   There are several features on the surface of the moon that resemble terrestrial mountain ranges. It is unlikely that any of them has an origin similar to that of the mountain ranges on the earth, most of which are formed by up-and-down slippage of immense blocks of the terrestrial crust. Instead, it appears that the mountain ranges of the moon are related to the formation of the larger maria, the circular basins apparently produced very early in the lunar history by the collisions of the large objects that were common in interplanetary space. Mountain ranges such as the lunar Alps and the lunar Apennines are apparently remnants of the high rims of these ancient craters. Much of the rim has since been worn away by further cratering, but the mountains remain; in some cases, their heights rival their terrestrial counterparts. Among the highest mountains on the moon are some near the south pole, where altitudes have been estimated to exceed 10,000 meters above the bases.

Lunar Volcanoes   Another kind of mountain recognized for certain only during the last 20 years is the lunar volcano. These volcanoes are small—very small by comparison with giant Martian volcanoes, but on the order of the size of terrestrial cinder cones—and fairly inconspicuous by comparison with the largest craters produced by impact. They have fairly gentle slopes and often show summit craters like typical terrestrial cases. Around some of them are lava tubes and lava channels produced by streams of flowing lava, some of which show solidified lava over parts of them, producing natural tunnels.

Fig. 4-10 Lunar mountains photographed by the Apollo 17 astronauts. *(NASA photograph.)*

Fig. 4-11 Mount Hadley, almost 5,000 meters (15,000 feet) above the plain, photographed on the Apollo 15 mission. *(NASA photograph.)*

Fig. 4-12 St. George's crater, formed in the side of a mountain. Antennae associated with the Apollo 15 mission are shown in the foreground. *(NASA photograph.)*

Fig. 4-13 Shröter's Valley, the Moon's largest sinuous rille. Notice the small mounds at the left, which are probably lunar volcanoes. Also in the lower left is a smaller rille, a lava channel. Volcanic peaks are visible in this view. *(NASA photograph.)*

Fig. 4-14 A close-up view of a section of Shröter's Valley, which is shown in Fig. 4-13. *(NASA photograph.)*

Fig. 4-15 The interior of a terrestrial lava tube. *(Photograph courtesy of P. B. Lucke.)*

Analysis of the moon rocks brought back to terrestrial laboratories by the Apollo and Luna programs has given a great deal of information about the age and history of the moon. By measuring the amounts of various radioactive and nonradioactive elements in the rocks, scientists can determine the ages of the various rocks accurately. Naturally radioactive materials, such as certain kinds of uranium, radium, rubidium, and potassium, gradually decay into other recognizable materials. The rate of decay is well known, so that by measuring the amount of the radioactive material and the amount of the "daughter" element, it is possible to tell how long this decay has been going on. In all cases, the amounts of material are exceedingly small, and very sensitive and complicated laboratory equipment must be used.

**4-3
AGE OF THE
MOON**

The oldest rocks found on the moon so far have ages of 4.0 billion years. These include some remarkable rocks that are unusually highly radioactive, called KREEP basalts. The name is derived from the fact that they are unusually rich in potassium (chemical symbol K), rare earth elements (REE), and phosphorus (chemical symbol P). They are recognized as basalts because of their similarity otherwise to common once-molton volcanic rocks on the earth called basalts.

The youngest rocks found on the moon are iron-rich basalts found in the maria, with ages of between 3.1 and 3.7 billion years.

**Ages of Lunar
Rocks**

Fig. 4-16 A lunar basaltic rock. This rock was collected by the Apollo 15 mission. *(NASA photograph.)*

Fig. 4-17 An example of a large-scale terrestrial basaltic lava flow. Much of eastern Oregon and Washington is covered with a series of layers of basalt. The layering can be seen in detail in this canyon, which was formed at the end of the last ice age by the swollen Columbia river.

From age measures and a wide variety of chemical and mineralogical evidence, the recent exploration of the lunar surface has resulted in a fairly complete understanding of lunar history. Many mysteries remain, but the general pattern is clear.

Lunar History

The moon apparently formed at the same time as Earth and the other planets, about 4.6 billion years ago. It most probably formed by gathering together many smaller bodies and space debris, as did the planets. During its initial 1.5 billion years, it underwent a complicated, active, and still somewhat mysterious evolution. Initially, much of the surface was apparently molten. Evidence about the interior during this phase is contradictory and it is not known whether it was hot and molten or cold and solid.

About 500 million years after its formation, the moon underwent another period of melting, when parts of the surface were molten and when the KREEP basalts were formed. Then, after a period of relative inaction, a widespread extrusion of molten rock filled most of the deep basins of the moon, forming the mare surfaces between 3.1 and 3.7 billion years ago. Since that event, little or nothing has happened that can be traced to the moon's internal thermal activity; it is likely that its internal heat sources died forever then.

All subsequent activity on the moon is due to the constant bombardment of its surface by large and small meteorites. Over such a long time, this constant cratering has continually "tilled" the lunar soil, breaking up rocks, redistributing them across and below the surface, and fusing some of them together to

Fig. 4-18 Lunar breccia and basalt from Apollo 15. Astronauts Scott (left) and Erwin (right) are examining it. *(NASA photograph.)*

Fig. 4-19 Lunar tilling of the soil.

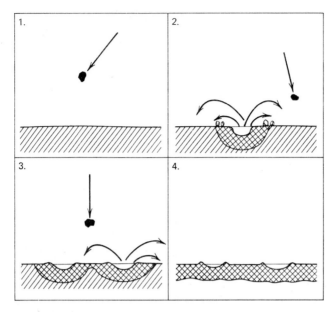

form new, complex composite rocks. The process is slow, but over the 3 billion years since lunar geology died, it has had time to alter the lunar surface effectively to depths of a meter or more.

4-4 OTHER SATELLITES

The other satellites of the solar system fall naturally into two different groups—those that appear to be captured minor planets, and those that seem to be natural satellites.

Captured Satellites

Some of the moons of the solar system are very small irregular bodies that apparently are related by origin to the asteroids (Chap. 5). It is likely that these objects in fact were asteroids that had been captured previously during an accidental close approach to the planet.

The two moons of Mars, Phobos and Deimos, may be examples of this class of satellite. Both these objects are very small; Phobos is an oval-shaped object 25 kilometers long, and Deimos has a major diameter of only 15 kilometers. Both the moons are irregular, as are many of the minor planets. Photographs of the surfaces of these two satellites were taken by Vikings 1 and 2; they show irregular and heavily cratered surfaces, with the density of craters so large that it is very likely that both satellites have surfaces so old that they have been worn away by collisions with meteorites since the solar system began. Both are much too small to have ever maintained a permanent atmosphere, because the

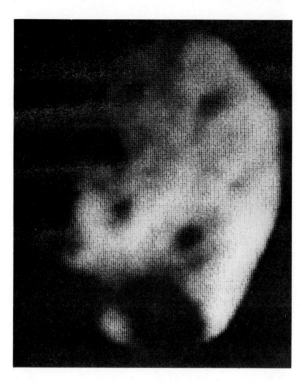

Fig. 4-20 The satellite Phobos of Mars in a Mariner 9 photograph. *(NASA.)*

masses are grossly inadequate for keeping gases from escaping out to space.

Other satellites of this same type include ten of the fourteen satellites of Jupiter, all of which are small, ranging in estimated diameter from 20 to 200 kilometers. The smallest satellites of the planet Saturn, which has a total of ten, are probably also captured satellites.

Natural Satellites

The other class of satellites in the solar system is typified by the larger moons of Jupiter and Saturn, as well as by the large satellites of the outer giant planets, Uranus and Neptune. These satellites are considerably bigger than the first type (up to 5,000 kilometers in diameter) and reach sizes approximately equivalent to those of the smallest of the inner planets, Mercury and Mars. They are most likely objects that formed at approximately the same time and during the same process as the planets around which they revolve. Titan, the largest satellite of Saturn, has an atmosphere consisting of heavy gases, including methane and ammonia. The larger satellites of Jupiter may also have thin atmospheres. Observations under exceedingly good conditions have shown markings on the larger Jovian satellites and on Titan that clearly indicate differences of brightness from one position to another on the surface. Maps of these markings

Fig. 4-21 Jupiter's four galilean satellites can be seen through small telescopes, and their motion around Jupiter can be followed from night to night. Charts such as this for each month are available in the *American Ephemeris*.

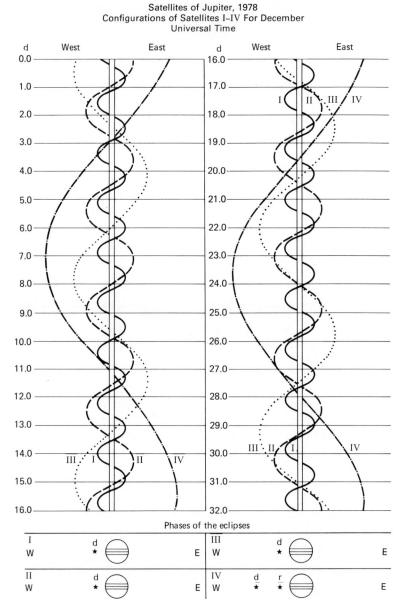

Satellites of Jupiter, 1978
Configurations of Satellites I–IV For December
Universal Time

have been drawn, and spectra of these large satellites suggest that some of the portions of the surfaces are covered with layers of frozen ices, including water ice and ices of other compounds of hydrogen.

One of the most remarkable satellites is Jupiter's second moon, Io. In 1973 it

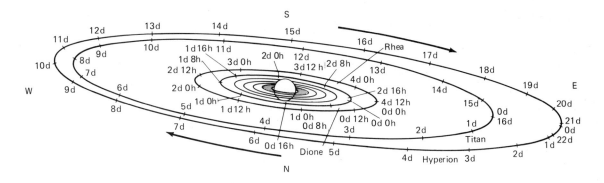

Fig. 4-22 Saturn's brighter satellites can be followed with small telescopes. Charts like this, which are published each year in the *American Ephemeris*, help to identify which is which.

was discovered by Pioneer 10 that Io has a thin atmosphere, about a billionth of the earth's atmosphere. Then in 1974 astronomers were astounded to find that it is surrounded by an envelope of yellow, glowing sodium vapor. Since sodium is commonly found in planetary materials in the form of salt (NaCl, sodium chloride), it has been concluded that Io must be covered with salt deposits that sometimes evaporate, perhaps when Io passes through Jupiter's radiation cloud, causing bright yellow flaming skies.

Further exploration by spacecraft will no doubt lead to more surprising discoveries among the Jovian satellites. They will make good, solid landing places and their surface deposits of ices, rock, and salts probably contain many unimagined clues to events in the history of our solar system.

Saturn's Rings

The remarkable system of rings that makes Saturn so unusual in appearance and so spectacular to view is made up of billions of chunks of ice or ice-covered rock, with each chunk a tiny satellite circling in an orbit close to the planet. The rings were probably formed by material that was too close to Saturn to form into a single larger satellite, as they are inside the limit of tidal breakup. Any large satellite that close to Saturn would be torn apart by the tides exerted on it by the planet, and thus it is unlikely that a close large satellite would ever have formed. Radar, infrared, and optical measures tell us that the chunks around Saturn are centimeters to meters in size, and covered with ice.

The Rings of Uranus

In 1977 the astronomical world was surprised by the discovery that Saturn is not the only ringed planet. Five narrow rings around the planet Uranus were discovered by groups of astronomers who had set out to measure Uranus' size, shape, and atmosphere by watching a faint star gradually being occulted by Uranus. To their surprise, the star was first dimmed by the otherwise invisible rings, which are too narrow and faint to see directly. The rings average only 10 kilometers in width (compared to 20,000 kilometers for Saturn's rings).

1. Make a graph showing the number of satellites for a planet versus its distance from the sun. Is there a good correlation? Plot the number of satellites versus the planet's rotation period, its mass, its density, and its temperature. Which gives the best correlation? Speculate on why correlations such as these might or might not be expected.

2. Draw a series of views of the moon showing the important events in its history as learned from the Apollo missions.

EXPERIMENTS AND OBSERVATIONS

1. Using one of the photographs in this book or an enlargement of a lunar photograph, derive a crater count for various crater sizes on the moon. A magnifying glass with a graduated reticle will be useful, or use a plain magnifying glass with a millimeter ruler. Calibrate the scale of the photograph and determine the average rates of creation of moon craters with different sizes. To calibrate the scale, use either data on the angular or on the linear extent of the photograph, or use a known feature.

2. Plot the extent of the maria of the moon as seen without a telescope, with binoculars, or with a small telescope (or use photographs in this book). Draw on graph paper so that you can count up the number of squares covered by the maria. Compare this with the surface area of the moon to find the percentage of the lunar near side that is mare surface.

3. Using a variety of media, such as sand, mud, flour, plaster of Paris, etc., carry out cratering experiments by shooting or dropping projectiles onto the

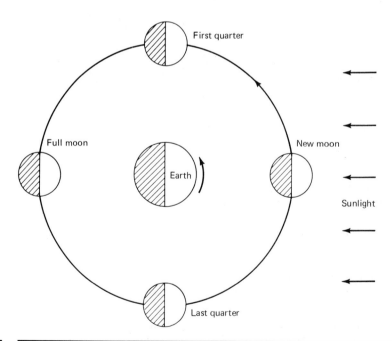

Fig. 4-23 The phases of the moon depend upon where it is in its orbit around the sun, and upon the direction toward the sun.

Fig. 4-24 This lunar chart can be used with small telescopes or binoculars to identify the more conspicuous lunar features.

surfaces. How does the crater shape (outline, depth, rim height) depend upon the nature of the surface and upon the velocity, size, shape, and angle of incidence of the projectile?

4. Using data from this book and/or the books listed in Appendix O, construct scale models of a terrestrial volcano and a large lunar crater. You can use cut-out layers of cardboard, clay, flour-salt plaster, or some other medium.

5. Using binoculars or a small telescope (20X or higher is best), observe as many moons of Jupiter and/or Saturn as you can make out. Identify them by reference to the *American Ephemeris, Sky and Telescope, Astronomy Magazine* or other source. Follow them for several hours. Can you detect any motion? Follow them for several nights and derive periods for as many as possible from your observations.

5

ASTEROIDS AND METEORITES

Meteorites and their parent bodies, the asteroids, give further evidence about the history and evolution of the solar system. In this chapter you will learn how, by bringing meteorites that fall to the earth into the laboratory, we can set the age of the solar system accurately and can deduce the cosmic abundance of many of the elements.

**5-0
GOALS**

On the night of the first day of the nineteenth century, January 1, 1801, the Italian astronomer Piazzi discovered a remarkable object in the sky. He had been mapping faint stars by means of optical observations through a telescope, and on this night he found in the star field that he had been studying a star that had not been there before. Excited by the possibility of a new discovery, he carefully measured its position with respect to the other stars and looked for it again the next night. He found that it had moved, and on succeeding nights he traced its motion as it moved slowly among the stars. Piazzi announced his discovery as that of a new planet, because it moved in a smooth and regular way like a planet. He named it Ceres, because he was Sicilian and Ceres was the Greek goddess assigned to Sicily.

**5-1
DISCOVERY**

Before it had been observed very extensively, the planet Ceres was lost because of its motion westward, which took it too close to the sun to be observed. Piazzi had observed it for several weeks before becoming ill for a period. When he recovered, the planet was gone, and he could not find it again.

The scientific world was elated by the discovery of Ceres and dismayed at the planet's eventual loss and lack of recovery. A young German scientist named Gauss, who was interested in mathematics and physics as well as astronomy, seized upon this problem as a challenge. He set to work attempting to determine the orbit of the planet based on the few observations of its position measured by Piazzi before it disappeared. Gauss was only 23 years old at the time, but was a brilliant and creative young man who in a short period was able

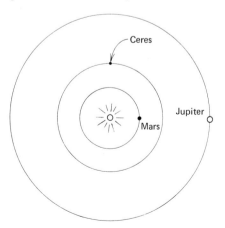

Fig. 5-1 The orbit of Ceres.

to invent a method for determining a planetary orbit from only three observations. When he applied his method to the observations of Ceres, he discovered that its orbit lies between the orbits of Mars and Jupiter. He was able to predict its position with enough accuracy so that it was rediscovered near the end of 1801, almost exactly where Gauss predicted it would be.

In addition to the remarkable mathematical feat of Gauss, a further feature of the orbit of Ceres was important—its location among the planets. For centuries astronomers had noted an unusually wide gap in the distribution of the planets between the orbits of Mars and Jupiter. Kepler had noticed this and had tried unsuccessfully to account for it. Succeeding astronomers had attempted to explain what appeared to be the absence of a planet, since otherwise planets had their distances distributed in a very regular way described by a mathematical formula called Bode's law (Chap. 2). Ceres was found to have an orbit almost exactly where Bode's law predicted, and therefore it was considered for a time to be the long lost "missing" planet.

However, its unique fame was short lived. In 1802 the astronomer Olbers, while searching for Ceres, found another planet almost identical in brightness. He named it Pallas, and its discovery was soon followed by Juno (1804) and Vesta (1807). All four of these objects were found to have rather similar orbits lying between Mars and Jupiter, and all four were very faint and small, virtually never visible to the naked eye. Because the objects were much smaller than the previously discovered planets, they were called *minor planets*. Another common term used for them shortly after their discovery was that invented by Sir William Herschel, who suggested the name *asteroid* (meaning starlike), since though definitely planets, through the telescope they look like stars because of their small size.

In conformity with the practice of classical civilization, the planets have traditionally been given the names of Greek and Roman gods and goddesses. Unlike the comets, which are named after their discoverers, the minor planets are named *by* their discoverers. Since too many of these minor planets are known to always use their names conveniently, asteroids have been numbered in the order in which their orbits have been determined. Many more asteroids have been discovered than numbered, because it has sometimes not been possible to determine the orbits accurately enough to give them an official number.

Traditionally the names of the minor planets are always feminine, except for those with peculiar orbits. For example, two groups of asteroids that follow Jupiter in its orbit (each 60° away from the major planet) in front of and behind it (as seen from the sun) are called Trojans, and these are given masculine names. Also, those planets that have orbits bringing them much closer to the sun than the average are given masculine names—for example Icarus, whose orbit brings it even closer to the sun than the planet Mercury.

Because there are now thousands of asteroid discoveries, astronomers long ago exhausted the list of Greek and Roman deities. The Heidelberg astronomer

Reinmuth, for example, discovered and named more than a thousand minor planets. Astronomers have had to look to other sources for names and a look at the list of asteroid names discloses that people have used the names of colleges, cities, wives, ocean steamers, and even pets, each of which now has a permanent place in the heavens like those reserved for the gods in classical times.

By the middle of the twentieth century, more than 1,500 of the minor planets discovered had had their orbits sufficiently well determined so that they could be kept track of permanently. These have been given official numbers. Estimates based on several plates taken by the world's largest telescopes have shown, however, that the total number of asteroids that we are presently capable of discovering is more nearly 100,000. The faintest would be among the faintest objects presently visible and would be hardly bigger than large meteorites.

Most of the orbits of the asteroids lie between Mars and Jupiter. The orbits are mostly nearly circular, like those of the major planets, with periods of between 4 and 7 years. Almost all orbits are nearly in the plane of the ecliptic, like the planets.

5-2 ORBITS

Although the majority of the asteroids behave very much like the larger planets in terms of their orbital properties, there are a few with unusual orbits. The Trojan asteroids (mentioned above) have periods of nearly 12 years, like that of Jupiter. A few asteroids that come close to the Earth and to the sun, including such objects as Hermes, Eros, and Icarus, also have unusual orbits. An example of a minor planet with a long period and large mean distance is Hidalgo, number 944, with a period of 14 years and a semimajor axis of 5.8 astronomical units. Hidalgo is also unusual in having a large inclination angle, with an orbit that lies 42° from the plane of the ecliptic. This suggests that Hidalgo has possibly suffered a close gravitational encounter with one of the

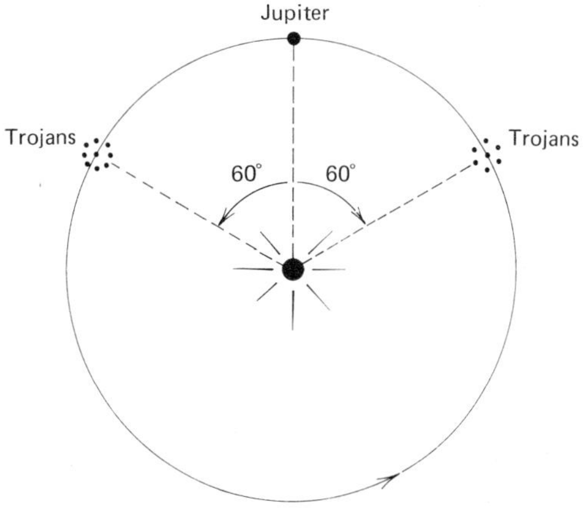

Fig. 5-2 The Trojan asteroids.

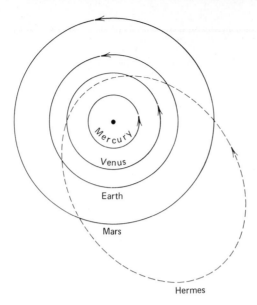

Fig. 5-3 The orbit of Hermes.

planets (probably Jupiter), and has had its orbit changed drastically from the typical orbit of an asteroid. Another unusual asteroid, Chiron, discovered in 1977, has an even larger orbit out between Saturn and Uranus.

Jupiter's Influence: Kirkwood's Gaps

One of the more interesting features of the orbits of the minor planets is the strong influence that the planet Jupiter has on them. When the motions of the asteroids and the characteristics of their orbits are examined, it is found that certain periods seem to be avoided. Many asteroids have periods either slightly bigger or slightly smaller, but virtually none have these periods. This effect produces gaps in the asteroid belt, called *Kirkwood's gaps* after their discover-

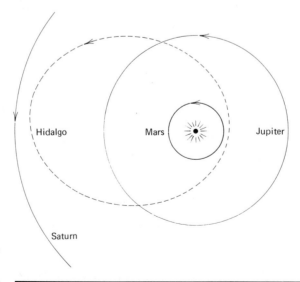

Fig. 5-4 The orbit of Hidalgo.

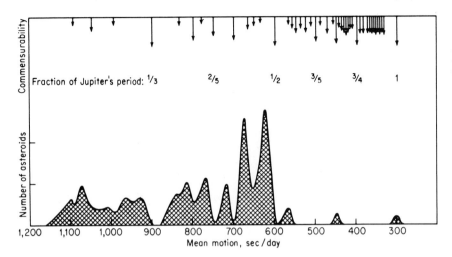

Fig. 5-5 The Kirkwood gaps. Notice that the number of asteroids is very small near orbits that are in simple rhythm with Jupiter's motion.

er. It is now known that these gaps are caused by the planet Jupiter. Mathematical calculations of gravitational perturbations on a small body when a large planet is nearby show that these perturbations are most severe for objects with periods that are a simple fraction of the large planet's period. The perturbations are sufficiently strong to throw the small body out of its orbit and into another orbit in a relatively short time. Therefore, when it was discovered that the Kirkwood gaps occurred for periods that were a simple fraction of Jupiter's period, namely $1/2$, $2/5$, $1/3$, and so forth, it was understood immediately that these gaps are due to perturbations on any asteroid that might have had such a period—causing it to move into a different orbit with a different period.

Earth-Grazing Orbits

Some of the asteroids with anomalous periods either have small enough orbits or large enough eccentricities in their orbit so that they can come quite close to the earth. An example is the asteroid Eros, which has come within 23 million kilometers of the earth. This is sufficiently close for astronomers to be able to make out its irregular shape through the telescope. Another example is Hermes,

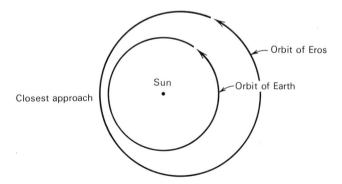

Fig. 5-6 The orbit of Eros.

which came even closer, within 800,000 kilometers. Asteroids have very likely come much closer than that over past history, and large meteorite craters on the earth are evidence that some of these actually collide with the earth.

5-3
SIZES

With the very largest telescopes, it is possible to determine the sizes of the four largest asteroids and of the smaller ones that come very close to the earth, such as Eros. The largest asteroid is Ceres (with a diameter of 950 kilometers); but most of the asteroids are smaller than 200 kilometers, with the overwhelming majority only a kilometer or so across. For the small asteroids, the only available method for estimating sizes is by means of measuring their brightness and calculating what their sizes would be on the basis of an assumed *albedo* (the measure of the fraction of sunlight reflected back from the asteroid to the earth). The faintest asteroids that can be photographed with the largest telescopes are estimated on this basis to have diameters of less than a kilometer. If one of these faint asteroids should collide with the earth, it would create a crater and an explosion that would potentially demolish a portion of the earth about the size of a large city.

The total amount of material in all the minor planets together, calculated by adding up estimated masses for all the asteroids that are calculated to exist in the solar system, turns out to be only about 1/1,000 the mass of the earth. Therefore, although it was once proposed that the asteroids resulted from the breakup of a large planet that once existed between Mars and Jupiter, the debris from such a hypothetical planet would include much more material than exists. It is much more likely, as commonly believed now, that the asteroids originated as many small objects (not necessarily very much bigger than Ceres). In the years since the formation of the solar system, they have occasionally collided with each other, breaking up into smaller and smaller pieces.

5-4
SURFACES

Because of their exceedingly small size, surface features on asteroids have never been seen with any degree of certainty. Even with the largest telescopes and under the best of conditions, it is normally just barely possible to make out that they are not stellar and that they have a definite disk. However, it is possible to estimate the albedo for those with known sizes, and therefore to determine at least some crude information about the nature of their surface material. For Ceres and Pallas, the two largest of the asteroids, the albedo is almost exactly like that of the moon and Mercury, indicating a very dark surface which reflects about 6 percent of the light. This is about the equivalent of the amount of light reflected from a nearly black, rough piece of paper. Therefore, the surfaces are probably dusty and black, like the surface of the moon.

For Juno and Vesta, however, the albedos appear to be larger, indicating that these two asteroids have a surface material that is brighter than the moon's. Possibly, in fact, they have a thin surface layer of icy material (perhaps frozen carbon dioxide), or a light fine dust.

Recently the infrared spectra (Appendix D) of asteroids have allowed us to find out more about what materials the asteroids are made of. When we compare the infrared light of asteroids with sunlight reflected from various kinds of meteorites, we find that about 80 percent of the asteroids match up with certain special types of meteorite called *carbonaceous chondrites* (Sec. 5-7). These are the most primitive, primordial meteorites, the ones that have remained nearly unchanged since first condensing out of the cloud that formed the sun and planets. The remaining 20 percent, mostly large asteroids of about 200 kilometers in diameter, have spectra that are like stony and iron meteorites.

These differences tell a great deal about the origin of the asteroids, and by analogy, something about the formation of Earth and the other planets. Those with the composition of carbonaceous chondrites cannot ever have been part of a large planet, because then their composition would have shown the results of change; the materials of low melting point would have been melted or even evaporated away, and the heavier elements would have been segregated from the lighter elements by sinking under gravity to the center of such a planet. But the 20 percent of asteroids with stony and iron composition must have once been part of such a planet. Studies of meteorites indicate that most likely there were several small planets originally formed in the asteroid belt, plus large numbers of small fragments that never accumulated together into a planet. The larger ones gradually built up sufficient mass so that their heavier materials separated out, with the iron sinking to the center. Then, because there were so many objects in such a narrow range of position, they began to collide with each other. Some were completely destroyed and others had their outer layers gradually chipped away, laying bare their iron-rich cores to our view.

This picture of how the asteroids came to be seems to fit all the facts and agrees well with contemporary ideas of how the planets formed and gradually acquired their present compositions. However, it will remain for the future to test these ideas, probably most convincingly when a spacecraft makes a rendezvous with an asteroid.

**5-5
ASTEROID
SHAPES**

Many minor planets vary in brightness in a regular way, with rotation periods of a few hours. This is understood if the asteroid either has one side brighter than the other, or if it is irregular in shape. In either case, rotation would cause its brightness as seen from earth to appear to change in a way characterized by the period of the rotation. Irregularity of shape is the most likely of these explanations, as seems to have been demonstrated in 1931 when the asteroid Eros made one of its close approaches to the earth (within 23 million kilometers). When observed through a large telescope, the image was definitely elongated, similar to the appearance of an unresolved double star. The astronomer van den Bos, for example, observed Eros through the large refracting telescope of the Lamont Hussey Observatory, which was located in a game reserve in South Africa. He found it to be spinning counterclockwise with a period of 5 hours and 17 minutes. Its estimated size, based on such measurements, is approximately 25 kilometers long and about 10 kilometers wide; it is shaped something like a slab of rock.

**Varying
Brightness**

From photometric measurements, Eros had already been found to vary in brightness with a period of 5 hours and 17 minutes. This was understood to be because of the fact that the minor planet alternatively showed us a larger and then a smaller total area as it rotated, reflecting first more and then less sunlight in our direction.

Other asteroids have similarly been found to vary in brightness in a regular way, with periods for the shortest on the order of 2 hours and the longest on the order of 10 hours. Sometimes the variation for a given asteroid is very marked, and at other times there is less change in brightness, but there is always the same period. The reason for these changes is the changing angle between us and the asteroid's rotation axis. When an asteroid rotates with its axis of rotation nearly pointing in our direction, very little variation in brightness will be seen; but when such an irregular asteroid rotates with its axis of rotation

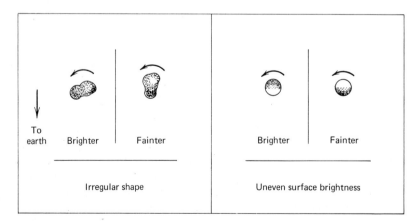

Fig. 5-8 Two ways to explain the variations in brightness of asteroids.

pointing at right angles to our line of vision, the amount of variability in brightness will be at a maximum. For Eros, for example, the variability is sometimes too small to be detected; at other times the brightness changes almost by a factor of 3 over its period.

Reasons for the Irregularity

Why are the asteroids so often irregular in shape while the major planets are so nearly spherical? This question can be answered easily in terms of the physical makeup of a body of rock. The earth's almost perfect spherical shape is due to the fact that the rocks of the earth are much less strong than the gravitational force that the earth's total mass exerts on its contents. Because gravity attracts bodies in a spherical way, and because the total gravitational force is proportional to the mass, the very massive earth pulls all its material in toward its center with a pull of gravity very much greater than the strength of the rocks. Any rocks that might have stuck out above the average surface at some early period in the earth's history have been pulled in by this strong gravitational force to make earth almost perfectly spherical.

On the other hand, if a planetary body is sufficiently small so that gravitational force on its materials due to its mass is smaller than the strength of its rocks, these rocks can withstand the gravitational force and can maintain whatever irregular distribution they might have. Calculations show that for typical materials of rocklike character, any object with a radius greater than about 150 kilometers is too massive to remain irregular; the gravitational force is greater than the strength of the rock. On the other hand, minor planets with diameters less than this amount can remain irregular in shape without being pulled into a sphere. Interestingly in this connection, the four largest minor planets (all of which are larger than 150 kilometers) have nearly constant brightness, while those asteroids with large variations in brightness are among the fainter, and therefore probably smaller, ones.

5-6 METEORITES

Several years ago a group of mountain climbers on the snowy slopes of Mt. St. Helens in the northwestern United States were startled by the appearance of an intensely bright light in the early dawn sky. They stopped and watched a luminous object as it moved in from the ocean toward them, crossing over the city of Portland, Oregon, and finally disappearing after a series of bright explosive outbursts. A day later a berry picker in Washougal, Washington, northeast of Portland, picked up a fresh meteorite. It was a small stonelike object, and it had dug a shallow crater in the ground where it had fallen.

This was a fairly typical example of a meteorite *fall*, an instance when at least one and usually many persons witness the brilliant passage of a meteorite body through the earth's atmosphere to its surface. The flash of light is called a *meteor*, and the object that reaches the ground is called a *meteorite*. If its descent is witnessed and reported, it is called a meteorite fall, but if a meteoritic

Fig. 5-9 A bolide, as bright as those that can produce a meteorite to land at the surface of the earth. This one happened to pass near to the Andromeda galaxy. *(Photograph courtesy J. Klepesta.)*

object is merely found on the ground without its passage through the atmosphere having been seen, it is called a meteorite *find*. There are many more meteorite finds than falls, because it is often possible to find a meteorite in or around the ground (even if it fell thousands of years ago), providing one knows what a meteorite looks like.

Meteorite Showers

Although meteorites most commonly fall as single objects, it is also possible to have many fragments fall, causing what is sometimes called a *meteorite shower*. Such a circumstance usually occurs when a large body in space is shattered and broken up during its passage through the atmosphere, which is a very energetic event because of the high velocity of a typical meteoric body in space. The velocity is usually on the order of at least 20 kilometers per second, about 40,000 miles per hour. At this speed, collision, even with such a thing as the tenuous gas in the earth's upper atmosphere, can be an explosive event that can melt and shatter the incoming body. For that reason, an object must be fairly large, about the size of an apple, for it to pass all the way through the atmosphere to the ground before being completely destroyed.

Examples of meteorite showers are fairly common. One of the most spectacular was the shower that appeared near Holbrook, Arizona in 1912. A railroadman was having dinner in a Santa Fe section house when suddenly it sounded as if hail were falling on the roof. Because the weather had been clear and sunny all day, he immediately went outside to see what peculiar change of

Fig. 5-10 The Lost City Meteorite, recovered as a result of a survey of the sky carried out by the Smithsonian Astrophysical Observatory's Prairie Network. *(Smithsonian Astrophysical Observatory.)*

weather could be occurring, and found himself surrounded by a rain of stones, which were later identified as meteorite fragments. More than 10,000 small meteorites were eventually picked up from the area.

Although astronomers have identified more than a dozen different chemical families of meteorites and have given them various mineralogical names, there are only two principal kinds of meteorites, commonly called *stones* and *irons.*

**5-7
KINDS OF
METEORITES**

Fig. 5-11 The Henbury meteorite craters of Australia. Fourteen separate craters were formed here.

Fig. 5-12 A stony meteorite. This one is one of the Allende carbonaceous chondrites which fell in Mexico.

Stones Stony meteorites are the most difficult for a nonscientist to recognize, and for that reason there are very few stony meteorites among the finds of the world. They look like ordinary rocks, with their most distinguishing features being a melted crust that is produced by the intense heating during their passage through the atmosphere at high velocity. Many have small spherical objects called *chondrules* imbedded in them. A great deal of chemical and mineralogical study of chondrules have convinced scientists that these are droplets formed during an early stage in the formation of the solar system, and it is

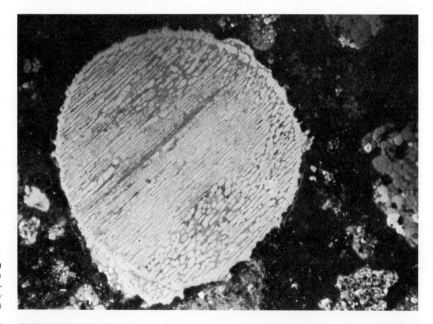

Fig. 5-13 A chondrule, a small spherical formation in a stony meteorite. *(Microphotograph courtesy D. Brownlee.)*

realized that the stony meteorites contain a large number of clues about conditions during that early period.

Iron meteorites are all of nearly the same composition, about 90 percent iron with the rest mostly nickel, a little cobalt, and smaller amounts of other metals. Because of their composition, iron meteorites are much easier to find, and about 90 percent of all meteorites that have been found on the ground without having been seen to fall are irons. Usually they show a characteristic pattern of erosion by melting that gives them an easily recognizable appearance, as shown in Fig. 5-14. Iron meteorites have a composition that is rather similar to the mix of metals that makes up stainless steel, which explains why they last for centuries without rusting. Many of the essential parts of tools for ancient peoples and primitive civilizations were made from meteorites, which were found to be both strong and impervious to rust. It is unlikely, of course, that any of these ancient peoples realized that their ax blades and hammer heads had been fashioned out of material from outer space.

Irons

The meteorites are important sources of information about the history of the solar system and the planets. Their importance comes from the fact that they are the most readily available and inexpensive samples of materials from space, and from the fact that they contain evidence about the conditions during the early history of the solar system. They show that the conditions during the formation of the solar system were very complicated, and involved a mix of atoms, molecules, dust, and gas, all heated by the early sun with its rapidly changing properties.

**5-8
HISTORY OF
THE
METEORITES**

Fig. 5-14 An iron meteorite, one of several collected from the neighborhood of the Arizona Meteorite Crater.

Fig. 5-15 Widmanstatten figures in a section of an iron meteorite. These crystal-like shapes tell something about the temperatures and pressures that existed when the meteorite formed. *(Photograph courtesy D. Brownlee.)*

Some Deductions   The details of the mineralogy and chemistry of meteorites tell enough about the history of these objects to fill many books. The following are some clues deduced about their origin and nature:

• Some meteorites show only metals (the irons), and other meteorites show only rocklike materials (called *silicates* because of the abundance of silicon). Since the only known way by which the heavy elements—the metals—can be separated from the silicates is by settling to the bottom of the hot interior of a planet, this differentiation suggests that at least some of the meteorites were once part of a moderately large body, such as a planet. However, the total deduced number of meteorites in the solar system is not sufficiently large to make up a planet-sized object, and therefore it is most likely that the differentiation occurred in a smaller object or objects, such as asteroids.
• In some meteorites, the metals and the silicates are mixed together smoothly, and these materials were apparently never in a planetary body.
• Droplet shapes with a very fine grain structure have been found. These show little difference in the location of kinds of materials that vaporize when heated (called *volatiles*) and *nonvolatiles*, suggesting that these meteorites were once

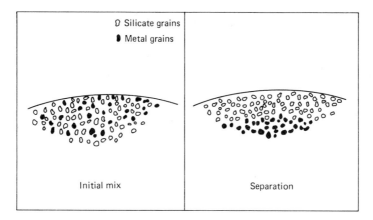

Fig. 5-16 How the heavy elements, particularly the metals, might have separated from the lighter elements, the silicates.

Silicate grains

Metal grains

Initial mix

Separation

heated rapidly (to produce the melted droplet shapes) and then cooled rapidly so that the materials that were volatile did not have time to escape. This rapid heating and cooling process seems characteristic of many of the meteorites.

• The breaking up of many minerals, which geologists call *brecciation* for earth rocks, indicates that meteorites were once exposed to violent shocks, as this broken character normally cannot be produced by other means.

• Some meteorites contain hydrocarbons, which cannot be found unless the temperature is less than 300° in a vacuum or less than 500° in the presence of 1 atmosphere of pressure. They have planetary abundances of hydrogen, rather than abundances like the interplanetary gas or the sun, and are believed to have been formed near the surface of a planetary body with an atmosphere made up of the gases found in the atmospheres of the outer planets (methane, ammonia, and water).

• Diamonds are found in meteorites, and they can be formed only under exceedingly great pressures. For example, if the temperature is approximately that of the melting point of iron, the pressure needed to produce a diamond is

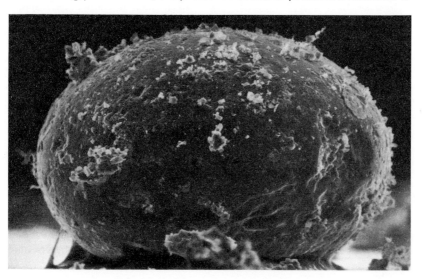

Fig. 5-17 This microscopic-sized sphere of glass was found in a stony meteorite, and was formed before the meteorite formed, 4 billion years ago. Tiny craters have been found in it. These craters were produced by impacts of the interplanetary dust of that time. *(Photograph courtesy D. Brownlee.)*

on the order of 40,000 times the atmospheric pressure of the earth. This implies that either the diamonds were formed near the center of a body that was at least a thousand kilometers in diameter (where such pressures could exist), or else the diamonds were formed by a shock process such as might occur when two interplanetary bodies collide. We have clear evidence that diamonds can be formed in meteorites by such collisions, because the Canyon Diablo meteorite fragments found near the Arizona meteorite crater show diamonds which were formed at the time of the collision of the meteorite with the earth, when the crater was formed.

From such considerations as these, astronomers have come to realize that meteorites show a very complicated but well-documented history. The original mix of materials, the primordial matter that eventually made up the planets and the sun, is represented by the *carbonaceous chondrites*. These peculiar and rather rare stony meteorites contain large amounts of carbon and hydrogen, which were among the most abundant elements in the presolar material. From evidence such as that above, it is clear that some of the original bodies made up of this material were heated long ago and that this caused a loss of the volatile materials (the carbon and hydrogen compounds). This heating could have occurred because of the accumulation of large amounts of matter to form a small planet, which would have a hot center because of natural radioactivity. The heating might also have occured as a result of collisions with other smaller bodies. For cases where the heating was quite extensive and the temperatures and pressures very high, the lighter materials (the rocklike silicates) would be lost, and all that would remain would be the heavy elements, primarily iron. Evidence about the ages and distribution of different kinds of meteorites suggests that this formation into small planets and their eventual demolition occurred several billion years ago.

Meteorite Orbits   Several meteorites have been photographed as they passed through the sky, and their paths have been determined with enough accuracy so that it has been possible to establish their orbits. It is found that they have orbits that are similar to the orbits of those asteroids which come close to the earth. This is evidence that the meteorites have their origin in the minor planet belt, and that they have been formed from the many collisions that, by our calculations, take place between the asteroids there. Because collisions between asteroids can cause the fragments to have orbits quite different from those of the colliding bodies, some meteorites have orbits sufficiently small to bring them close enough to collide in turn with the earth.

Meteorite Ages   The ages of meteorites can be measured in the same way that the ages of the lunar rocks are measured. Most meteorites are found to have ages of about 4½

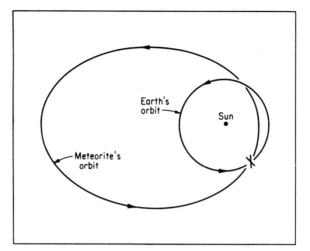

Fig. 5-18 A meteorite orbit, determined from photographs taken of the Pribam meteorite in Czechoslovakia.

billion years, which is now considered the best estimate of the solar system's age. By comparison, the ages of the oldest rocks on the moon are about 4 billion years, and recent discoveries on the earth have shown that rocks there also go back almost that far.

When very large meteorites strike the earth, they form distinctive craters rather similar in shape to the lunar craters. Recognized meteorite craters on the earth range in size from 6-foot holes in the ground, produced by football-sized meteorites, to immense circular formations about 100 kilometers across. Their shape and the mechanics of formation are similar to the lunar case.

**5-9
METEORITE
CRATERS**

One of the best-known meteorite craters is the Arizona meteorite crater, near Canyon Diablo in Arizona. About 1,300 meters (4,000 feet) across and 100 meters deep, this crater was formed by a meteorite estimated to have had a mass of over 10,000 tons. Fragments of iron meteorite have been found scattered over the ground surrounding the crater, and small meteoritic particles exist in the desert soil to distances of 10 kilometers out from the crater.

Even larger craters exist, in Australia for example, where the Wolf Creek Crater is perhaps the largest recently formed crater on the earth's surface.

Meteorite craters are recognized and distinguished from other kinds of craters by the presence of meteoritic material, by the existence of a "basement" of shattered rocks, by the presence of uptilted layers in the crater rim, by the occasional presence of *shatter cones* (rock formations formed by the shock wave produced by the cratering collision), and by the existence of certain minerals found only where extremely explosive events take place.

These criteria have demonstrated the existence of many very old meteorite

Fig. 5-19 Henbury Meteorite Crater number 4, photographed from the floor of the crater. This Australian meteorite crater is 47 meters in diameter.

crater formations around the world; for instance, the Sierra Madera formation in West Texas, the Vredefort formation in South Africa, and many circular lakes and depressions in Northern Canada.

Tektites, small and not yet entirely understood fragments of glass, are

Fig. 5-20 The Wolf Creek meteorite crater in northwest Australia. (Photograph courtesy Australian National Mapping Service.)

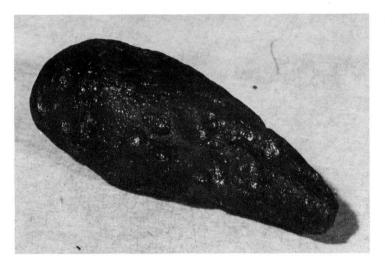

Fig. 5-21 A tektite from the Philippine Islands.

probably the result of large crater-producing meteorite impacts. Tektites have rather strange chemical compositions and a variety of peculiar shapes. They are found only in certain locations on earth. For some of these sites a known meteorite crater can be identified, but not for others. Since large meteorite craters are formed only rarely (every 50,000 years or so), we have little opportunity to witness the process, whatever it might be, that can lead to the production of these rare and odd-shaped pieces of glass.

QUESTIONS

1. What other solar system objects are so small that they might be expected to vary in brightness for the same reasons that asteroids do? What is this reason? Do they?
2. A faint asteroid is discovered to be about 10,000 times fainter than Ceres. Assuming that it has the same albedo as Ceres, what is its approximate diameter?
3. What is the period of Hungaria (No. 434) if its semimajor axis is 1.94 astronomical units? Is its orbit unusual?
4. If the asteroid Ceres has the same density as a stony meteorite (about 3.5 grams per cubic centimeter), what is its total mass?
5.* Hidalgo's orbit brings it as close to the sun as 2.0 astronomical units. What is its greatest distance from the sun? What planets is it "near" at its least and greatest distances from the sun?
6.* About two meteorites per day fall over the whole earth. Calculate the likelihood that any one person will be hit by a falling meteorite during his or her lifetime.
7. Could the meteorites be the remains of a large planet like Jupiter that was somehow destroyed?

8. What characteristics of meteorites indicate that they are associated with the asteroids?

OBSERVATIONS AND EXPERIMENTS

1. Using the *American Ephemeris*, *Sky and Telescope*, the current *Observer's Handbook* (of the Royal Astronomical Society of Canada), or a similar source, plot the path of one of the bright asteroids for its next close approach to the earth. Using binoculars (for Ceres) or a telescope (for the others), find the asteroid and compare its night-to-night motion with your plot.

2. Using variously shaped rocks held against a black background, determine their light curves of rotation, either by comparison with an array of "standard" nearly spherical rocks (look at them through the wrong end of a telescope to make them all small), or by using a photographic exposure meter. How irregular must they be before you can detect the light variations? How irregular must they be to produce variations of a factor of 2 or 3?

3. Visit a museum, if one is nearby, and note the ratio of iron to stony meteorites in the display (if any). Can you explain this ratio?

4. Small meteorites, meteorite fragments, and tektites can be purchased from several sources, such as the American Meteorite Laboratory in Denver, Colorado. Obtain one and list the differences that you can see and measure in various ways between the meteorite and the native rock in your area. Would you recognize a meteorite in the ground if you happened upon one?

5. From the museum samples you see, the meteorite you bought, or from photographs in books, can you deduce the direction of motion of a meteorite as it passes through the atmosphere? Meteorites heat up and *ablate* (lose molten material) from their surfaces during those few seconds of passage through the air to the ground.

COMETS AND METEORS

6-0
GOALS

Comets apparently formed at great distances from the sun, in the outer solar system. Most meteors are fragments of cometary material that has broken lose and chanced to collide with the earth. Thus these two kinds of objects can tell us about conditions in the early solar system far from earth; otherwise, few clues exist.

6-1
COMETS

There is a good chance that as you read these lines, there is a comet (probably a newly discovered one) in the night sky available for your observation with a small telescope. There is seldom any period of time when there is not at least one such comet in the sky, and every year or two, one is usually close enough to the earth to be seen with the naked eye. Whether through a telescope or with the unaided eye, a bright comet is a memorable object to observe. Bright comets have played a conspicuous part throughout the recorded history of mankind, and their unusual and awesome appearance led many prescientific people to the conclusion that they were omens heralding some important aspect of heavenly policy. Their unexpected appearance in the sky, their relatively fast motion (moving across the background of stars in a period of a month or so), and their remarkably long tails all indicated a considerable divergence from the otherwise orderly behavior of the heavenly bodies.

More than a thousand comets have been discovered so far; some 500 of

Fig. 6-1. Comet Arend-Roland, with its anomalous "forward spike," which was produced by dust in its orbital plane, seen edge-on. *(Lick Observatory.)*

Fig. 6-2. Comet Ikeya. *(Smithsonian Astrophysical Observatory.)*

these have been sufficiently bright to allow their discovery before the use of telescopes. Most comets that have been found in recent years are too faint to be seen well without a telescope. At present, the rate of discovery of comets is approximately eight per year. Many of these are discovered by amateur astronomers who use special comet telescopes with wide fields of view, and who spend immense amounts of time scanning the sky for new discoveries.

Comet Names

Comets are usually named after their discoverers. For example, one of the comets visible at the time these pages are being written is Comet Sandage, which was named after an American astronomer who happened upon it while observing galaxies. For comets that are very bright, that return periodically, and that were discovered before recorded history, the given name is that of the astronomer who first identified it and calculated its orbit. An example of this is Halley's Comet—one of the brightest of the comets seen periodically from the earth, which had its orbit first determined by the Astronomer Royal, Sir Edmund Halley, in the seventeenth century. Halley was able to show that the orbit that he calculated from its appearance in 1682 was almost exactly the same orbit as those calculated for two previous comets—a bright comet in 1607 and one seen in 1531—both of which were observed carefully by astronomers of those times. Checking the records further, Halley found that there had also been a bright comet in 1456, in 1301, in 1145, and in 1066 (a year famous for another event in English history as well). The nearly equal intervals between the appearances of all of these recorded bright comets led Halley to the conclusion

Fig. 6-3. The orbit of Halley's Comet. It was at aphelion in 1948. Perihelion is in 1986.

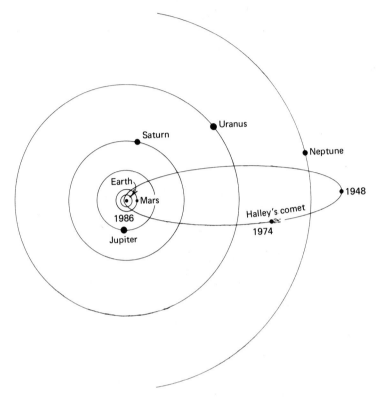

Fig. 6-3. The orbit of Halley's Comet. It was at aphelion in 1948. Perihelion is in 1986.

that they were all one and the same object, a comet with a period of about 75 years. The slight discrepancy in the periods, Halley reasoned, could easily be explained by perturbations of the comet's orbit by the planets Jupiter and Saturn. On the basis of these calculations, Halley predicted that the comet would return in early 1759. It did make a spectacular appearance in that year, at one point displaying a tail more than 1 million kilometers long.

6-2 COMETARY ORBITS

Most of the comets with known orbits are found to move in nearly parabolic paths. A *parabola* is like an ellipse, except that it is open at one end. If the cometary orbits are truly parabolic, then comets fall into the solar system, pass close to the sun, and then go out again never to return. However, as Fig. 6-4 shows, a very large elliptical orbit is virtually identical, as seen from the earth, to a parabolic orbit. Therefore, it is very possible that all of the apparent parabolic orbits for comets are actually elliptical orbits of exceedingly large size. A few orbits are found to be hyperbolic, but these apparently are all instances of comets that have approached sufficiently close to one of the major planets (usually Jupiter) to cause their speed to be increased anomalously, thus

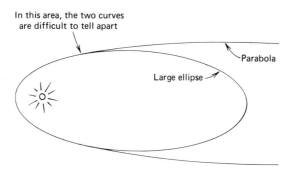

In this area, the two curves are difficult to tell apart

Parabola

Large ellipse

Fig. 6-4. Parabolas and ellipses of large size are difficult to distinguish from each other.

throwing them into a hyperbolic orbit from what was previously perhaps a very large elliptical orbit.

Sizes

For most of the comets observed from the earth, the perihelion distance is about 1 astronomical unit. A few comets come much closer to the sun at perihelion, and some of these are referred to as *the sun-grazing comets*. They are very interesting objects, because their very close approach to the sun causes a noticeable destruction of their outer layers and allows astronomers to determine their chemical makeup more completely.

Shapes

Although most comets have orbits that are very large, a few do have fairly small elliptical orbits; two comets even have orbits that are nearly circular, somewhat like planetary orbits. It is found from the statistics of cometary orbits that it is likely that those comets with small orbits, including the circular ones, were thrown into such orbits as a result of encounters with Jupiter or Saturn. As shown later in this chapter, a comet that has a perihelion as close to the sun as 1 astronomical unit cannot last very long, because of solar heating. Therefore, the events that put such comets into relatively small orbits must have occurred fairly recently (within the last few thousand or million years).

Direction

Only about half the comets travel around the sun in the same direction as the planets. One of the remarkable facts about the other bodies of the solar system is their commonality in direction of motion; they all travel in what are called *direct orbits*. Only half the comets share this direction of motion; the other half move in the opposite direction, thus having *retrograde* orbits. Halley's comet, for example, moves in a retrograde direction, clockwise as viewed from an outpost far above the north pole of the sun. This is an important fact with regard to the origin of these objects and their relationship to the origin of the solar system.

Inclinations   The orbits are found to be almost random in their angles of inclination, whereas all the planets and asteroids have orbits strongly concentrated to a single plane, the ecliptic. Comets have orbits that appear in any plane with almost equal probability.

Thus the two most important orderings of motions for the bodies in the solar system are not shared by the comets. It is concluded that the conditions during which the planetary bodies were formed are not common to the conditions during which the comets were formed, and the implications of this fact for our understanding of the origin of the comets are discussed in Sect. 6-7.

Jupiter's Family   There is a large group of comets that has orbits that are all somewhat similar,
of Comets   and these are now known to have been captured by the planet Jupiter. They are called Jupiter's family of comets. They were "captured" by the influence of Jupiter, which caused their orbits to have short periods with aphelia close to the orbit of Jupiter. It has been calculated that whenever a comet comes by chance within about 300,000 kilometers of Jupiter, it can be thrown into a new orbit with an aphelion of about 5 astronomical units and with a period of between 5 and 7 years. About 50 such comets are now known, and almost all have direct orbits.

6-3   Comet structure consists of three important features: the nucleus, the coma,
STRUCTURE   and the tail.
OF COMETS

The Nucleus   The most basic physical feature is the nucleus of the comet, which is usually exceedingly small and often very bright. Measurements show that nuclei have diameters on the order of a few kilometers. The nucleus is the hard body of the comet, and the other two features are merely produced by the emanation of gas and dust from the nucleus. At very large distances from the sun, where sunlight

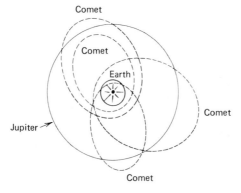

Fig. 6-5. Jupiter's family of comets is a group of short-period comets that have been captured by Jupiter and perturbed into small, Jupiterlike orbits.

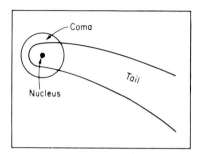

Fig. 6-6. The structure of a comet, defining the principal parts.

does not cause gas and dust to be ejected from the nucleus, a comet consists of nothing but this tiny hard body. Its smallness indicates that it is a relatively insignificant object in mass. If a comet should collide with the earth, as apparently happened in Siberia in 1908, its small size means that a relatively small amount of damage would be caused. (The Siberian event destroyed a forest, but produced no crater.)

The Coma

The portion of a comet most easily seen from the earth is called the *coma*, which consists of a bright spherical envelope around the nucleus, with a size on the order of 100,000 kilometers. The coma's size depends very much on the distance of the comet from the sun. It is very small for comets when they are more than 4 astronomical units from the sun, it reaches a maximum size between 1 and 2 astronomical units from the sun, and then contracts in size again if the comet comes closer to the sun than 1 astronomical unit.

The coma is made up of gases that have been emitted from the nucleus because of that body's warming by the sun. Its small size at large solar distances is due to the small amount of warming possible at such distances, while its large size at 1 or 2 astronomical units is due to the fact that at this distance the solar heating is sufficient to evaporate a considerable amount of material. The coma then becomes smaller if a comet goes closer to the sun, because the solar radiation pressure pushes more of the gas away from the comet, and therefore reduces the amount in the gaseous coma.

The Tail

The third important structural component of a comet is its *tail*. The tail is usually long, often curved, and diffuse in appearance. Sometimes the tail is remarkably structured, with visible knots and wisps. Its length varies from 0 to 10 million kilometers, the length depending among other things on the distance of the comet from the sun. When a distant comet first becomes visible in its path inward toward the sun, it often has no visible tail at all; as it gets closer the tail begins to form and grow. The tail is at its longest when closest to the sun; then after perhelion passage, the tail again begins to diminish in size.

The tail does not follow a comet, as one might guess from photographs, but rather always points away from the sun. A comet approaching the sun will have its tail approximately following it, but then when the comet has passed the sun and is returning out toward deep space, its tail will be in front—it will be moving tail first. The antisolar direction of the tail is caused by the fact that the tail is made of gas and dust that have been pushed away from the coma by the pressure of solar radiation and solar particles. This pressure, of course, always pushes the material away from its source, the sun.

Some comets have two tails: one somewhat curved, very diffuse, and slightly reddish in color; and the other more straight, often structured, and more bluish in color. The straight tail is found to be made up of gases blown away by the sun's radiation, and the slightly curved, diffuse tail is made up of dust. The dust is not pushed away so rapidly as the gas, and therefore it forms a curve made up of the orbits of many small particles.

6-4
MASSES OF
COMETS

Because cometary nuclei are so small in size, their masses are also small. Tests have shown that the masses are probably on the order of 10^{16} grams or so, which is only an exceedingly small fraction of the mass of the earth. Two comets have passed directly within Jupiter's system of satellites, with no effect whatever on the motions of these satellites. This fact led astronomers many years ago to the conclusion that cometary masses were certainly much less than a billionth of the mass of the sun. One of these, Comet Brooks II, passed even closer to the planet Jupiter than its innermost satellite, approaching to only about 75,000 kilometers of the surface of the planet. No effect was found on any members of Jupiter's satellite system, but the comet's orbit changed drastically from a period of 29 years to a period of only 7 years.

Mass Loss

The tails of comets show that material is continuously being lost due to the outward pressure from the sun. It is estimated that between 100 and 1,000 tons

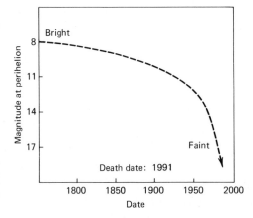

Fig. 6-7. The disintegration of Comet Encke, which is expected to disappear completely in 1991. *(After F. Whipple.)*

of material per second are ejected by the comet when it is close to the sun. This large rate, over many years of approaches, clearly results in the gradual disintegration of comets. Astronomers have found, in fact, that periodic comets that have approached the sun several times during recorded history have definitely shown a systematic fainting of brightness, and this has led to the realization that these comets are gradually being worn away by their solar encounters. Comet Encke, for example, has been calculated on the basis of its continuously fainter and fainter brightness to have a life expectancy of only 20 more years or so; it will probably disintegrate completely at approximately the turn of the twenty-first century.

Spectroscopic measurements show that comets are primarily made up of dust and ices. The coma and tail consist of gases such as cyanogen, carbon, methane, water, and other combinations of the elements hydrogen, oxygen, nitrogen, and carbon. Sun-grazing comets have occasionally shown other elements as well, such as iron, but these show up only if the comet is sufficiently close to the sun to melt and vaporize these heavy elements. The actual molecules and ions that show up in a given comet depend on its distance from the sun, and this can be understood in terms of the amounts of solar heating necessary to produce a particular ion or to vaporize a particular molecule.

**6-5
CONTENT
OF COMETS**

All the various aspects of the observed properties and structure of comets can be understood in terms of the nature of the cometary nucleus. No one has yet sampled a comet by direct means, although plans are presently underway to send an unmanned space probe toward a comet in the 1980s for such direct sampling. In the meantime, it is necessary to build a theoretical model of the cometary nucleus based on the observed facts. The most widely accepted model is that developed about 1950 by the astronomer Fred Whipple, who based his reasoning on the many then-puzzling facts about cometary structure and behavior. Whipple's model is called the *icy conglomerate model* of the cometary nucleus, and it basically consists of a body made up almost entirely of ices of substances such as water, carbon dioxide, methane, and ammonia. Mixed up with these ices are particles of dust consisting of heavier elements such as silicon, magnesium, aluminum, iron, and various compounds of these materials. When such a body of ices and dust moves toward the sun, the solar radiation upon it warms up the outer surface to the extent that the ices sublimate, changing directly from solids to gases. The gases thus formed escape to form the spherical coma. The brightness of the coma is due to the fact that the gases within it fluoresce because of solar radiation, involving absorption of higher energy (short wavelength) solar radiation by the atoms and molecules, and subsequent reradiation of this energy in characteristic wavelengths (many

**6-6
THE NATURE OF
THE COMETARY
NUCLEUS**

Fig. 6-8 Model of a cometary nucleus, based
on Whipple's cometary
theory.

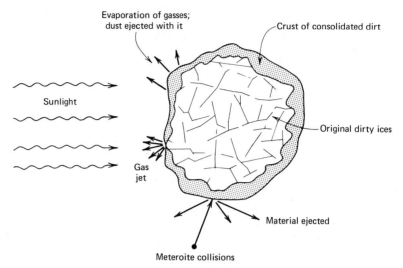

of which are in the visible part of the spectrum). Some of the light is also sunlight reflected from the dust particles that are freed from the frozen conglomerate nucleus by the solar action on the surface.

A rather clear confirmation of the Whipple icy conglomerate model of comets is given by the erratic and otherwise unexplainable behavior of certain comets, particularly of Comet Encke. This comet, which has a short period of a little over 3 years, is sometimes observed to speed up erratically and sometimes to slow down, with no apparent gravitational cause. Comet Encke's strange accelerations and decelerations occur most often when it is closest to the sun, and therefore Whipple reasoned that the peculiar behavior had something to do with evaporation of the cometary material. He showed that a comet that has made many close passages to the sun (like comet Encke) should have a crust of material of higher than average density; this crust would be left over from extensive evaporation of the ice. The dusty crust would normally slow down the evaporation process because it would protect the ices underneath from extensive solar heating. Whipple then realized that whenever there is a weak spot in the crust of such a cometary nucleus, solar evaporation would again cause the gases to escape rapidly. But in this case, they would escape preferentially in one direction. The resulting jet effect would give the comet a small push in the opposite direction; the comet would be temporarily "jet-propelled" by this effect, and therefore irregularities in its motion as observed from the earth would result.

**6-7
THE ORIGIN
OF COMETS**

From the abundance of the elements in comets, it is clear that comets are made up of materials in a mix similar to that of the sun. They seem to be primarily hydrogen, hydrogen compounds, and other common light elements. The fact that most comets have nearly parabolic orbits further indicates that comets spend the vast majority of their time at very great distances from the sun. Comets probably come close to the earth only when perturbations on their orbits (perhaps caused by each other) change the orbits drastically; otherwise, most of them probably have large orbits that never bring them in as close as the planets. It is also relevant that cometary orbits are nearly random in direction and in orbital inclination.

All these facts have been brought together in the hypothesis that the comets were formed very early in the history of the solar system and at very great distances from the sun. Their random orientations could be due to the conditions in the outer portions of the presolar gas and dust cloud. The randomizing of motions that is caused by encounters of one comet with another in distant space might also cause their random orientations. According to this hypothesis, comets exist in vast numbers (millions or billions) at distances of hundreds · or thousands of times the distance of Pluto. Few examples are ever seen, and these are due to perturbations that bring them close to the sun from the dark and otherwise empty regions of the outer solar system.

On a clear night away from city lights, it is possible to see several meteors during an hour's patient observing. Under good conditions, four or five meteors can be witnessed every hour during the evening hours and as many as fifteen per hour in the darkness of the predawn hours. The meteors will be seen as bright streaks of light, appearing to come from any given direction and having a variety of apparent brightnesses and velocities. Sometimes very bright meteors will end suddenly in a flash of brillance, or possibly split into several pieces that continue on, fading together. All will begin as faint objects that brighten rapidly and then fade again. If a meteor shower is occurring the number of meteors seen may be ten or a hundred times the normal number, and most of these will appear to originate from a common point in the sky.

The term *meteor* is used to apply to the visible phenomenon that occurs when an object collides with the earth's atmosphere and a streak of light is produced. The light itself is caused by the heating of the atmosphere just in front of the incoming body, which causes the air and any gas from the evaporated surface of the body to glow and to emit characteristic radiation. The term *meteorite* (Chap. 5) is used to refer to a body that survives the collision with the atmosphere and arrives at the surface of the earth, where it can be found and studied. The term *fireball* is used to refer to a meteor that is excessively bright, brighter than the brightest planets.

In addition to these official definitions, there is a more complicated relationship between these terms. It is now known that most meteors are objects very different from most meteorites in their physical properties. With only a few exceptions, meteors appear to be low-density, fragile objects that

6-8 PIECES OF COMETS: METEORS AND DUST

Fig. 6-9. A bright meteor, photographed by a telescope that remained still while the stars formed trails. The meteor's trail was interrupted by a rotating shutter. *(Harvard Observatory photograph.)*

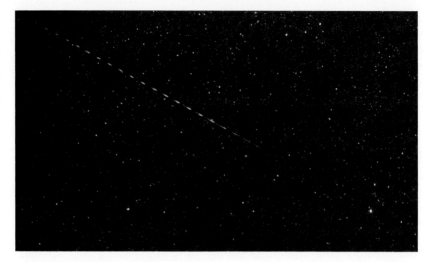

originate from the disintegration of comets, while meteorites appear to be high-density chunks of rocks and iron that originate among the minor planets in the asteroid belt. Meteorites, when they are seen to fall, usually produce fireballs in the sky. Although not all fireballs are due to meteorites that eventually land on the ground, it is believed that a good fraction of these exceedingly bright meteors are made up of hard, meteoritic rocklike materials rather than cometary fragments.

Diurnal Rates   Although on an average dark night it is possible for a person to see four or five meteors every hour in the evening sky, she or he might see as many as ten or fifteen per hour in the early morning. The hourly rate of meteors changes during the night, with a much larger number of meteors visible after 2 A.M.

This phenomenon is called the diurnal (from the latin word for day) variation

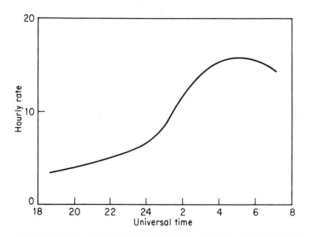

Fig. 6-10. The rate of meteors changes depending upon the time of day.

in the rate of meteor encounters. It results primarily from the fact that in the morning hours, the earth's motion in its orbit is in the direction that is more or less straight up in the sky, while the evening side of the earth faces back in the direction from which the earth has come. Therefore, the morning side encounters more meteors because it is the "front" of the moving earth, in somewhat the same way that the front windshield of a car encounters more raindrops than the back window. In the case of meteors, the orbits are found to be almost (but not exactly) randomly distributed, so that the earth's motion plays the dominant part in determining the diurnal rate of encounters.

Seasonal Rates

There is also a difference in the rate of meteors according to the different seasons. This is due to the fact that meteor orbits are not completely uniformly distributed around the sun. There is a greater number of meteors seen from July to November than during the rest of the year, and this is apparently related to the fact that meteors are found to have a preponderance of orbits in the space that the earth encounters around the sun during the northern hemisphere's late summer months. The periodic comets also show a preponderance of close approaches to the earth during this same period—a result of the connection between comets and most meteors.

There is also a wide variation in the rate of meteors on certain dates, which are the occasion of the phenomena called *meteor showers.*

6-9 METEOR SHOWERS

On October 10, 1946, the world witnessed a spectacular display of celestial fireworks. During that night, meteors were seen so frequently that they almost formed a continuous uninterrupted pattern. As many as 4,000 meteors per hour were witnessed at locations where the night sky was dark. All the meteors seemed to emanate from a particular point in the sky in the constellation Draco.

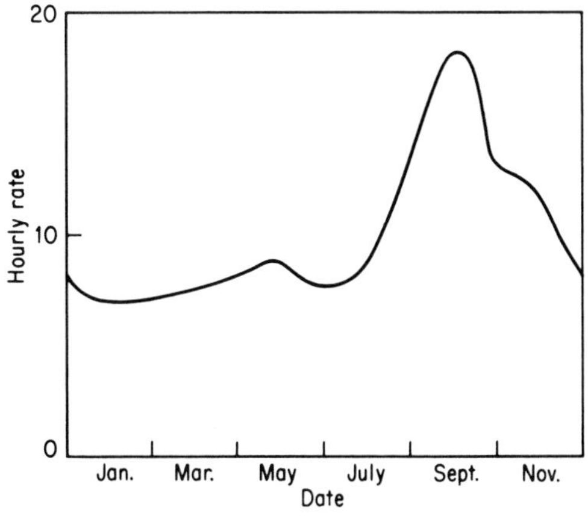

Fig. 6-11. Seasonal meteor rates differ because of the uneven distribution of meteor orbits around the sun.

Fig. 6-12. The Draconids of 1946 produced a spectacular shower, filling the sky.

Fig. 6-12. The Draconids of 1946 produced a spectacular shower, filling the sky.

This specatcular event was an instance of an unusually rich meteor shower. Astronomers knew ahead of time that the earth on that day would be passing close to Comet Giacobini (comet 1900 III), crossing its orbit just 15 days after the comet's passage. The earth thus collided with particles of dust and ices that the comet left behind in its path. On some previous occasions, however, astronomers had not been able to make predictions of spectatcular meteor showers, and the result was sometimes widespread fear and panic. In 1833, the Leonids were so spectacular that many witnesses were convinced that the world was ending, because the sky was ablaze with bright meteors, falling at the rate of about 200 per minute.

On about ten occasions every year, there are meteor showers that are conspicuous and well worth watching, though smaller in scale than the occasional spectacular displays. The richest is the Perseid shower which shows

TABLE 6-1. MAJOR METEOR SHOWERS*

Name	Date of maximum	Type
Quadrantids	January 3	Permanent
Lyrids	April 21	Permanent
η Aquarids	May 5	Permanent
δ Aquarids	July 28	Permanent
Summer daytime showers	May-July	Permanent and periodic
Perseids	August 12	Permanent
Orionids	October 21	Permanent
Taurids	November 7	Permanent
Leonids	November 16	Periodic
Geminids	December 12	Permanent
Ursids	December 22	Permanent

*From Brandt & Hodge, "Solar System Astrophysics," McGraw-Hill Book Company, New York, 1964. By permission of the publishers.

a maximum of some 50 per hour. Table 6-1 lists the names and dates of some of the showers visible at night each year and separates showers into two categories—the *periodic showers* and the *permanent showers*. The permanent showers remain much the same every year, whereas the periodic showers are usually visible only occasionally, with a recurrence period of several years. The Perseids are a permanent shower because the number of meteors seen is roughly the same every year, while the Draconids are a periodic shower because only every 13 years or so is a bright display witnessed.

Astronomers now understand the reason for the phenomena observed with meteor showers. Meteor showers are found to occur when the earth passes close to the path of a comet. Permanent showers like the Perseids owe their regular occurrence to the fact that the cometary debris from the comet whose orbit the earth intersects is now spread fairly uniformly over the entire orbit. The meteor bodies are particles ejected from the comet that now travel around the sun in nearly the same orbit as the comet. For many of the permanent showers, the comet itself appears to have completely distintegrated, and all of its material is now in the form of particles sharing its orbit.

The Cause of Meteor Showers

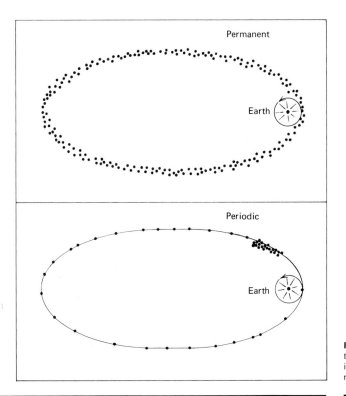

Fig. 6-13. The distribution in the orbit of debris in periodic and permanent meteor showers.

Periodic showers, on the other hand, occur in cases where the cometary debris is more highly clumped around the position of the comet in its orbit. Thus, we witness a conspicuous shower only when both the earth and the comet and its surrounding halo of material both arrive at the same time at the intersection of the orbits. For many of the periodic showers, the comet is still visible. The evidence suggests that the periodic showers are associated with young comets, while the permanent showers belong to older (in some cases dead) comets.

6-10 METEOR ORBITS   Orbits of individual meteors can be determined if their exact path through the atmosphere is known and their velocities are mesasured. Both can be determined by careful reduction of meteor photographs if at least two well-separated meteor cameras succeed in photographing a meteor. From the observed path, it must be deduced to what extent the earth's gravitational pull has altered the meteor's motion, as the earth will have caused the direction to change inward toward the earth and the velocity to increase toward the earth. As both of these changes can be calculated from gravitational theory, these effects can be allowed for, and the original direction of motion and velocity can be found, thus giving the orbit.

Observed Velocities   Meteors are found to have velocities in the atmosphere ranging between 11 and about 74 kilometers per second (from 25,000 to 170,000 miles per hour). It is readily understood why these velocities are the observed limits. No meteor can have a velocity less than 11 kilometers per second, because that is the velocity given by the earth's gravitational pull to any body that comes near the earth. Even if a meteor has no velocity at all with respect to the earth (that is, if its orbit is the same as the earth's), then it will still acquire an 11-kilometer-per-second velocity if it falls to the earth.

The upper limit of about 74 kilometers per second is the result of the fact that practically all meteors must belong to the solar system. At the earth's distance from the sun, no body may have a velocity of more than 42 kilometers per second without leaving the solar system forever. With any higher velocity, the object would have too much energy for the sun's gravitational pull to hold it, and it would escape from the solar system. Thus, any natural object seen to have a greater space velocity than this cannot belong to the solar system, as it has too much energy and must be a high-speed interloper from interstellar space. Therefore, the highest *observed* velocity for a solar system object will be the maximum possible space velocity (42 kilometers per second), plus the earth's orbital velocity (30 kilometers per second), plus another kilometer per second or so due to the earth's gravity—totalling some 73 or 74 kilometers per second. This maximum velocity can be observed only when there is a head-on collision.

Studies of hundreds of meteor orbits show that they are arranged nearly randomly in space. They are not limited to the plane of the ecliptic, nor are they moving exclusively in direct orbits. Nearly half move in the opposite direction to that of the planets. Periods are almost all long—greater than 20 years—and maximum distances from the sun are usually more than 10 astronomical units. In all respects, the orbits are arranged similarly to cometary orbits.

Orbits of Ordinary Meteors

As described in Sec. 6-9, shower meteors have orbits that are shared by specific comets. The Draconid shower, for example, has an orbit like that of Comet Giacobini, and the Northern Taurids have orbits like that of Encke's comet. For most shower meteors, the inner parts of the orbits agree best with each other and with the parent comet (when one is known), while the outer parts of the orbits are at times different, because of various perturbations and uncertainties.

Orbits of Meteor Showers

Because meteorites that land on the ground are fundamentally different kinds of objects from most meteors (Chap. 5), information on the physical and chemical nature of meteor bodies (sometimes called *meteoroids*) must come from other sources. The best information comes from detailed analysis of the trails of meteors on photographs and from meteor spectra. The physical properties (density, size, shape, and physical strength) of a meteoroid determine how bright it will be at different times during its passage, how rapidly it slows down, and how and where it disintegrates. Typically, meteors are found to brighten and decelerate rapidly, and the data can be explained only if the meteoric bodies are very small, highly fragile, and of low density (less dense than water). This evidence suggests strongly that they are chunks of ices.

**6-11
THE NATURE OF
METEOR BODIES**

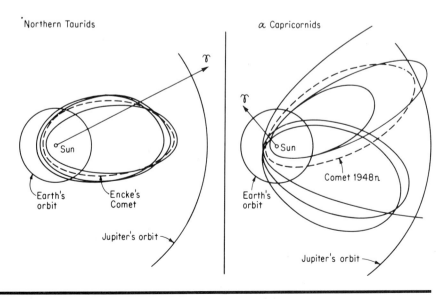

Fig. 6-14. Orbits of the northern Taurid meteors and the Alpha Capricornid meteors.

Meteor spectra primarily show lines of oxygen and nitrogen (largely due to the atmosphere heated up by the meteor) and lines of hydrogen, sodium, magnesium, silicon, aluminum, iron, and related elements (due to the material in the meteoroid). It is likely that the spectra can be explained if meteoric bodies consist of ices (water ice, dry ice, etc.) mixed with dust of a rocklike composition. Meteors, like comets, seem to be describable principally as chunks of "dirty ice."

**6-12
INTERPLANETARY
DUST**

The existence of dust between the planets can be inferred from many different kinds of evidence, even without detecting it directly. Measurements of the sizes of meteors have shown that the most abundant are the smallest, and there is reason to expect that this trend continues down into the size range of the interplanetary dust. Measurements of the dimensions of meteorites shows a similar effect, with the smallest meteorites being much more abundant than the larger. Similarly, therefore, it would be expected that the smallest meteorites of all (of dust size) would be the most abundant. Also, of course, the continual disintegration of comets is observed for those that come relatively close to the sun, and it is thus indicated that small particles are continuously being provided from comets to the interplanetary medium. All these lines of evidence suggest that there should be a considerable quantity of dust particles (microscopic in size) in the space between the planets, and recent measurements by space probes have given us a great deal of information about these objects.

**The Zodiacal
Light**

On a very dark night, away from city lights, it is possible to see a faint glow in the sky during the hours immediately after sunset and immediately before sunrise. This glow extends fairly high up into the sky from the horizon, with its brightest portions near the horizon. It is centered along the line of the ecliptic, among the constellations of the zodiac, and is called the *zodiacal light*. It is now understood to be the result of the reflection of sunlight from the particles that make up the interplanetary dust cloud. Observations of the zodiacal light made with telescopes designed specially for the purpose have shown that the light increases in intensity toward the sun, and that it is chiefly restricted to the ecliptic, extending only 20 or 30° from the plane of the earth's orbit.

Particle Density

From analysis of the zodiacal light intensities and colors, it is possible to calculate the approximate number of particles in space that produce the observed effect. The analysis is difficult, and assumptions must be made about the particle shapes and optical properties. The best estimates based on zodiacal light measurements give fairly reasonable results when compared with modern satellite data. The calculated density of particles in space near the earth is found to be approximately 200 particles per cubic kilometer (for particles of all sizes). This is equivalent to a mass density of approximately 10^{-24} grams per

Fig. 6-15. The zodiacal light, photographed from the observatory on Haleakala. *(Photograph courtesy J. Weinberg.)*

cubic centimeter—several hundred times the mass density of the solar wind. It is nevertheless an exceedingly small density.

Sizes of Particles

It is very difficult to determine the sizes of the particles by analysis of the zodiacal light because of the many assumptions that must be made about other properties of the particles. Calculations indicate that most of the particles must be something on the order of 10 microns in size. (A micron is 1 millionth of a meter). It is clear that particles very much larger than this cannot be dominant because if the particles were bigger than 100 microns or so, the reflected light would be considerably reddened, while color measurements of the zodiacal light indicate a color very similar to that of the sun. Similarly, the particles cannot be exceedingly small because if they were smaller than a micron or so, the scattered light from them would be noticeably blue.

The Distribution of Particles in Space

From the zodiacal light measurement, it is possible to conclude that most of the particles of the interplanetary dust exist in a cloud that is restricted somewhat to the plane of the planetary orbits. The cloud is much thicker than the differences between the planes of the various planetary orbits, extending up to heights above and below the planetary orbital planes of several tenths of an astronomical unit. The particles are very much denser near the sun than in the outer parts of the solar system, and the zodiacal light suggests that the density decreases rapidly outward in a smooth fashion.

Space Probe Measurements

The most accurate determination of interplanetary dust properties has come from measurements made by artificial satellites, space probes, and lunar experiments. The density of particles, their sizes, their orbits, and even in some cases their physical and chemical properties have been determined by these means. The prime difficulty in studying the interplanetary dust by any of these techniques results from the destructive nature of any collection of this material. No matter how the dust might be collected, it involves an impact of at least several kilometers per second, and such an impact with any reasonable surface material inevitably destroys the dust particle almost completely. Therefore, study has primarily been limited to the impact events (the craters left by the collision). To collect an interplanetary dust particle unharmed is an exceedingly difficult technological problem that has only been solved in a few instances by using the earth's upper atmosphere as a protective cushion (Sec. 6-13). The measurements have provided a determination of the number of interplanetary dust particles in space near the earth that is more precise than that inferred from the zodiacal light. The results are in fairly close agreement with the zodiacal light measurements, but give a great deal more information on the number of particles of different sizes. Results are usually presented in terms of a *flux*, which is a measure of the number of particles passing through a given area of space. If one were to hold up a normal-sized window frame (with an area of roughly 1 square meter) and then count the number of particles passing through the frame in space near the earth, one could count approximately 1 particle per day with a diameter of 10 microns, about 1,000 per day with a size of 1 micron, and only 1 particle in 300 years larger than 1,000 microns (1 millimeter, or approximately the size of a grain of sand).

Space experiments devised in the early 1970s were sufficiently sophisticated so that orbits of some of the particles could be measured, even though the particles were eventually destroyed by collision. This feat was accomplished by having a volume exposed in space that consisted of several layers of very thin film, which were coded electrically. If a particle should penetrate a film at any position, that position would be recorded and signaled back to the earth. By this means, both the velocity and direction of motion for the particle can be determined as it passes through the various thin films until its eventual destruction. Results so far show a preponderance of orbits that are similar to cometary orbits and meteor orbits, with most of the differences explainable in terms of the various sources acting on the interplanetary dust particles in space.

6-13 DUST AND THE EARTH

The most effective means of nondestructively collecting interplanetary dust appears to be to utilize the earth's atmosphere. Because the atmosphere decreases gradually outward in density, a particle encountering it from outer space can be gradually decelerated from its space velocity without being destroyed. This process is effective only for very small particles that have a relatively large surface compared to their volume, so that they can radiate away

Fig. 6-16. This interplanetary dust collector is taken to very high altitudes by a large balloon (in background) to get above the terrestrial contamination of the lower atmosphere. *(Courtesy D. Brownlee.)*

energy comparatively rapidly. Such small particles, if they come into the earth's atmosphere and reach the ground without being destroyed, are often called *micrometeorites.*

The earth's atmosphere has a disadvantage for this kind of work in that it is very heavily contaminated by terrestrial dust from various sources, including volcanoes, wind-blown soil, other natural particles, and human pollution. Terrestrial dust is carried high into the atmosphere by winds and is so abundant that it is impossible to distinguish interplanetary dust in any part of the lower atmosphere of the earth. For that reason, the most successful collections of interplanetary dust have been made by experiments carried very high into the atmosphere by balloons, which reach altitudes as great as 140,000 feet (about 40 kilometers), and high-flying aircraft. Two kinds of particles are found at these altitudes, and they are most likely identified either as micrometeorites that have been virtually unaltered by their passage in the upper atmosphere, or as ablation products made up of material melted or fragmented away from larger bodies that had been destroyed in the atmosphere. The micrometeorites are quite small (less than 10 microns in diameter) and irregular in shape, with composition similar to that of carbonaceous chondrites (Chap. 5). These are probably cometary particles, the ices of which have been lost by the heating in

Fig. 6-17. A particle collected from the upper atmosphere by a balloon-borne micrometeorite collector. *(Courtesy D. Brownlee.)*

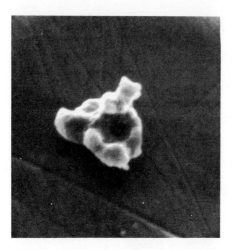

the atmosphere. The ablation products are normally spheres, which indicates that they were formed by the melting of the surface of a larger body. Their composition ranges from that of iron meteorites to that of carbonaceous chondrites.

**6-14
THE SOLAR
WIND**

The dust is not the only inhabitant of the space between the planets. Spread throughout the solar system is tenuous gas, primarily made up of material lost from the sun, and called the *solar wind*.

Magnetic Storms

The first indication of the presence of gas particles in the solar system was the discovery of occasional sudden disturbances of the earth's magnetic field. Normally, the earth's field is steady in shape and constant in strength. A

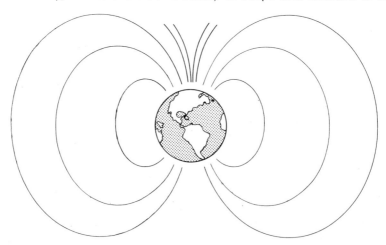

Fig. 6-18. The earth's magnetic field.

Fig. 6-19. An aurora, photographed behind one of the buildings of the Harvard College Observatory.

magnetic compass points steadily toward the magnetic pole of the earth because the magnet of the compass needle always attempts to line itself up with the direction of the earth's total magnetic field.

Occasionally, but infrequently, the magnetic field of the earth suffers a disruptive disturbance. Called *geomagnetic storms*, these are usually accompanied by a disturbance in the earth's upper atmosphere as measured by radio instrumentation. Furthermore, there is often a considerable amount of auroral activity during these times. In the 1930s it was suggested that the geomagnetic storms, as well as the accompanying phenomena, are probably due to streams of charged particles (protons and electrons) emitted by the sun. A large number of charged particles would be capable of distorting the earth's magnetic field because of the fact that there exist strong interactions between magnetic fields and electric fields. If charged particles in a cloud encounter the earth, they would be expected not only to disturb the magnetic field of the earth, but also to disrupt the upper atmosphere by altering the conditions in the ionosphere, sometimes penetrating deep enough into the upper atmosphere to produce the bright lights of the aurora.

It was found that this hypothesis not only explained these otherwise inexplicable phenomena, but it also fit well with developing theories of events on the surface of the sun. *Solar flares*, which are intensely energetic and hot disturbances on the solar surface, were found to occur just a few days before geomagnetic storms on the earth. Calculations showed that the flares were responsible for the ejection of bursts of charged particles that traverse the distance between the sun and the earth in 2 or 3 days.

Fig. 6-20. The sequence of events, as seen from the earth, after a solar flare.

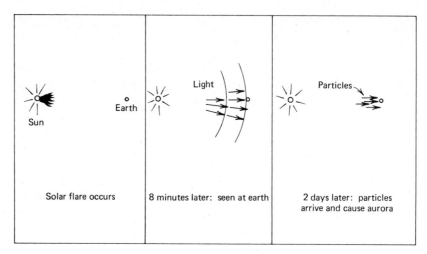

Sun / Earth

Light

Particles

Solar flare occurs | 8 minutes later: seen at earth | 2 days later: particles arrive and cause aurora

Comet Tails   Although it was clear from the magnetic storm evidence that charged particles (making up an interplanetary gas plasma) existed at times of intense solar flare activity, it was not known originally whether or not a steady stream of gas particles exists in the solar system. The first indication that such must be the case came from a study of comet tails. The comets often show two different kinds of tails, a gas tail and a dust tail (Sec. 6-3). During the 1950s a study of the behavior of the cometary gas tails showed a somewhat remarkable effect. Gas tails have distinguishable structure in many cases, and it is possible to measure the rate at which the gas moves away from the comet down the tail by studying this structure day after day. It was found that the gas molecules are continuously accelerated away from the sun by a greater amount than can be explained simply on the basis of the sun's radiation. Therefore, it was realized that there must be a continuous ejection from the sun of particles that can cause the observed accelerations. These particles must be mostly hydrogen nuclei (protons) and electrons, because the sun is made up mostly of hydrogen. They must interact with the molecules and molecular ions of the comet tail, thus accelerating them away from the sun at an anomalously high rate.

Space Measurements   Although a considerable amount of information about the solar wind came from a study of magnetic storms and comet tails, the best data on its properties have been the result of concentrated effort on the part of space vehicles. Early satellites only reached altitudes of a few hundred kilometers above the surface of the earth, and their measurement of the gas medium there, beyond the earth's atmosphere, showed a remarkable and certainly not typical situation. Among the first discoveries were the van Allen Belts, immense pockets of trapped charged particles held in place by the outer portions of the earth's

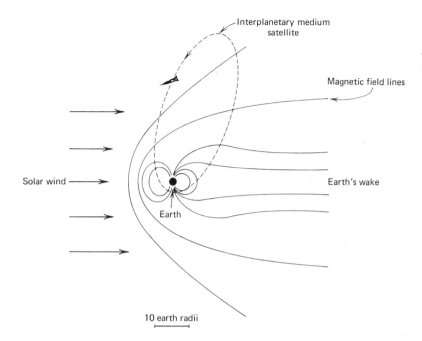

Interplanetary medium
satellite

Magnetic field lines

Solar wind

Earth's wake

Earth

10 earth radii

Fig. 6-21. Space probes have made important contributions to our understanding of the interplanetary medium.

magnetic field. The early satellites quickly demonstrated that at altitudes of only a few hundred kilometers above the earth, the interplanetary medium as such does not exist. The earth's magnetic field and attendant phenomena exert considerably too much influence on the material. However, as satellites were sent into larger and larger orbits, and as space vehicles were sent to the moon and to other planets, it became possible to probe the solar wind at sufficiently large distances from the earth to get a realistic measure of its properties.

The number of particles in the solar wind is exceedingly small. The space probes have shown that at the earth's distance from the sun, there are only about two particles per cubic centimeter on the average, which is the equivalent of a matter density of about 2×10^{-24} grams per cubic centimeter. This is incredibly thinner than the best vacuums obtainable in terrestrial laboratories and represents a very low mass concentration. The density is not constant throughout the solar system; it is hundreds of times larger near the sun and also somewhat larger exterior to the earth's orbit.

The direction of motion for the particles of the solar wind is in all cases directly outward from the sun. Some minor effects due to disturbances in the large scale solar magnetic field are noticeable, but these are only important in the more distant parts of the solar system. Measured velocities for the particles themselves average approximately 500 kilometers per second (1 million miles per hour). The velocity is somewhat less at greater distances from the sun than the earth's distance, reaching values of only 10 kilometers per second or so in the outer solar system, where the solar wind gradually diffuses into the interstellar medium.

By studying the relative motions of the particles in the solar wind, it is possible to derive a temperature for the gas. Measurements from space probes lead to the conclusion that the temperature must be on the order of 200,000 K. This is, of course, an exceedingly high temperature, but even so the solar wind does not have much effect on the solar system objects that it encounters. The reason for the lack of any heating effect is the exceedingly small density of the gas, which means that although each particle contains a very large amount of energy, this energy (when added up for all the particles) is insignificant because of their small number.

Needless to say, if the solar wind should be greatly increased in density, the effect on solar system objects could be considerable. It has been hypothesized that in the early history of the solar system, the solar wind might have been occasionally much more substantial than at present. The possible effects of this hypothesized intense solar wind are involved in some theories of the formation of the solar system. For example, the relatively high densities of the inner planets, especially of the planet Mercury, are possibly explained in terms of the sweeping away of all its low density materials by intense bursts of solar wind early in the history of the solar system.

The reason for the existence of the solar wind was not understood until the late 1950s. At that time, it was pointed out that the steady, continuous solar wind (not the bursts) was the natural extension of what is called the *solar corona*. The corona of the sun is a very thin outer atmosphere that extends outward to several times the solar radius from the surface of the sun. It is made up of exceedingly hot gas, with temperatures as high as 2 million degrees, very much hotter than the temperature of the sun's surface (6,000 K). The corona is produced by matter ejected from the surface of the sun, both by violent motions during the solar surface storms and by the general circulation (called *convection*) of the layers of the sun that are just beneath the surface.

Studies of the structure of this corona show that it is not a stable configuration as the earth's atmosphere is. The solar corona, with its observed properties, cannot remain around the sun in a steady and fixed state. Theoretical calculations show that it must instead eject gas out into the solar system continuously at high velocities. These calculations predict that there will be a steady stream from the corona of particles with properties very similar to what is observed by space probes for the solar wind. Therefore, it is now understood that the solar wind's particles are ejected by the solar corona because of its inherent instability. Material is continuously lost from the solar surface into its thin outer atmosphere, the solar corona, and then particles are eventually lost from the corona out into the solar system in the form of the solar wind. The sun in this way continuously loses this matter, but the rate of loss is exceedingly small and has no important effect on the long-term evolution of the sun.

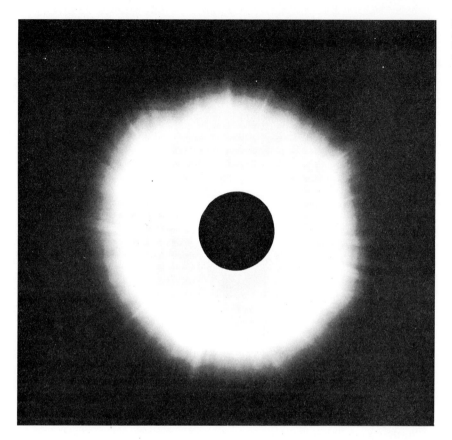

Fig. 6-22. At times of a solar eclipse it is possible to see the solar corona. *(Yerkes Observatory.)*

Besides, it is also gaining material from the other component of the interplanetary medium—the interplanetary dust.

1. At the present rate, how many *new* comets will be discovered before 2000 A.D.? What does this tell you about the probable number of comets in the solar system?
2. Calculate the year of the next appearance of Halley's Comet.
3. Why is it important to astronomy to know whether cometary orbits are parabolic, hyperbolic, or large ellipses?
4. In a table contrast the characteristics of the orbits of comets with those of the planets.
5. If a comet loses 100 metric tons per second while near the sun, how long would a typical comet last if it had a circular orbit near the sun? Why do comets actually seem to last much longer than this?
6. Windows of spacecraft are pitted by collisions with interplanetary dust. Estimate the number of approximately 1-micron-sized pits that would be

formed on a 100-square centimeter spacecraft window on a 5-day mission to the moon and back.

7. Calculate a comparison between the density of the solar wind and the density of the earth's atmosphere at sea level.

8.* Why can we not sample and study the solar wind at the earth's surface? At the top of the earth's atmosphere?

9. Approximately how long would it take a solar wind particle to reach the earth's distance from the sun, if its velocity were uniformly that given in the text?

10.* How many solar wind particles would collide with an astronaut in space near the earth during a 10-minute space walk?

11.* What is the approximate rate of mass loss to the sun due to the solar wind?

OBSERVATIONS AND EXPERIMENTS

1. If a bright comet is in the sky at this time, observe it with binoculars or a telescope. Measure the size of its coma and its tail, and from its orbit (often given in astronomy magazines) calculate their intrinsic size. Keep track of its position with respect to nearby stars, and plot its position day by day on a star chart (which you can make). Astronomers can calculate the orbit of a comet from only three such measurements (if they are precise). Can you estimate its brightness by comparison with nearby stars?

2. Build a "working comet model" by mixing snow or refrigerator frost with sand or dirt. Allow sunlight (or a sunlamp) to shine on it while it is in a cold room or outdoors on a cold day. The conditions will be roughly similar to those for cometary heating in space; the solar radiation will cause evaporation of the ice, and gradually the dirt will form a compacted skin, inhibiting further evaporation.

3. On a clear dark night (no moon), count meteors over the sky for 30 or 60 minutes. This is best done lying down, facing straight up, dressed warmly. Compare your rates with those in this chapter. If possible, repeat the counts at from 2–4 A.M. and compare results. Can you estimate the approximate brightness of the meteors by comparisons with stars? Are there more bright meteors or faint meteors?

4. On a night when a permanent meteor shower is expected, repeat these counts and compare. Do you detect a common point of apparent origin for the shower meteors? Plot the meteor paths on a star map and determine the origin from the intersections of lines extended from the paths. Is the radiant (the apparent common point of origin) in the expected constellation?

5. Choose a dark, moonless night away from city lights and attempt to see the zodiacal light, either about an hour after sunset or an hour before sunrise. Why choose those times? Draw the zodiacal light on a star map. Does the brightest part lie on or near the ecliptic (the zodiacal zone)? How wide is the zodiacal light? Making the crude assumption that it is on the average about 1 astronomical unit from you in space, calculate its thickness.

6. Leave a pan of water out on the roof of a building and/or in a remote place for 24 hours. Using a millipore filter (obtain one from your chemistry or biology department), filter the water and examine the residue with a microscope or a high-power magnifying glass. There is a possibility that some of the dust you will collect is interplanetary dust, either micrometeorites, meteor ablation fragments, or droplets. But the overwhelming majority of the particles will be terrestrial. How would you determine which are extraterrestrial?

7

LIFE ON OTHER PLANETS

The purpose of this chapter is to provide you with food for thought about our uniqueness in the universe. Are there other habitable planets? Are some of them already inhabited?

In the last 10 years or so, scientists have made a great deal of progress toward understanding some of the steps in the process of the origin of life on earth. Most of their work has dealt with the possible ways by which earthly materials, obeying the laws of physics and chemistry and having plenty of time, could have evolved naturally from inanimate objects into living things. This involves intricate arguments and ingenious experiments in organic chemistry and biochemistry, and the astronomical input is only important at the beginning. After the first or second stage of the process, astronomy steps out and big molecules with big names take over.

**7-1
THE ORIGIN
OF LIFE**

In the following paragraphs, some of the ways by which life may have formed are reviewed. The geological evidence for the time scale of biogenesis is described, the seesawing arguments about the earth's early atmosphere are outlined, and the possible sources of prebiotic organic molecules are given.

The earliest direct evidence about the origin of life on earth comes from the record preserved in the oldest terrestrial rocks. Modern age-dating methods make it possible to establish ages for the rocks and rock formations involved, and thus it is possible to have a reliable idea of how many years ago the beginnings of life occurred.

**The Geological
Evidence**

The oldest rocks found on the earth are only 3 billion years old. This has been noted by some geologists as a rather curious fact, since the earth is known to be considerably older, approximately 4.6 billion years old, which agrees with the ages of other objects in the solar system, such as the meteorites. It is not known whether the relatively young age for the oldest terrestrial rocks is due to a long period when the earth's crust was not stable and rocks could not persist, or whether it is due to some other cause. It may be that we simply have not yet found older rocks, which may exist in some unexpected part of the earth's surface.

Among the earliest fossil evidences of the existence of life on earth are the

10 μm

Fig. 7-1. Tiny fossils of primitive microscopic life as found in 3 billion-year-old rocks in Swaziland. This series of sketches shows a few selected examples that demonstrate that these life forms reproduced by simple cell divison, *(After Barghoorn and Kroll.)*

small microscopic spheres found in the Figtree series of rocks in South Africa, which have been dated geologically as being about 3 billion years old. These objects are not clearly fossils of living organisms, but are believed by their discoverers, Barghoorn and Schops, to be similar to the tiny proteinoid spheres produced and studied by Fox and Young. These artificial microspheres were produced in the laboratory under primitive conditions. In experiments with heated mixtures of various amino acids, Fox and Young found that long chain *peptides* and *dipeptides* formed in time, and these fairly complex organic molecules appear to be similar chemically to proteins in living things. Furthermore, they formed microscopic spherules that were approximately the same size as bacterial cells in living things. They showed characteristics that were almost like properties denoting life. Two could combine to form a single somewhat larger sphere, or one of them could split apart. They showed the properties of osmosis, and their size could be regulated by the amount of salt present in the solution. Although it is not known for certain that the fossil spheres are the same kinds of objects as those studied in the laboratory, their appearance is very suggestive.

The oldest clearly living fossils that are well established are from the Gunflint formation, an iron-rich rock found in Southern Ontario. Here the age is between 1.6 and 2 billion years, and the fossil evidence of microscopic plant life is completely convincing. Both fungi and algae exist, and because algae operate by means of the complicated and late-developing process of photosynthesis, it is believed that primitive life must certainly have begun long before the flowering of the Gunflint flora.

The Earth's Primitive Atmosphere

Much of the present theory for the development of life on the earth depends critically on the chemical composition of the earth's atmosphere in its early history. It is generally believed by biologists that life could never have developed on the earth if the atmosphere had always had the same composi-

Fig. 7-2. An example of the rock in which early life forms are found. This is a *chert*, a metamorphosed rock originally consisting of sediments from a lake or sea in which microscopic life forms were trapped. *(David Azose photograph.)*

tion as the present terrestrial atmosphere. Specifically, the evidence indicates that naturally occurring events in the development of prebiotic organic chemicals could never have occurred if free oxygen had been present in the atmosphere. Of the hundreds of experiments that have occurred in laboratories since the pioneering experiments in the early 1950s, success in producing organic compounds out of simple elements has only been achieved in an atmosphere devoid of oxygen. Therefore it is believed that the predecessor molecules to life could not have formed in the presence of an oxygen atmosphere, and unless these molecules came from elsewhere, life could never have gotten started.

The astronomers Henry Norris Russell and Donald Menzel showed many years ago that the earth's primary atmosphere must have been completely lost in the early stages of the earth's development. This is based on the argument that in comparison with the sun there are no noble gases, and if the abundance of elements in the material forming the earth were similar to those in the sun, which otherwise is roughly the case, then these noble gases must have been lost at a time when the temperature was high, on the order of 5000 to 8000° absolute.

By analogy with the major planets, Jupiter, Saturn, Uranus, and Neptune, it is believed that the primitive first atmosphere of Earth must have been rich in hydrogen molecules, including molecular hydrogen, methane, and ammonia. This mixture also included appropriate amounts of the noble gases. From geological evidence it appears that any ammonia must have been minimal, because the oldest rocks of sedimentary origin on the earth are found in what geologists call cherts, which would not naturally form in the presence of ammonia. The environment in which these sedimentary rocks were laid down was highly acidic, and such conditions argue against anything more than very small amounts of ammonia, either in the atmosphere or in the seas and

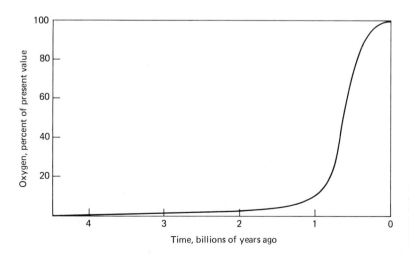

Fig. 7-3. Geological evidence indicates that the oxygen of the earth's atmosphere reached its present abundance only recently.

oceans, as early as 3.4 billion years ago. Thus, the primitive atmosphere must have been lost by the earth before 3.4 billion years ago.

It is still uncertain whether the above facts are mutually consistent; it was apparently necessary to produce the early predecessors of living molecules in the presence of a reducing atmosphere which must have contained hydrogen in molecular form. The hydrogen, however, must have left the earth very quickly, because of its light weight and the relative warmth of the terrestrial position in the solar system. Whether enough hydrogen was left in the earth's atmosphere for the prebiotic molecules to form before the earth's atmosphere changed from its primitive reducing nature to its secondary volcanic-gas composition is still unknown. It has been shown chemically that ammonia and methane are unstable without the presence of molecular hydrogen, so whenever molecular hydrogen was lost, these compounds also must have left.

One final kind of evidence against the presence of oxygen throughout the early history of the earth's atmosphere comes from biological evidence itself. The present way in which metabolism occurs in cells suggests that early simple organisms must have gone through their development without the benefit of free oxygen. They seem to have developed remarkable and ingenious ways of achieving metabolism without using oxygen, and these ways continue to the present. Examples include *fermentation*, which is a pathway of metabolism that biologists call the hexosemonophosphate cycle, and *photosynthesis*. Particularly if oxygen had been present, biologists argue, many of these pathways would not have been necessary. Respiration, which uses oxygen fully, must in this view have been a much later development, after sufficient oxygen existed in the atmosphere.

It has also been shown that if molecular hydrogen had been present, the hexosemonophosphate cycle would not have developed. This apparent contradiction suggests that much of the early development of energy-generating methods by life forms occurred after the hydrogen atmosphere of the earth had been lost, but before oxygen was produced in significant quantities by life itself.

Sources of Organic Molecules

In the modern view of the development of life, there were several stages of increasing complexity. First it was necessary for simple organic molecules to be present in reasonable abundance, and then various processes led to more and more lifelike objects.

Biologists have shown that if single organic molecules were available under proper conditions and were given long periods of time, they could have joined together to form very large *macromolecules*. In time, these combined together to form what are called *molecular aggregates*, which are highly complex mixtures. At this time, as first proposed by A. I. Oparin, it is believed that because of surface tension the aggregates developed simple membranes around them, which separated them from the liquid medium out of which they had formed. One end of the fatty acid chain is attracted to water, and thus any

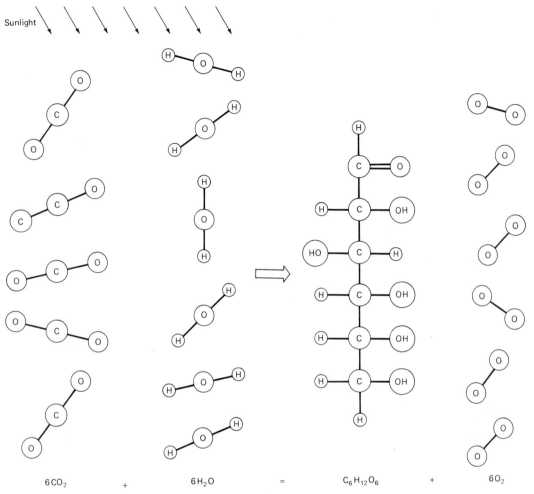

$$6\,CO_2 \quad + \quad 6\,H_2O \quad = \quad C_6H_{12}O_6 \quad + \quad 6\,O_2$$

fatty acids in a macromolecular aggregate could collect at the surface to produce such a membrane. Nucleic acids then began, somehow, to organize these aggregates, as they do now in life, and thus ensured the ability of these aggregates to split into two similarly organized parts. Following this, evolution began, and only those aggregates that could use energy most efficiently survived. It is believed that these objects depended upon outside molecules as sources of energy, and developed a means of storing energy as phosphates. As time went on these objects, which were depending upon ready-made food in their environment, found their source of food diminishing, and evolution is therefore believed to have favored any ability to produce energy internally. Photosynthesis is such a process, and it is believed that it must have developed at this point. With the introduction of photosynthesis, life became independent of its surroundings to some extent, and large-scale production of oxygen began.

Fig. 7-4. The essential features of photosynthesis involve the formation of glucose (chemical formula: $C_6H_{12}O_6$) and oxygen, out of carbon dioxide (CO_2) and water (H_2O).

Of course, it should be remembered that much of this outline for the development of life is sketchy, and details at several points are not yet understood completely.

From the astronomical point of view, the most important question raised by biogenesis is the question of the source for the organic molecules that must have been present before life could form. There are two possibilities that are presently considered reasonable. The first and most generally accepted is that the organic molecules were formed in a primitive terrestrial atmosphere, out of which they accumulated in the oceans, and were perhaps concentrated in abnormally high densities on the beaches. Many experimenters, starting with Stanley Miller and Harold Urey in 1951, have shown that fairly complicated organic molecules, including various amino acids, purines, sugars, and lipids, and even some more complicated molecules, may have been formed. Usually three molecules are prevalent, hydrogen cyanide (HCN), formate (COOH), and formaldyhyde. The most commonly assumed primitive atmosphere is one rich in methane, ammonia and hydrogen, but the geological evidence cited above raises the question of whether methane and ammonia were likely components of the earth's atmosphere. Scientists, however, have produced the same organic compounds by experimenting with an assumed atmosphere of carbon dioxide and nitrogen, as well as other mixtures of this sort, including water.

The simulation experiments involve the introduction of a source of energy into a mixture of gases in the laboratory. Included among different sources of energy in recent experiments have been heating, electrical discharges, ultraviolet light, and x-rays. Almost without fail, the resulting mixture includes a reasonable number of organic molecules.

A second possible source for the organic molecules that preceded the formation of life is an extraterrestrial one. This became realized as a possibility only after it was shown that the carbonaceous chondrites, until recently a little-known type of stony meteorite, were found to be much more common than originally thought. It is thought quite possible that the carbonaceous chondrites are the most abundant type of meteorite in outer space, and only their fragile physical nature makes them rare on earth, except in the form of fine dust. If they do not become completely destroyed by their passage through the atmosphere, which is probably usually the case, they are quickly broken down on the earth's surface and lost. If they are among the most abundant kinds of meteorites encountering earth, it is significant that they are rich in carbon and carbon compounds. In fact, recent research shows that the carbonaceous chondrites contain all the more stable elements in the same proportions as in the sun, and about 6 percent of their cosmic proportion of carbon, making it an abundant component. Even more interesting is the fact that much of this carbon exists in compounds, some of which are complex organic compounds. A great deal of controversy once surrounded the subject of the organic molecules found in carbonaceous chondrites, but much of this controversy is now

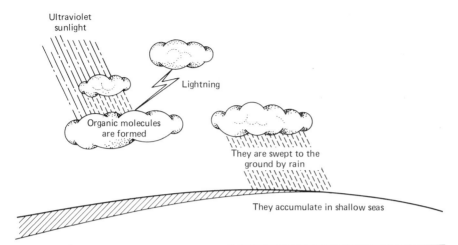

Ultraviolet
sunlight

Lightning

Organic molecules
are formed

They are swept to the
ground by rain

They accumulate in shallow seas

Fig. 7-5. The simple organic molecules that formed the basis for more complex molecular systems may have been caused by the action of the sun's untraviolet light or lightning, working on the earth's atmosphere before oxygen was present. Rain carried these molecules to the ground, where they could be concentrated in shallow seas.

resolved. During the early 1960s there seemed to be evidence of life forms in the carbonaceous chondrites, but these were found to have been introduced on the earth itself after the meteorites had already fallen. There is no controversy, however, about the presence of fairly complex organic molecules in freshly fallen carbonaceous chondrites, and these are generally recognized to be prebiotic molecules that have no relationship to actual living organisms.

Carbonaceous chondrite meteorites usually contain several amino acids, and of those that have been studied in detail, so far, many are common in proteins of earth organisms. But there are also many amino acids that do not occur in large quantities in terrestrial biological materials. Besides the fairly common and relatively simple organic molecules, the carbonaceous chondrites also contain molecules called *nitrogenous bases*, such as purines and pyrimidines. These nitrogenous bases are not the ones that are common in biological samples, and it is curious to note that the laboratory experiments with primitive earth atmospheres have produced neither those nitrogenous bases that are common in biological samples, nor those found in the carbonaceous chondrites.

It is also noteworthy that there are certain differences between the distribution of the different kinds of organic molecules found in the meteorites and those found in terrestrial life. It is not possible to say that the meteoritic compounds are all foreign to life, because many of them are not. Some are fairly common, and others are found in low concentrations in living matter. However, the difference in general is quite striking. In life, there are approximately twenty amino acids that are common to proteins. All these have one peculiarity in common. They are characterized by having an amino group and a carboxylic acid group that are attached to the same carbon atom. Under these conditions, organic chemists call these alpha-amino acids. In the carbonaceous chondrites, especially in the Murchison and Murray meteorites, which are among those most completely studied, the amino acids do not show this preponderance of alpha types. Instead, all possible isomers of amino acids that contain two or three carbon atoms are present. This feature suggests that the carbonaceous chondrites contain amino acids that were formed by a random and unbiased process, while those in life forms on earth have been formed selectively.

Another, rather similar difference between the meteorite compounds and those in terrestrial life has to do with the way in which the amino acids rotate plane-polarized light. This in turn is the result of the preferential orientations of each individual part, which can be either "left-handed" or "right-handed." In the meteorites, there is an approximately equal number of amino acids of one type or the other, but in life amino acids are all of one orientation.

A further difference between the organic compounds in life and in the meteorites has to do with the structure of the hydrocarbons. Those in the Murchison carbonaceous chondrite, for example, are almost all double ring structures, whereas in life hydrocarbons are most often long, linear molecules.

These important differences between the meteorites and life might at first be considered evidence against the hypothesis that life originated from organic material that had been brought from outer space in the meteorites. However, such a conclusion is unwarranted. In the first place, life has a very selective way of development, and it is unlikely that the raw materials, the prebiotic organic molecules, would have been maintained in their original proportions through the long, selective process of biological evolution. In the second place, most of the differences between the meteoritic organic molecules and biological ones also exist between those organic molecules that are produced in experiments with primitive terrestrial atmospheres and biological materials. Therefore, neither hypothesis has preference over the other in this regard.

Biologists have suggested that a necessary concentration of molecules may have appeared on the beaches of the earth's primitive oceans. The hypothesis

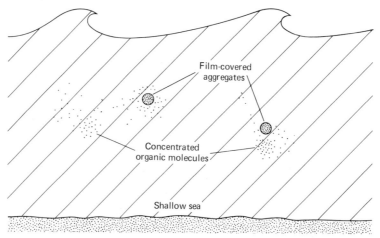

Fig. 7-6. The organic molecules may have collected together under water to form tiny film-covered aggregates, looking almost like simple cells (Fig. 7-1).

that the organic molecules were formed in the primitive atmosphere relies on just such a concentration process, as does the hypothesis that they came in as small meteorites or as dust from interplanetary space. These possibilities both require some process for concentration of the molecules in smaller areas, where their density can be sufficiently great for them to coalesce into aggregates. At this time any of these sources of organic molecules can be considered equally likely, since we do not have good numerical information on the numbers of such molecules that might have been produced by ultraviolet light, x-rays, or discharges in the atmosphere. The primitive-atmosphere hypothesis may turn out to be preferable from the point of view of the numbers of organic molecules that might have formed, but the interplanetary dust and meteorite hypothesis has the advantage of providing a continuous source that kept on going regardless of the constituents in the earth's atmosphere and regardless of other, perhaps inhospitable, conditions on its surface. It may well turn out that these sources were both involved, and then it will remain only to determine which was the more important.

With our present ideas about how life came about on earth, we can ask: Where else in the solar system are there similarly favorable conditions for life to begin and flourish? We have seen in previous chapters that there is very little to choose from. Venus has an atmosphere, but no liquid water and much too much heat. Mercury has no atmosphere at all. Jupiter and the outer planets are too cold at their visible surfaces and probably don't even have solid surfaces. Beneath Jupiter's clouds is a monstrous sea of molecular hydrogen, not water. Though some biologists have suggested that some simple life forms might, at least temporarily, float in Jupiter's atmosphere, most agree that conditions are just too formidable. Titan, a satellite of Saturn is also a possibility; it has an atmosphere not too unlike earth's early one, but its temperature is too low for life to develop in the same way as it did on earth.

**7-2
LIFE
ELSEWHERE IN
THE SOLAR
SYSTEM**

This leaves Mars, long considered the planet most likely to have some form of life like earth's. Searching for such life was, in fact, a prime motivation behind the Viking 1 and 2 space probes. These landers, which set down on the Martian surface in 1976, had three biological experiments on board, as well as a camera that photographed the surroundings periodically and would have shown any large Martian life forms that happened to be nearby. The experiments involved tests made on Martian soil samples—tests in which the soil was cooked to see whether water would be emitted, as would be the case with terrestrial microscope life. The soil was also fed a broth ("chicken soup"-like) and then checked for excess production of O_2, and given "labeled" carbon (detectable because of its radioactivity) to see whether the Martian "bugs" would eat it and release it later. The results from these experiments were disappointing—not so much because nothing happened—but because something happened, but we don't know what. The first experiment said "no life;"

Fig. 7-7. This Martian vista, photographed by Viking 2 and showing desolate wind-blown dunes and black volcanic basaltic rocks, is similar in many respects to photographs of deserts on the earth, as in Fig. 7-8. *(NASA.)*

no water was released until the temperature was quite high, 350°C for the Viking 1 site and 200°C for a sample taken from under a rock near Viking 2. This means that if there are any organic compounds, they must be scarcer than one part per billion in the soil. The other experiments, however, gave perplexing results. Large amounts of oxygen were given off after giving the soil some water, its broth and its radioactive carbon, but the behavior was not like

Fig. 7-8. A desolate California desert scene with wind-blown sand and black volcanic basaltic rocks, plus one element conspicuously lacking in Fig. 7-7, i.e.,life.

Fig. 7-9. At its more favored locations, life forms dominate the earth's landscape and render it unlike anything to be seen on the other planets. How many kinds of things (including various life forms) show up here that would not be seen on Mars? What things could be seen on Mars?

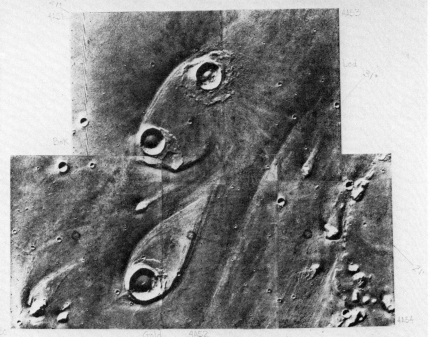

Fig. 7-10. Mars once had flowing water, as this Viking orbiter photograph shows. The dry channel of a once-flooded river cuts through the Chryse region, having left islands around the upturned rims of craters. Life might have been able to develop in such places if liquid water had existed for a long enough period of time *(NASA.)*

that of life, especially since the soil continued to give off oxygen even after being heated to so high a temperature that life should have been killed. This puzzling ·result means that something is happening to cause oxygen to be released in the presence of water. A few biologists think it might be due to some exotic form of microscopic life, but most believe that ordinary chemical reactions are responsible. Unfortunately, to be sure we will probably have to go to Mars again, and this time we must be equipped to do even more exhaustive tests.

Apparently our earth is the only abode of life in our solar system. In this part of space, at least, humanity is probably alone.

7-3
PLANETS OF
OTHER STARS

Our best hope of finding life like ours lies with the distant stars. How probable is it that somewhere in our Galaxy there is a star like the sun, with a planetary system that includes one planet like earth, and where conditions for life to form have been favorable? This is one of the most intriguing current questions in astronomy.

From our present understanding, our solar system appears to be a common sort of thing. Very probably, at least some planets are formed whenever a star forms. Since there are several hundred billion stars in our Galaxy, it is likely that some kind of planet revolves around many of them, perhaps a hundred billion. Among these hundred billion, only a fraction will have the right kind of star to favor life, presumably something like our sun, and only some fraction of these will have planets at the right distance and with the right mass to have an earthlike atmosphere and liquid water on the surface. These should, if our present ideas are correct, be suitable for life, and life most likely has developed on them. Even if only $1/100$ of the stars have suitable planets, it means that there could be a billion planets in our Galaxy harboring life. This is a very rough estimate, of course, and could be very wrong, but it does indicate that our present understanding says that life on other planets may be very common.

Life, no matter how common, does not imply intelligence. On earth, among the millions of life forms, only one species has developed sufficient skill and intelligence to read this book. But the human record of technological competence is extremely short compared to the age of our Galaxy. If the age of the earth is likened to a mile's walk, then the period of humanity's technological history is $1/1000$ of an inch at the end of that walk. There are an estimated 1 billion earthlike planets in our Galaxy. If a billion people were all to start out walking a mile, it is unlikely that they would all arrive at the last $1/1000$ inch at precisely the same moment. We have achieved space travel, and with our radio telescopes we can send or receive signals for great distances. Any intelligent beings on one of those other 1 billion planets could easily have passed that point millions of years ago, and could now be unimaginably more competent than we are.

But where are they? With all that competence, why haven't they detected our first steps in space, our first attempts to hear them? Why haven't they come here and introduced themselves? Scientists who have pondered this point out these possibilities:

1. We are alone. No other life exists in the Galaxy. Ours is one in a billion.
2. We are first. Happy chance puts us out ahead. This seems unlikely, of course, if there are a billion other planets to compete with. But it's not possible to say it has a small probability. If there is only one planet at the present with intelligent, technologically advanced life on it, then the probability that we happen to be on it is large!
3. They've come and gone. No convincing evidence of such a visit in the geologic records exists. More important, perhaps, is the lack of any recognized garbage on the moon. In recent years we have littered the lunar surface, but that surface was found by us to be unviolated by anyone else's visit. It has apparently lain just as it is for 3 billion years.
4. Civilizations never survive beyond a certain point. About the time that they have radio telescopes and begin space flights, they also have the potential to blow themselves up and always promptly do so.

If the last choice is correct, then it is the duty of astronomy to bring the lessons of the stars to earth, to let people see what we'll be missing if we follow that pattern, to give us a chance to be the exception, to be for our Galaxy the intelligent life that jumps that barrier, that learns to control itself and its primitive instincts before it is too late. Astronomy tells us that billions of other worlds lie out there in space waiting to be explored, and that billions of years, the whole lifetime of the sun, lie out there in the future waiting to be lived.

QUESTIONS

1. Describe the earth's early atmosphere.
2. What evidence tells us that our atmosphere couldn't have always been as it is now?
3. Why was the earth's early atmosphere lost?
4. Why are carbonaceous chondrites, rather than iron meteorites, likely sources of material for life as it was forming on the early earth?
5. How do the organic compounds in meteorites differ from those in terrestrial life?
6. Do you think that we should attempt to communicate with intelligent beings on other planets? Why or why not?

CONCEPTS OF STELLAR EVOLUTION

THE SETTING

Before beginning the story of the birth and death of stars, this chapter provides you with some necessary background on distances, brightnesses, and motions of stars.

**8-0
GOALS**

The stars are very distant objects. The nearest is vastly more distant than the most distant planet, and the farthest is so distant that even the world's largest telescopes cannot detect it. Like the planets, moon, and sun, the stars appear to rise in the east and set in the west, but this motion is entirely due to the rotation of the earth and is only apparent. Unless special instruments are used, it is not possible to detect any true motion for the stars. Their relative positions in the sky, which are seen from each seasonal angle as the earth revolves in its yearly path around the sun, remain the same from year to year and almost the same from century to century.

    Ancient astronomers had no accurate means for determining the distances to the stars. Most ancient peoples seem to have believed that the stars were distant lights or holes in a remote sphere that turned around the stationary earth. Some of those few Greek astronomers who correctly surmised that it was the earth's turning that made the stars appear to move also correctly guessed that the stars were exceedingly distant objects.

**8-1
DISTANCES**

Even without instrumentation, it is possible to obtain an estimate of the distances to the stars if their true nature is realized. If we assume that the stars

**Apparent
Brightness**

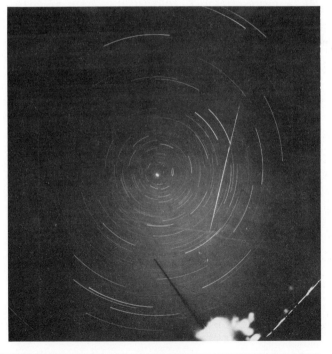

Fig. 8-1. Circumpolar star trails. When a camera is pointed north and is left open for a time exposure, the turning of the earth causes the stars to appear to trace out circles, with the north celestial pole at their center. The straight-line streak in this photograph is a bright meteor. *(Smithsonian Astrophysical Observatory.)*

are distant but very luminous glowing objects that are similar to the sun, and further assume that all have about the same real or intrinsic brightness as the sun, then it is possible to obtain an estimate of their distances by comparing their apparent brightnesses with that of the sun. This assumption is not strictly correct, as we know (from measurements of distances more precisely determined) that some stars are thousands of times more luminous than the sun, and others are thousands of times fainter. But the sun is approximately average in brightness, and we can therefore obtain a reasonable answer when its luminosity is used as typical.

The light from the brightest stars contains energy that can be measured as arriving at the surface of the earth at the rate of about 5×10^{-5} ergs per square centimeter per second. An *erg* is a very small amount of energy, approximately the amount that an insect expends if it climbs onto a piece of cardboard (see Fig. 8-2). By comparison, sunlight arrives at the earth's surface with an energy of about 1.5×10^6 ergs per square centimeter per second. Comparing these two figures shows that the sun has an apparent brightness that is 30 billion (3×10^{10}) times greater than the apparent brightness of the brightest stars. Since apparent brightness depends on the square of the ratio of the distances, the nearest stars must be approximately the square root of this ratio more distant than the sun, or $\sqrt{3 \times 10^{10}} = 1.7 \times 10^5$. Thus, it is calculated that the nearest stars must be at a distance of about 170,000 astronomical units, which is 2.6 light years—only somewhat less than the true distance found by triangulation for the nearest star (4.3 light years).

1 erg

A beetle climbing onto a piece of cardboard

1 billion ergs

Fig. 8-2. The erg is a very small unit of energy.

A man climbing up one step

Although at present the above method isn't used for determining the distances of ordinary stars, the principle of measuring distances by means of apparent luminosities enters into many of the ways that astronomers determine distances to specific kinds of stars. For example, it is found that stars spend most of their evolutionary lifetime as stable objects of a given temperature that depends almost exclusively on their total luminosity. Both measurements of individual stars and theoretical calculations show that stars of different temperatures have certain specific luminosities. Therefore, if it is possible to measure the temperatures of stars during their normal stable phase, it is possible to infer their intrinsic luminosities. A measurement of their apparent luminosities immediately leads to a determination of the distance by the method employed in the above paragraph. Astronomers use spectroscopic methods to determine stars' temperatures and to ascertain whether or not the stars are in the appropriate phase of their life cycle. They then can determine distances to stars on the basis of what are called *spectroscopic parallaxes*, distances obtained utilizing this method. (See Appendix D.)

Similarly, the measurement of a large number of stars in star clusters can allow the determination of the stage of development for individual stars as well

Fig. 8-3. Two star clusters, one large and faint, the other smaller but made up of brighter stars. *(Harvard College Observatory.)*

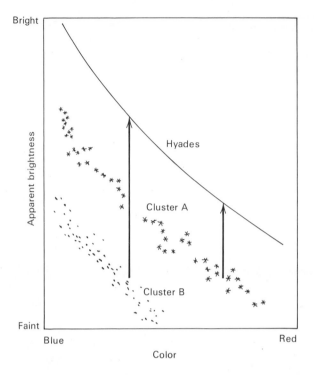

as their temperatures; therefore, measuring the apparent luminosity of stars in clusters can lead to an accurate determination of the distance to the cluster.

A final example is the determination of the distance to certain kinds of stars, called *variable stars*, that are unstable and that consequently pulsate in size and brightness (Chap. 12). Some of these pulsating stars are found to be close enough to the sun for their distances to be measured by other means, and it is found that for many of them there is a characteristic luminosity that depends on their period of pulsation. For the bright objects called *cepheid variables* (named after a bright example, the star Delta Cephei), there is a well-determined relationship between the luminosity and the period. It is therefore possible to determine the distance to any cepheid variable if only its period and its apparent luminosity are measured. Cepheid variables form the cornerstone of the distance scale among other galaxies and are an important tool in the determination of the size of the universe.

Triangulation   The most fundamental determination of the accurate distances to the nearby stars is based on *triangulation*, the method of measurement used by surveyors to determine the distance to an unattainable distant point. The principle of the method is very simple. The object for which measurement of the distance is desired is made one point of a triangle; the opposite side and the opposite angles are measured precisely. For instance, a surveyor who wishes to measure the width of a river can select two trees on his or her side of the river and one tree on the unattainable opposite side, thus making a triangle of these three landmarks. From the first tree, the surveyor can measure the angle between the distant tree on the opposite side and the other nearby tree. Then the idea is to pace off the distance between the two trees on the near side of the river and measure the angle between the distant tree and the first tree (as seen from the second). Then a triangle can be drawn on a piece of graph paper, with these angles marked and with a scaled-down representation of the distance between the two nearby trees. The intersection of the two other sides of the triangle on the graph will then give the scaled-down position of the far shore—and the river's width.

Astronomers do a very similar thing when they are determining the distance to the nearest stars. The triangle has as its side of known length the distance across the earth's orbit. The two angles are measured by determining the relative position of the nearby star to be measured and the very faint, very distant stars in the background. This angle is measured once from one side of the earth's orbit and then 6 months later is measured from the other side of the orbit. The very small angles thereby determined indicate the very large distances to stars. The nearest star has a measured angle p, called *parallax*, of less than one second of arc. From the triangle shown in Fig. 8-6, it is seen that half the baseline is just 1 astronomical unit, about 150 million kilometers (93 million miles). An angle of one second would indicate a distance of 3.3 light years. This distance is defined as a *parsec*. For the nearest known star, Proxima Centauri, the angle measured is 0.765 seconds of arc, thus indicating that its distance is 4.3 light years. Light travels at approximately 300,000 kilometers per second, and therefore a light year, which is defined as the distance light travels in one year, is a distance of nearly 10^{13} (10,000 billion) kilometers, or 6,000 billion miles.

The nearest stars are all at least 4.3 light years away. They are clearly very much more distant than the edges of the solar system, as Pluto is only $5\frac{1}{2}$ light

— 3.27 light yr = 1 parsec.

Distance = $\dfrac{3.27 \text{ light years}}{\text{parallax in sec. of arc}}$

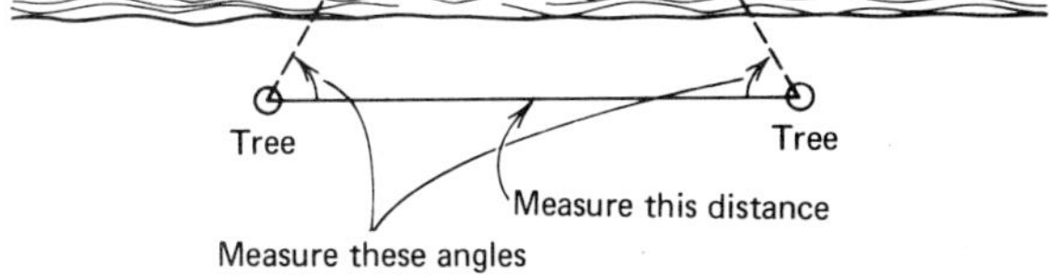

Fig. 8-5. Determining the distance across a river by triangulation.

Tree

River

Tree   Tree

Measure this distance

Measure these angles

Fig. 8-6. Determining the distance to a star by triangulation. The distant stars are so far away that they can be taken to be a fixed background.

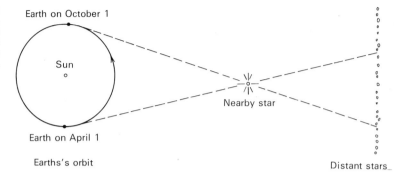

Earth on October 1

Sun

Nearby star

Earth on April 1

Earths's orbit

Distant stars

hours from the sun. With present technology, it takes only a few days to reach the moon by spaceship and only a few years to reach the planets. By comparison, with a spaceship capable of these velocities, it would take more than 10,000 years to reach the very nearest of the stars.

Separations of Stars

The vast distance between the sun and the nearest star is fairly typical of the separation of stars in our local Galaxy. Stars are so exceedingly small compared to the huge distances between them, that if a spaceship were sent off in a random direction, the probability that it would ever collide with or even have a close encounter with a star is very small. Most of the stellar universe is empty space, containing only the very thin interstellar medium of gas and dust (Chap. 9).

Our system of stars can be thought of as approximately similar in arrangement to a swarm of bees. It is made up of many small objects (stars) that are concentrated toward the center and more spread out in the outer parts, similar to a swarm of bees. However, if we compare the two systems by mentally

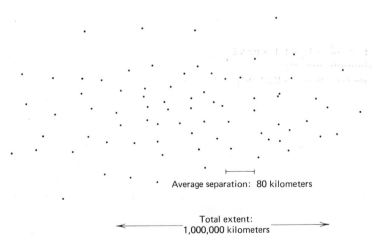

Average separation: 80 kilometers

Total extent:
1,000,000 kilometers

Fig. 8-7. The swarm of bees analogy.

scaling down our Galaxy so that an individual typical star is approximately the size of an individual bee, the difference between these two clusterings of objects becomes conspicuous. The average separation between the bees (or bee-sized stars) would be approximately 80 kilometers (50 miles), and the entire size of the swarm would be more than 1 million kilometers.

The apparent luminosity of a star is a measure of the amount of energy in the form of light that reaches the earth from the star. It was shown above that the apparent luminosities of the nearest stars are about 30 billion times fainter than the apparent luminosity of the sun. The faintest stars visible without a telescope are about a thousand times fainter than the brightest stars visible in the sky, and the largest telescopes in the world can see stars that are a million times fainter yet.

**Apparent
Luminosities**

The true or intrinsic luminosity of a star is called its *absolute luminosity*. This is the brightness that the star would have when compared with other stars if they were all seen at the same distance. The sun has approximately average absolute luminosity, but other stars have greatly different values. The brightest stars known in our Galaxy have absolute luminosities approximately 1 million times greater than the sun's; these objects are called *supergiant stars* because of their great brightness and large size. The faintest known normal stars have absolute luminosities that are nearly 1 million times fainter than the sun; these are called *red dwarfs*, as they are not only faint but are also cool (red) and very small.

**Absolute
Luminosities**

Astronomers use a rather unusual system for numerically describing the brightnesses, both apparent and absolute, of stars and other astronomical bodies. This system originated with the Greeks and was based on the kind of response that the human eye has to various brightnesses of images. The Greek astronomers divided the stars into six different groups according to their apparent brightness. They called these different groups *magnitudes*, and the brightest stars were considered to be of the first magnitude—the faintest being of the sixth magnitude. To the human eye, these magnitude intervals looked approximately like equal divisions, and so they apparently made up a natural system of measurement.

Magnitudes

When more precise instruments for measuring the brightnesses of stars became available, it was found that the magnitude intervals are not divisions of equal luminosity. Instead, it was found that the human eye's response is logarithmic rather than linear, and the magnitude system is found therefore to have a logarithmic base. This means that a proper calculation of the magnitude

Fig. 8-8. The apparent magnitude of a star depends upon its absolute magnitude and its distance.

Small distance
Faint absolute magnitude
Bright apparent magnitude

Large distance
Bright absolute magnitude
Faint apparent magnitude

Earth

for stars and of the magnitude differences between them must be made by using logarithms of numbers. By convention, the interval between stars of five magnitudes in brightness is taken to be exactly a factor of 100 in luminosity. Therefore, a star of the first magnitude is exactly 100 times more luminous than a star of the sixth magnitude, which is the faintest that can be seen without a telescope. A first magnitude star is 100×100 or 10,000 times more luminous than a star of eleventh magnitude, which is about the faintest star that can be seen with a 6-inch telescope. It is $10^2 \times 10^2 \times 10^2 \times 10^2 = 10^8$ (100 million) times brighter than a twenty-first magnitude star, about the faintest than can be photographed with a 60-inch telescope.

Apparent Magnitude

The *apparent magnitude* of a star is defined according to convention established by astronomers many years ago; i.e., several of the brightest stars in the sky are defined as standards of magnitude. All other stars can then be referred to these standards, and luminosity differences from the standards lead to the determination of apparent magnitudes for them.

All apparent magnitudes depend on both the intrinsic luminosities of the stars and their distances. A star of very bright absolute luminosity will have a bright apparent magnitude if close to us, but it could have a very faint apparent magnitude if it is very distant. For example, the two apparently brightest stars in the sky are Sirius and Canopus. While Sirius is a rather normal star in terms of absolute brightness, Canopus is an exceedingly bright star in terms of absolute brightness—almost 10,000 times brighter than the sun. They both appear to be roughly the same brightness because Sirius is relatively near to us (about 8 light years), while Canopus is very much more distant (500 light years away).

Among the twenty brightest stars (Appendix K) are some that are brighter than first magnitude, the arbitrarily chosen brightest interval of the ancients.

TABLE 8-1. ASTRONOMICAL MAGNITUDES

If the magnitude of Star A is this much smaller than Star B,	Then Star A is this many times more luminous than Star B
1	2.5×
2	6.3×
3	16 ×
4	40 ×
5	100 ×
10	10,000 ×
15	1,000,000 ×
20	100,000,000 ×

Stars brighter than this are either 0 magnitude, or if even brighter, they are assigned a negative magnitude. Sirius, for example, has an apparent magnitude of −1.4, meaning that it is 1.4 magnitudes brighter than a star of zero magnitude, or 2.4 magnitudes brighter than a first magnitude star. The faintest stars that can be photographed with the world's largest telescopes have apparent magnitudes of about 24, and they are therefore over 25 magnitudes fainter than Sirius. In apparent luminosity, this difference is the equivalent of 10^{10} times, or 10 billion.

Absolute Magnitudes

In order to compare the intrinsic, true luminosities of different stars, astronomers have defined the absolute magnitude of a star as the apparent magnitude that it would have if it were at a distance of 10 parsecs (33 light years). The absolute magnitude of the sun is the apparent magnitude we would measure if the sun were that distance away from us rather than being vastly closer; this has been found to be 4.8 magnitude. This means that if the sun were removed to a distance of 10 parsecs, it would be among the fainter stars that we could see. The absolute magnitude of Sirius is +1.5, three magnitudes fainter than its apparent magnitude, due to the fact that it is nearer than the standard distance used to calculate absolute magnitudes. The bright star Deneb, the brightest star in the constellation Cygnus, is of first magnitude in apparent brightness, but it has an absolute magnitude so bright that it is one of the brightest stars in our portion of the Galaxy, with a value of −7.0. If it were moved from its distance of some 1,000 or 2,000 light years in to the standard distance of 10 parsecs, it would have an apparent magnitude of −7, thus making it the brightest object in the sky, excluding the moon and sun. It would be bright enough to cast faint shadows at night.

The Stellar Luminosity Function

Astronomers define the natural distribution of absolute magnitudes of stars in terms of what is called a *luminosity function*. This is a table or graph that tells how many stars in a unit volume of space are included for each different magnitude interval. For example, Fig. 8-9 shows the luminosity function for stars in the solar neighborhood of our Galaxy and illustrates the fact that the most luminous stars are exceedingly rare compared to the faint ones. There are about 100,000 times as many stars like the sun than stars of absolute magnitude −5, and about 4 times as many stars of absolute magnitude +15 than stars like the sun. The most common type of stars are the faint red dwarfs, roughly 10,000 times fainter than the sun.

The luminosity function for the solar neighborhood is found by determining the distances of large numbers of stars near the sun and measuring their apparent magnitudes. Luminosity functions for other parts of the universe, for star clusters for example, or for the central region of our Galaxy, or for other galaxies, can be determined similarly and are found to differ in various ways from the luminosity function of local space. For example, for some star clusters,

Fig. 8-9. The stellar luminosity function tells how many stars in space are of various intrinsic brightnesses.

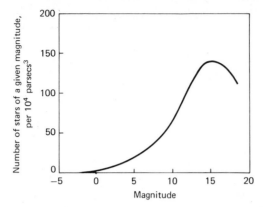

the most common stars are intrinsically bright, with only very few stars having faint absolute magnitudes. For other clusters, there is a considerable depletion of the brighter stars. These various differences can be understood in terms of the evolution of stars. The first kind of cluster, with very few faint stars, is known to be a very young object for which enough time has not been available for the faint stars to have formed as yet (see Chap. 10). For the second kind of cluster, which is greatly depleted in stars brighter than the sun, it is understood that this depletion is due to the fact that this kind of cluster is exceedingly old, and its bright stars have all long ago decayed (see Chap. 14).

8-3 STELLAR MOTIONS

Though to the unaided eye it appears that the only motions of the stars are the apparent motions produced by the rotation and revolution of the earth, precise measures of star positions have shown that all stars have real motions of their own. However, because of their exceedingly great distances, these motions are almost imperceptible. The effect of the distance of an object on its apparent

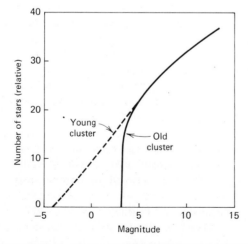

Fig. 8-10. Luminosity functions compared for a young and an old cluster. The young cluster still has many of its brighter stars, which have already died in the old cluster.

motion is simply understood. For example, a very distant airplane near the horizon will have a very much slower apparent motion than will a car passing by on the road immediately in front of an observer. Though the motion of the car may be only 30 miles per hour and the velocity of the airplane 600 miles per hour, the effect of the much greater distance of the airplane gives it a much slower apparent motion as seen by a single observer. Similarly, the lights of an airplane take only a few minutes to cross the sky at night. As seen against the background of stars, it has a very much greater apparent motion than do the background stars, which move so apparently slowly that they would take centuries to detect by eye. In this case, the airplane's true velocity may be 600 miles per hour, while the typical stars in the background have true velocities of about 50,000 miles per hour.

Proper Motions

The amount of angular change of position measured for a given star is called its *proper motion*. It is measured in terms of seconds of arc and is typically much less than 1 second of arc per year. To determine the proper motions of stars, it is necessary to plan far ahead. Normally, a measurement consists of two sets of photographic plates taken many years apart in time—at least 10 years, and in some cases as many as 50 years. Measurements with precise measuring instruments that can detect the very smallest shifts of the nearby stars with respect to faint background stars on the plate lead to determinations of these proper motions. Because of the distance effect described in the above paragraph, the nearest stars have larger proper motions than more distant objects. For example, Sirius, one of the nearer stars, has a motion that carries it across the sky at the rate of 1.3 seconds of arc per year. On the other hand, Canopus, the second brightest star in the sky, at its greater distance of about 500 light years, has a proper motion of only 0.02 seconds of arc per year. Deneb, at about three times greater distance yet, has an exceedingly small proper motion, only 0.004 seconds of arc per year. It is clear that many years must elapse between photographs to detect this when it is remembered that a single photograph of a star field with a large telescope has a resolution seldom better than about 1 second of arc.

Proper motions of stars depend not only on their distances from the sun, but also on their true space motions. We only see one component of this space

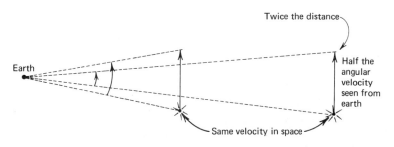

Fig. 8-11. The relationship between apparent motion and distance.

motion when we observe the proper motion. For example, if a star should be moving directly toward or away from us, we would not detect any proper motion at all, even though its space motion could be very large. Therefore, when astronomers make use of proper motions to determine the properties of star motions in the neighborhood of the sun, they must make statistical allowances for the fact that only one component of the motion is observed.

Radial Velocities

Spectroscopic utilization of the Doppler shift (see Chap. 2) allows astronomers to determine one more component of the star's motion, the component in the line of sight. This is called the *radial velocity* of a star, and it can be measured quite accurately from stellar spectra. Unlike proper motions, radial velocities are the true velocities of the objects (the line of sight component) and do not depend on the distance. Therefore, the radial velocity for a star measured near the sun would be the same as that measured for that same star even it if were moved a thousand times farther away.

Typical radial velocities for stars are on the order of 20 kilometers per second. Some stars are found to have much greater velocities (up to 200 or 300 kilometers per second), but these are fairly rare. The star Sirius has a radial velocity of −8 kilometers per second, which means that it is moving toward us with that rather low velocity. Among the other brightest stars, the star with the largest radial velocity is Aldebaran, the brightest star in the constellation Taurus; it has a radial velocity of +54 kilometers per second, indicating that it is rapidly moving away from us. In both of these cases, the velocities quoted are in the radial direction only and do not include the motion of these stars sideways, at right angles to the line of sight, which is called the *transverse* or *tangential* motion.

Space Motions

If it is possible to measure both the proper motion of a star and its radial velocity, then it is possible to combine these two measurements to determine the total motion of that star through space and its direction of travel. This is called the *space motion* of a star, and it is known for large numbers of the nearest stars. Sirius, for example, has a radial velocity of −8 kilometers per second and a proper motion which, at its distance, is equivalent to a tangential

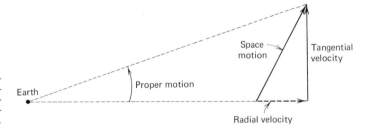

Fig. 8-12. The relationship between proper motion, space motions, radial velocity, and tangential velocity.

velocity of 17 kilometers per second. Its space motion can be calculated from these two figures to be 18.7 kilometers per second in a direction given by the direction of its proper motion across the sky, which is tilted about 30° toward us.

The space motions of most of the nearby stars are approximately parallel and average about 20 kilometers per second in a direction toward the constellation Eridanus (the River). This systematic motion is taken to indicate that the sun itself is moving amongst the nearby stars with this velocity in the opposite direction, toward the constellation Hercules. Like observers in a car who apparently see the trees and telephone poles all moving past them with a space velocity of 55 miles per hour, our viewpoint from the solar system shows us only an apparent systematic motion of the nearby stars, which is in truth a systematic motion of the sun with respect to its background.

Peculiar Motions

Actual space motions of stars with respect to some given background or reference frame are called the *peculiar motions* of the stars. A car on a freeway has a peculiar motion of about 55 miles per hour as measured with respect to the background of the road and surrounding countryside. However, as in the case of stars, the peculiar motion of this car can be seen to be something quite different if a different background is considered. For example, if we examine the peculiar motion of that car with respect to the sun, then we must take into account the fact that the earth is moving at 30 kilometers per second (66,000 miles per hour) in its orbit around the sun; therefore, the peculiar motion of the car in that reference frame is vastly greater than its occupants probably realize. Similarly, the peculiar motion of the sun with respect to the nearby stars is 20 kilometers per second. However, if instead of the nearby stars, the background of the entire Galaxy is considered as a frame of reference, it is found that the peculiar motion of the sun is about 250 kilometers per second. This large value is explained in Chap. 19, where it is shown that most of the stars of our local Galaxy are revolving around the center of the Galaxy in a way analogous to the earth's revolution around the sun.

QUESTIONS

1. Using the crude method of comparing distances from apparent brightness that was used in Sec. 8-1, calculate the approximate distance to the faintest star that is visible through a 3-inch telescope, i.e., those that are about 10,000 times fainter than the brightest stars in the sky.
2. What is the distance to the star Altair (α Aql.)? Its parallax is 0.20 seconds of arc. Check your answer with that given in Appendix K.
3. In about 10 billion years, during its advanced phases of evolution (Chap. 13), the sun will at one point be 100 times its present luminosity. What will its *absolute* magnitude be?
4. How many more stars are there in nearby space that are 100 times less luminous than the sun than there are solar-type stars?

5. From the ground, a 747 or other "jumbo jet" appears to move more slowly than smaller jets? Why?

6. How far from its present position will Sirius be in 100 years?

EXPERIMENT　To see what kinds of effects can influence stellar luminosity functions, construct an analogous "age function" for the members of your class. Compare this to the age function of the public at large (by sampling, by guess, or by library research). Estimate curves that would describe the age function for professors, for astronauts, for *Who's Who*, etc., and discuss the effects that make each of these different. What analogies can you find with stellar luminosity functions?

INGREDIENTS FOR FORMING STARS

**9-0
GOALS**

Each of us is made up of material that was once spread out between the stars in the form of gas and dust. The purpose of this chapter is to explore the nature of our elemental ancestry. What is this material that eventually forms into stars and planetary systems?

**9-1
THE GAS
BETWEEN THE
STARS**

This chapter and the ones that follow trace the evolution of stars from their birth to their death. A star can be born only if conditions are right and if the right materials are present. We know from observations of the chemical abundances in stars and from measurements of activity occurring in certain regions of our Galaxy that star formation occurs where large amounts of gas and dust are accumulated in complex clouds. This chapter is concerned with the properties of these clouds and of the other interstellar materials that make up the ingredients for star formation.

The interstellar material consists of both gas and dust, arranged in a general, fairly smooth interstellar medium and in pockets of relatively high-density gas-and-dust clouds. Most of the gas is hydrogen, the most common element in the universe, and the hydrogen exists in three different forms: as neutral atoms, as molecules, and as ionized atoms.

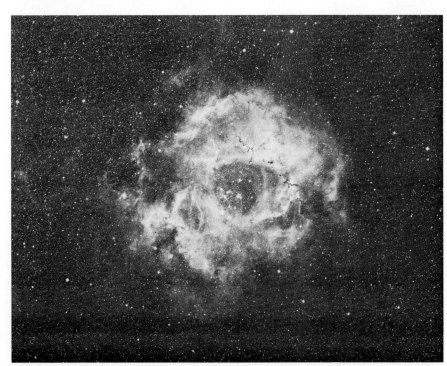

Fig. 9-1. The Milky Way galaxy is studded with bright, glowing clouds of gas like this one, The Rosette Nebula. *(Kitt Peak National Observatory.)*

The interstellar neutral hydrogen gas is completely invisible to optical telescopes, and therefore remained unobserved for many years. In 1944 the Dutch astronomer H. C. van de Hulst computed that the hydrogen atom should be capable of producing radio waves of 21-centimeter wavelength. He based his calculations on certain characteristics of the components of the hydrogen atoms. Consisting of a single proton with a single electron orbiting around it, the hydrogen atom normally can emit radiation only if the electron loses energy by changing from a more energetic to a less energetic orbit (see below). However, atomic scientists realized in the 1940s that the particles, the proton and the electron, have a property that is analogous to rotation, called *spin*. The proton and the electron can both be spinning in the same direction, or else they can be spinning in opposite directions. The situation is somewhat analogous to the earth-sun system, where the earth's direction of rotation is the same as the direction of the sun's rotation. However, in the case of the hydrogen atom, the stability of the configuration is greatest when the spin direction of the electron is opposite to the spin direction of the proton. Therefore, whenever a hydrogen atom possesses spins that are in the same direction for both particles, it has a certain tendency to reverse this situation; the electron flips over so as to become opposite in spin direction, and in the process it looses a small amount

Neutral Hydrogen

Fig. 9-2. A close-up of the center of the Rosette Nebula, taken with a larger telescope than Fig. 9-1, and showing the mixture of bright, glowing gas and dark dust clouds. *(Kitt Peak National Observatory.)*

Fig. 9-3. How the 21-centimeter line of neutral hydrogen is formed.

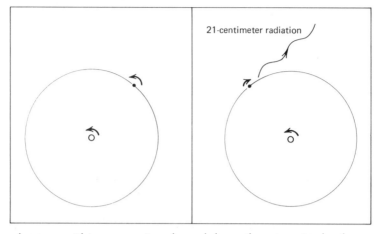

of energy. This energy is released from the atom in the form of radiation. Because of the extremely small amount of energy involved, the radiation is of long wavelength, 21 centimeters. The energy of a burst of radiation (a quantum of light) is inversely proportional to the wavelength. For large amounts of energy emission, the wavelength of the emitted light will be short (visible light, x-ray, etc.), while for very small amounts of energy emission, the wavelength is long (radio wavelengths).

Fig. 9-4. Radio telescopes at Owens Valley in California, used together as an interferometer. *(California Institute of Technology.)*

When van de Hulst calculated the properties of the neutral hydrogen radiation, he noted that the probability of the spontaneous emission of 21-centimeter radiation by neutral hydrogen atoms would be very small. This is because of the very long time it takes for an electron with parallel spin to flip naturally into the more stable configuration of antiparallel spin. Calculations indicate that on the average, a hydrogen atom will take 11 million years before achieving this reversal, and therefore the emission of 21-centimeter radiation by hydrogen is an exceedingly slow and weak process. Nevertheless, in the vast open spaces between the stars, there is room enough for hydrogen atoms to exist undisturbed for long periods of time and in large numbers—sufficiently large so that the 21-centimeter radiation from neutral hydrogen can be detected by sensitive radio receivers on the earth. Detection was first accomplished in 1951 by E. M. Purcell and H. I. Ewen at Harvard, who found the 21-centimeter radiation to be emitted most intensely from the plane of our Galaxy, the Milky Way.

Neutral hydrogen is now observed extensively by large telescopes at radio observatories all over the world. The study of the interstellar medium has

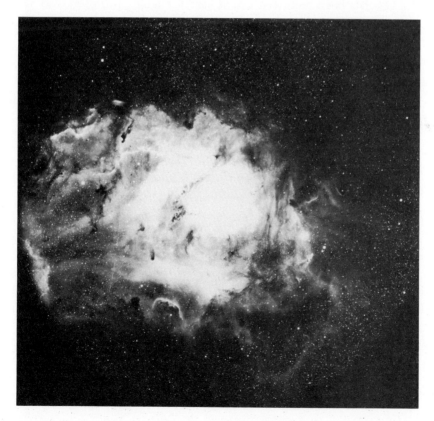

Fig. 9-5. A complex of gas and dust in Sagittarius, popularly called the Lagoon Nebula, photographed by the 4-meter telescope. *(Kitt Peak National Observatory.)*

advanced tremendously since the discovery of the 21-centimeter neutral hydrogen radiation, and it has provided us for the first time with a comprehensive picture of the nature of our own Galaxy. Unlike optical radiation, the long-wavelength radiation from neutral hydrogen can penetrate through the dark, dusty regions of our Galaxy from any position in the Milky Way. This allows us to observe the properties of distant and otherwise unobservable regions of our stellar system.

Because of the very long wavelength of neutral-hydrogen radiation, it is necessary to have large telescopes to observe it well. As explained in Appendix E, the resolution of a telescope depends on the wavelength of the radiation observed. In order to have the kind of resolution achieved at optical wavelengths by large telescopes, it would be necessary to build an immense telescope for neutral-hydrogen study. The wavelength of neutral hydrogen is approximately 500,000 times greater than the wavelength of optical light, and therefore one would have to build a telescope 2,500 kilometers in diameter in order to have the same resolving power that the Mount Palomar 200-inch telescope has at optical wavelengths. Of course, no single radio telescope of that size has been built. The largest used for neutral hydrogen work is 1,000 feet in diameter and located in Puerto Rico.

By radio interferometry (Appendix E), it is possible to achieve resolutions at long wavelengths that are as good as or better than that of the Palomar telescope, by using radio telescopes that are separated across long distances on the earth, across continents or oceans, for example. The most common configuration for neutral hydrogen studies involves smaller base-line radio interferometry, with separations of a few kilometers or so. These give resolutions of the order of 10 seconds of arc, which is adequate for most current studies of the distribution of neutral hydrogen in our Galaxy and in nearby galaxies.

Neutral hydrogen in our local Galaxy is found to be concentrated in long,

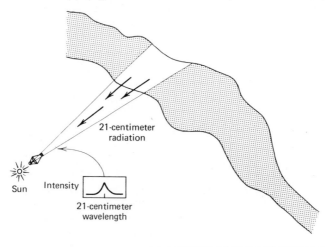

Fig. 9-6. Measuring the arms of the Galaxy using the 21-centimeter line of neutral hydrogen.

Figure 9-7. The Cambridge radio interferometer in England. *(Cambridge University.)*

narrow features that are identified as spiral-arm segments. The sun lies near the edge of one of these features, which are found to extend over the entire face of our Galaxy. The general pattern of these large complexes of neutral hydrogen is quite complicated and difficult to establish in detail, because of the difficulty radio astronomers have in determining distances to the sources of neutral-hydrogen radiation. Scientists are forced to assume distances based on the motions observed, which can be accurately determined from the observed wavelengths by using the Doppler shift to measure the radial velocities of the gas clouds. The densities in these large complexes of neutral hydrogen are extremely small, averaging about 2×10^{-25} grams per cubic centimeter, or about 1 hydrogen atom per 10 cubic centimeters. This represents only about 2 or 3 percent of the total mass density in the solar neighborhood, with most of the mass concentrated in the stars themselves.

One of the important recent discoveries about the interstellar medium is the amount of hydrogen in the form of molecules. Hydrogen molecules consist of two hydrogen atoms joined together, rather loosely, by what are called molecular bonds. These molecules cannot exist in stars (except in the very outer parts of cool stars), because the high stellar temperatures cause the atoms in the molecule to break apart. The molecular bonds are not strong enough to withstand the constant high-energy bombardments of particles at high temper-

Molecular Hydrogen

Fig. 9-8. The relation between the energy and the wavelength of light.

Long wavelength: each photon has small amount of energy

Short wavelength: each photon is highly energetic

ature. Hydrogen molecules have not been detected from ground-based observatories because their observable spectral features are located deep in the infrared spectrum, at wavelengths that are completely blocked out by the earth's atmosphere. Therefore, it has been possible to measure the molecular-hydrogen concentration in interstellar space only by space telescopes. The first detections, made in 1972 by the *Copernicus* satellite, showed a large concentration of hydrogen molecules, but only in very dusty regions of space. It is

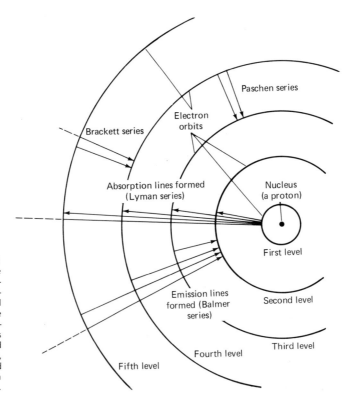

Fig. 9-9. An early and simple model of the atom, which was developed by Niels Bohr, pictured the protons (and neutrons, if any) in the center of the atom, surrounded by the electrons which revolved around the nucleus in orbits, much like planets around the sun. This diagram shows a hydrogen atom.

found that dust must be mixed in with the gas in order to allow the hydrogen atoms to join together, as they form best on the surfaces of grains of dust.

**Excited and
Ionized Hydrogen**

In areas where the neutral-hydrogen density is unusually high, it is found that some of the hydrogen emits optical radiation. Typically, such gas clouds are very much hotter than the general interstellar medium because of the presence of one or more exceedingly luminous and hot stars. The stars are there because they have formed out of the interstellar material in the gas cloud and, as pointed out in the following chapters, young, very recently formed stars generally include some with very high temperatures.

According to the blackbody laws, stars of high temperature emit large amounts of short-wavelength, high-energy radiation (see Chap. 3). This radiation generally heats up the surrounding gas to temperatures of 8,000 to 10,000 K. The high-temperature hydrogen atoms, because of the energy imparted by the short-wavelength radiation from the central stars, are excited or ionized. An *excited* hydrogen atom is one for which the electron has been given energy so as to place it in an orbit of higher energy than the base or ground state (Fig. 9-10). Because of this extra energy, the atom can emit light spontaneously when the electron then "falls" down to an orbit of less energy. Each time an electron returns to an orbit of less energy, a *photon* (a bundle of

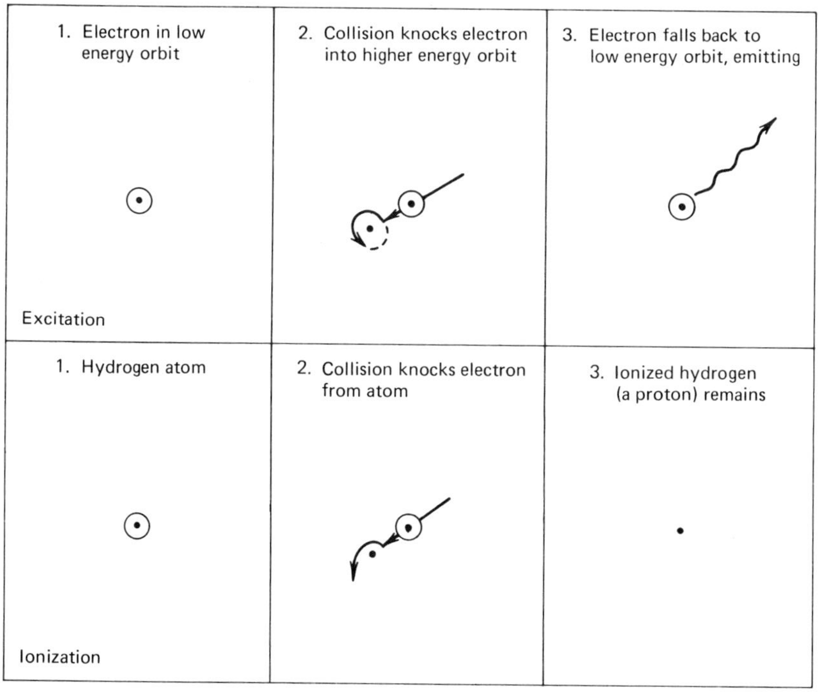

1. Electron in low energy orbit	2. Collision knocks electron into higher energy orbit	3. Electron falls back to low energy orbit, emitting
⊙		
Excitation		
1. Hydrogen atom	2. Collision knocks electron from atom	3. Ionized hydrogen (a proton) remains
⊙		•
Ionization		

Fig. 9-10. Excitation and ionization of hydrogen atoms.

light) is emitted by the atom. The wavelength of the emitted photon depends on the amount of energy difference between the initial orbit and the orbit to which the photon falls; in many cases, the amount of energy is appropriate for the light to be optically visible. Thus, an excited hydrogen atom is capable of emitting light that can be detected by optical telescopes on the earth.

Ionization of hydrogen atoms in gas clouds occurs when the energy imparted to the electron is sufficiently great so that it escapes entirely from the atom, leaving only the proton. In the space near the bright hot stars, where the energy density from the stars is highest, the hydrogen atoms are all ionized most of the time. The frequent ionization and recombination make this region glow brightly at optical and radio wavelengths. The sphere of ionized hydrogen is called a *Stromgren sphere*, after the Danish astronomer B. Stromgren, who first calculated the properties of such objects.

The visible portions of the gas cloud are referred to by astronomers as H II regions, while exterior to them, where ionization and excitation is not going on, the neutral hydrogen regions are called H I regions. H II regions are found to have typical sizes of about 15 light years and typical total masses of a few hundred times the mass of the sun. Densitites in H II regions are ten to a hundred times the average density of neutral hydrogen, with a typical value on the order of 10 atoms per cubic centimeter.

Other Elements and Molecules in Space

Although hydrogen is the most abundant element in interstellar space, measurements have shown the presence, especially in H II regions, of other elements, as well as of molecules and free radicals. (A *free radical* is a combination of two or more elements which form a portion of a molecule, but not a complete stable molecule). At optical wavelengths, it is possible to detect helium, oxygen, nitrogen, carbon, and sometimes other elements because of the presence of emission lines emitted by these elements (by the same process described above for hydrogen). The abundance of the elements can be determined if the general properties of the H II regions are measured. They are found to be hundreds of times less abundant than hydrogen, with the exception of helium, which is always found to be approximately $1/10$ as abundant as hydrogen.

At radio wavelengths, it has become possible in recent years to detect several molecules and free radicals in space, particularly in large and complex H II regions. The most conspicuous is the radical OH, which is an incomplete version of the water molecule. It is remarkably abundant in the denser portions of H II regions in various parts of our Galaxy. Complex physical and optical properties of the material in these dust and gas clouds, which involve processes called *optical pumping*, lead to an enhancement of the radiation from the OH radical.

Similarly, the water molecule (H_2O) is detected in some H II regions, as are almost fifty other molecules, including, for example, ammonia (NH_3) and

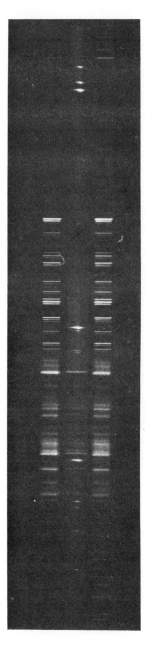

Fig. 9-11. The emission spectrum of a gas cloud.

Fig. 9-12. The Orion Nebula, a complex of gas, mostly hydrogen, which shines because of the excitation of the atoms caused by stars embedded in the gas. *(Manastash Observatory.)*

Fig. 9-13. A Stromgren sphere is a volume of space where the hydrogen is ionized and excited by the ultraviolet light emitted by a central hot star.

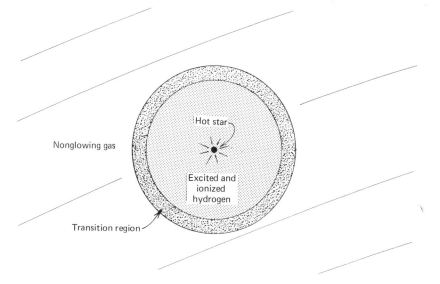

Hot star

Nonglowing gas

Excited and ionized hydrogen

Transition region

Fig. 9-14. Several large emission regions near the center of the Large Magellanic Cloud. *(Cerro Tololo Interamerican Observatory.)*

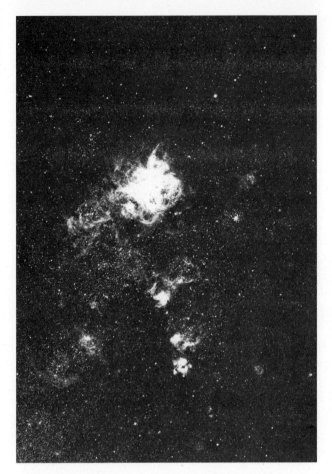

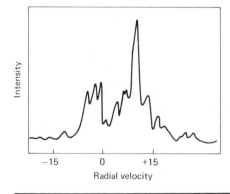

Fig. 9-15. The radio spectrum of H_2O (water vapor) from a giant gas cloud.

Radial velocity

formaldehyde (HCHO). These molecules are apparently formed by the interaction of radiation, gas, and dust, all of which are found to be important elements in the eventual formation process of stars as well.

**9-2
INTERSTELLAR
DUST**

**General
Reddening**

More than 30 years ago, astronomers recognized the effects of a layer of widespread interstellar dust in which the sun and most stars are imbedded. The dust was detected by its effect on the light of distant stars. It was found that stars of a similar spectral type, and therefore of a similar intrinsic color, showed different observational colors according to their distances. The farther away the star, the redder its measured colors. The most likely cause of this effect is the absorption of light by interstellar dust. Small dust particles tend to absorb and scatter short-wavelength radiation more than long-wavelength radiation, and so light coming from a distant star will have much of its blue radiation lost, but less of its red radiation. This gives it an anomalously redder color, as measured from the earth.

At the same time that a star's light is made redder by interstellar dust, it is also made dimmer. Tests show that stars lose about one-half of their light to dust for every 1,000 parsecs (3.3 thousand light years) of distance from the earth. In the astronomical magnitude system, this is the equivalent of a change of 0.75 magnitudes per 1,000 parsecs. This value is found to be the average for the solar neighborhood, but is much smaller than the amount of general absorption in some parts of our Galaxy, especially in the direction toward the galactic center, where stars are completely obscured from view at distances of only a few thousand light years from us. The center of our Galaxy itself, which is about 10,000 parsecs (33,000 light years) from the earth, shows an absorption of light amounting to something on the order of 20 magnitudes, which is the equivalent of a factor of 100 million. Thus, only about one-millionth of one percent of the light from a star at the center of our Galaxy reaches as far as the distance of the earth; the rest is absorbed and scattered by interstellar dust between us. Other distant sources of light in our Galaxy also cannot be seen at optical wavelengths because of the effect of interstellar dust, and therefore our information on them comes only from infrared and radio measurements.

Polarization

Appendix D describes the various features characterizing the nature of light and related radiation. In classical physics, light waves can be thought of as consisting of vibrations, like ocean waves. These vibrations occur in all directions (unlike ocean waves), as a light wave moves along through space. Light is polarized when some of the directions of a light-wave vibration are removed from the wave. Polarization of natural light occurs when light is

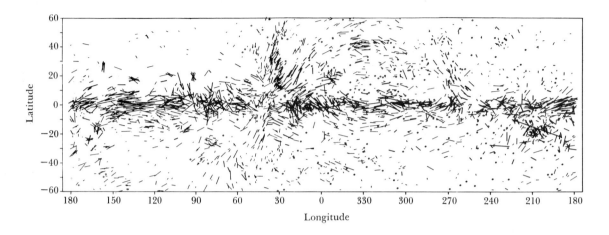

Fig. 9-16. Polarization of stars near the Milky Way, which is caused by interstellar dust that is apparently aligned by the magnetic field of the Galaxy. *(From D.S. Mathewson.)*

reflected off certain surfaces, such as the surface of a lake, when it passes through a sheet of polarizing material, such as the mineral mica, or when it passes through a dust cloud in which the grains of the dust are all aligned in approximately the same direction. Polaroid sunglasses are effective because the direction of sunlight polarization reflecting off a freeway, for example, is approximately perpendicular to the direction of light polarization passing through the sunglasses. Therefore, sunlight reflected off such a surface is greatly reduced in brightness, while natural unpolarized light coming from the surrounding scenery is much less reduced in brightness.

In interstellar space, it is found that starlight in some directions is measurably polarized, and that the amount of polarization depends not only on the direction but also on the distance of the star. From such evidence, astronomers infer that the polarization is due to the interstellar dust layer, which therefore must be made up of grains of dust that are irregular in shape and largely lined up in a parallel direction, with their major axis approximately parallel to the plane of the disc in our Galaxy. This feature of the interstellar dust is an indication of the presence there of a large-scale (but weak) magnetic field, which has recently been measured to be approximately $1/100,000$ the strength of the magnetic field of the earth (measured at its surface). Even this weak field is enough to align most of the grains of the interstellar dust medium along its lines of force, thus producing the polarizations that are measured for distant stars.

Distortion of the Galactic View

The general interstellar reddening produces a very decided distortion of our view of the local star system. Because of the absorption of distant starlight, which becomes virtually undetectable at distances that are much less than the size of our star system, the optical view of our local Galaxy is very limited. Before the general interstellar reddening was discovered, astronomers inferred that the thinning out of stars at distances on the order of a thousand light years

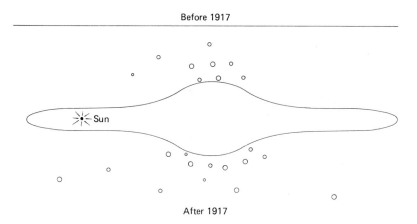

Before 1917

After 1917

Fig. 9-17. The Galaxy as recognized before and after Shapley's study of globular clusters.

was due to the fact that the edge of our star system was being reached. They therefore inferred not only much too small a size for our system, but were also led to the erroneous conclusion that the sun was near the center of the Galaxy. These errors were only recognized in 1917 when Harlow Shapley examined the distribution of objects that lie above the dust layer in our Galaxy, the globular star clusters (Chap. 19), and showed that our galactic system was in fact about 10 times bigger than was originally thought, and that the sun was much nearer to the edge of the system than to its center.

Dust Clouds

In addition to the general, fairly smoothly distributed interstellar dust, there are large regions in the galaxy where the dust is much more dense than average, and forms dust clouds and dust complexes. These can be detected as large dark areas against the Milky Way, behind which few if any stars can be seen. Measurements of the properties for these dust clouds can be made by determination of the colors and luminosities of the stars in their direction. This allows a separation of those stars that lie in front of the dust cloud from those that lie behind, and a determination from the latter of the optical properties of the cloud itself. Typically, the dust clouds are on the order of 25 light years across, and have total masses of dust grains estimated to be approximately 10 times the mass of the sun. The average density in a dust cloud is about 3×10^{-24} grams per cubic centimeter for the dust, which is about $1/10$ the density of gas in a typical H II region. The largest dust clouds have higher densitites than this and range in size up to several hundred light years, with a hundred or more

Fig. 9-18. A large gas cloud known as the Trifid Nebula, photographed with the Kitt Peak Observatory 158-inch telescope. Notice that the gas cloud is apparently embedded in a dust cloud, which obscures the stars surrounding it. *(Kitt Peak Observatory.)*

solar mass equivalents in total mass content. The smallest are almost unresolved objects, a light year or less across, called *globules*. These were discovered and studied extensively by the astronomer Bart Bok, who showed that they are probably related to star formation. Sometimes dust clouds are associated with large H II regions and complex mixtures of dust and gas; it is found, in fact, that the areas where interstellar molecules are detected are always rich in dust in the form of high-density dust clouds.

Circumstellar Dust

A third form in which dust is detected is that of an envelope of circumstellar dust around individual stars. Occasionally, a star with an extreme amount of reddening is found, thus indicating that the star itself must be imbedded in a very thick dust envelope. Such stars often have the appearance from earth of being "infrared" stars, because of the fact that they are frequently so highly reddened by the dust that they are only conspicuous at infrared wavelengths.

Fig. 9-19. One of the most familiar examples of a *dark nebula* or dust cloud is the Horsehead Nebula in Orion. *(Kitt Peak National Observatory.)*

Some of these objects have measured temperatures of only a few hundred degrees, rather than the typical thousands of degrees for stars; however, this temperature is inferred to be that of the circumstellar dust cloud rather than of the star imbedded in it.

Reflection Nebulae

A further source of information on the interstellar dust comes from the few cases where dust clouds are illuminated by nearby stars, producing a reflection nebula. (*Nebula* comes from the latin word for cloud). Reflecton nebulae look something like H II regions on direct photographs, but their differences are very striking when examined spectroscopically. An H II region has a spectrum that is made up almost exclusively of emission lines that are produced by the atoms of the gas, which are excited by the stars embedded in them. Reflection nebulae, on the other hand, show spectra that are identical to that of the star that illuminates the dust. The dust is merely reflecting the starlight toward us in the same way that planets reflect sunlight, without emitting light of their own. Densitites of reflection nebulae are roughly one particle for every 10^8 cubic centimeters of space. This is the equivalent of about 1 grain of dust in a volume the size of a large room.

The Nature and Origin of the Dust Grains

Extensive study of the properties of the dust can lead to some conclusions about its physical character and its chemistry. Most of the important information comes from the way in which the dust absorbs light, especially in the ultraviolet. Recent measurements from space telescopes have extended our knowledge of the interstellar absorption of dust far into the otherwise invisible ultraviolet portion of the spectrum. It is found from such studies that the dust is probably made up of fairly common elements, such as hydrogen, carbon, nitrogen, and oxygen. It is still not clear exactly what the detailed makeup of the dust must be, but it is possible that it consists partly of carbon atoms that have grown together to form graphite, perhaps mixed with ices or with an icy coating made up of other common·elements.

The dust in interstellar space has an origin that also remains somewhat obscure. It is found that dust can be formed around very cool stars (Chap. 13), and it may be that much of the dust that we see has been formed in

Fig. 9-21. A small dark nebula, known as Barnard 335, one of the very opaque objects called Bok globules. *(Steward Observatory, photograph by B. J. Bok.)*

circumstellar envelopes which have subsequently been dispersed. Alternatively, the dust may gradually grow by collisions of dust particles in interstellar space, some of which stick to each other, thus gradually building up a particle out of individual atoms and molecules. It is also seen from observing exploding galaxies (Chap. 20) that vast amounts of dust are formed in the process of these explosions by a means that is still completely unknown. A radio galaxy like M 82, which may have experienced an immense explosive event several million years ago, is filled with large amounts of dust whose origin was probably somehow related to the violent event.

Both the nature and the origin of the interstellar dust remains uncertain, and a great deal more research is necessary before this important element in the interstellar medium is understood. It is also clear that the interstellar dust is an essential ingredient for star formation, because it acts as a catalyst for the condensation of gas clouds into stars.

Fig. 9-22. Infrared telescope used by Leighton and others to map the infrared sources in the sky. *(Courtesy California Institute of Technology.)*

1. Compare the interstellar gas density with that of the solar wind. Which is greater?.

2. How could a large gas cloud be detected if there were no hot stars imbedded in it? Could it be detected optically if there were no dust mixed with the gas? If there were dust mixed in? How?

3. Calculate the mass of the *Lagoon Nebula* if its diameter is 100 parsecs and its average density is 3 atoms per cubic centimeter.

4. The bright star Deneb in Cygnus is at a distance of 1,000 parsecs. Assuming interstellar dust in its direction to be typical in density, find how much of Deneb's light is lost because of dust before it reaches us.

5.* The "Gulf of Mexico" is part of a large dust cloud associated with the "North America Nebula." Its distance is about 200 parsecs and its diameter about 1°. Using a scale drawing (or trigonometry), calculate its size in parsecs. Assuming a typical density, calculate its mass.

The objects discussed in this chapter are among the most interesting to look at with small telescopes. Charts to help you find the following are at the back of the book.

1. There are several bright H II regions (gas clouds) visible through binoculars or small telescopes in the Milky Way region.

WINTER OBJECTS

    a. *Orion Nebula* (M42, NGC 1976). About 30 minutes in apparent diameter, and about 460 parsecs distant, the Orion Nebula has a total mass of gas of about 700 suns. Four bright central stars, called the Triangulum, provide much of the ultraviolet light to the nebula. A large cluster and association of stars is imbedded in the region, with an age measured to be about 10^6 years. Several stars are found to be still forming.

    b. M43 (NGC 1982). Near the Orion Nebula, this gas cloud is part of the same gas-star complex. It is only about 2 minutes in diameter.

SUMMER OBJECTS

    c. *Lagoon Nebula* (M8, NGC 6523). In Sagittarius and nearly 25 minutes in diameter, this bright H II region is 1,500 parsecs distant. Its mass is about 3,000 suns. Very luminous O stars are involved in the gas, and an association of young stars surrounds it.

    d. *Trifid Nebula* (M20, NGC 6514). About 10 minutes in diameter, this object is at about the same distance as the Lagoon Nebula. Its mass is about 300 suns. Dark lanes of dust can be seen crossing it. (In Sagittarius).

    e. *Horseshoe Nebula* (M17, NGC 6618). At a distance of 1,800 parsecs, this gas cloud has an apparent maximum diameter of about 10 minutes and is long and narrow. Its mass is measured to be about 1,000 suns. (In Sagittarius).

2. Dust clouds can also be seen through binoculars, with small telescopes, and some even with the unaided eye. Some are visible as dark blank areas in the Milky Way, and some show up as bright nebulae, which are illuminated by the reflection of starlight from nearby stars.

WINTER OBJECTS

    a. *M78* (NGC 2068). A bright reflection nebula in Orion. About 5 minutes in diameter and 500 parsecs distant, M78 is illuminated by a nearby hot star. It is faint enough to require a dark sky and at least a 3-inch telescope.

    b. *Horsehead Nebula* (IC 434). A difficult object to see visually, except with a very dark sky and wide-field telescope, this complex of bright and dark nebulosity is only about 300 parsecs distant. (In Orion).

    c. *S Monocerotis Cloud.* A small dark nebula, conspicuous on deep photographs, near the star S Monocerotis. It is $1\frac{1}{2}°$ in diameter and 600 parsecs distant.

    d. *The Taurus Dark Cloud.* This well-studied dust cloud, best seen on photographs, is 2° in diameter and 120 parsecs distant. The total amount of dust is the equivalent of 5 suns.

SUMMER OBJECTS

    e. *North American Nebula* (NGC 7000). A mixture of gaseous and dust clouds that looks like North America on photographs, this complicated object is about 200 parsecs distant for the nearest dust cloud and up to 800 parsecs distant for the most distant clouds. A dark sky and wide field will show its outline. (In Cygnus).

f. *Cygnus Rift.* A large dark cloud complex visible with the naked eye against the Milky Way in Cygnus, about 5° across and 30 to 40° long. There are several smaller discrete clouds within it, with typical absorptions of some 1 to 3 magnitudes and distances of 200 to 600 parsecs.

g. *Sagittarius and Scutum Clouds.* This part of the Milky Way is also cut visibly by intervening dust clouds as seen without a telescope. One well-studied small cloud in Scutum, 3° in diameter, is 220 parsecs distant and shows a total of 3 magnitudes absorption.

3. All the above objects can be seen and studied better by photography. If your college has an equatorially mounted telescope, you can photograph the larger objects with an ordinary camera that is mounted firmly on the telescope. Follow the object, exposing for an hour or so, using the telescope as a guider to track accurately. If the telescope has a drive, try mounting film at the focus and taking a photograph using the telescope as the camera. You will have to guide with an attached telescope to keep the images round. By using red and then blue filters, the effects of absorption by dust can be discovered.

10

STAR FORMATION

10-0
GOALS

This chapter is designed to give you insight into what causes a star to form. An important conceptual device, called the *color-magnitude diagram*, is introduced; once you see how it works, it makes stellar formation and evolution easier to understand.

10-1
HOW HOT AND
HOW BRIGHT

Before a star can form, a cloud of gas and dust must in some way separate itself from the interstellar medium. It must contract to a dimension small enough so that it can achieve a gravitational identity of its own (separate from the surrounding medium), from which it eventually will collapse to stellar dimensions. Theoretical calculations have shown astronomers how this process can occur. It is found that star condensation depends upon certain conditions that are not generally prevalent throughout our Galaxy, but which do occur in certain areas. However, before it is possible to describe the development of a star, it is necessary to become acquainted with a conceptual device used by astronomers to describe the essential properties of a stellar object. This is the diagram, expressed in various units, that shows the relationship between the temperature and the luminosity of the star or stars.

The four most significant properties of a star are its surface temperature, its total luminosity, its mass, and its chemical composition. If these properties of a star are known, then its other properties can be calculated. For example, if the temperature and luminosity of a star are known, it is possible to calculate the radius of the star by using known physical laws that govern the conditions in the interior of the star. Other characteristics of a star—such as its rotation, its magnetic field, and its surface activity—have only minor effects on the large scale structure of a star.

The two quantities, temperature and luminosity, are the most convenient ones for separating stars into different kinds and for understanding stellar evolution. Both the temperature and the luminosity of a star change in time, whereas neither the mass nor the chemical composition changes markedly for a given star, at least not until its very late stages. Therefore, astronomers commonly plot the temperatures and the luminosities of stars in diagrams, both to show differences between stars of different kinds and to illustrate the changes undergone in individual stars as they evolve. The temperature-luminosity diagram in its various forms is the most used and most useful graph in stellar astronomy, and it provides a background against which most of the events in the life history of a star can be followed.

Temperature-
Luminosity Plots

There are three principal sets of units used for plotting the temperatures and luminosities of stars. The most fundamental involves the plot of the total luminosity of a star, either in units of the sun's luminosity or in energy units (such as ergs per second), and its temperature, as measured for its surface. Figure 10-1 illustrates this manner of plotting the data for stars of various kinds.

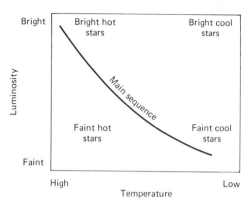

Fig. 10-1. The temperature-luminosity diagram for stars separates the normal stars (in the main sequence) from those with unusual conditions.

The plotted quantities—the total luminosities and the surface temperatures—are not directly measurable, but are quantities derived from theoretical calculations of stellar structure; they are, therefore, the quantities used most often by theoreticians.

A second set of units commonly used in astronomy involves observational parameters—the *absolute magnitude* of a star, which is related to its luminosity, and the *color* of a star, which depends on its temperature. Color-magnitude diagrams are illustrated in Figs. 10-2 and 10-3. These are observational quantities that are derived from data obtained directly at the telescope, and are therefore reliably known for actual stars. However, in order to compare observations with theory, it is necessary to convert absolute magnitudes into luminosities and colors into temperatures (or vice versa). With both these conversions, certain difficulties arise. For the conversion of absolute magnitude into luminosities, correction to the magnitudes must be made to account for the fact that much of the luminosity for many stars exists at wavelengths that are not measured by the typical magnitude systems used observationally. For example, an exceedingly hot star, with a surface temperature of 25,000 K, radiates most of

Color-Magnitude Diagrams

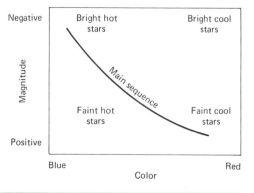

Fig. 10-2. The color-magnitude diagram provides a means of plotting the same kind of information as in Fig. 10-1, but in directly measurable quantities.

Fig. 10-3. The color-magnitude diagram for stars in the solar neighborhood shows a conspicuous *main sequence* of normal stars, and a few stars off the main sequence.

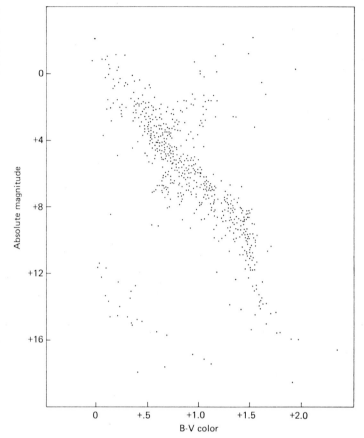

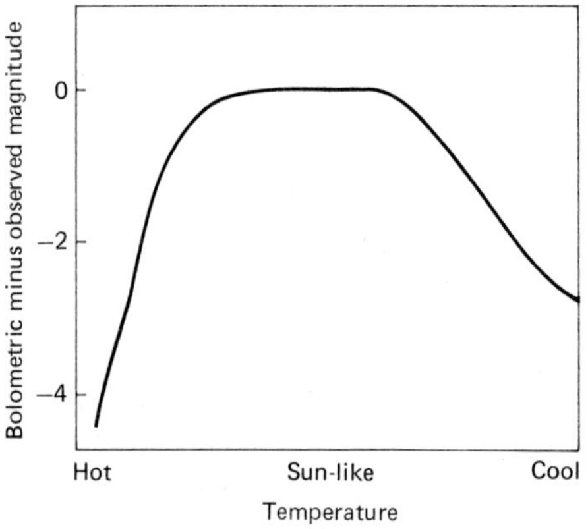

Fig. 10-4. Apparent vs. bolometric magnitudes. The bolometric correction is largest for stars of the lowest and highest temperatures, because the bulk of their radiation is not in the visual wavelength band.

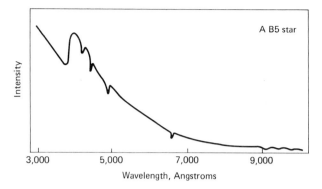

Fig. 10-5. The energy distribution of a high-temperature star shows that the intensity is highest in the ultraviolet (left side), and probably even higher beyond the limit (3000 Angstroms) of earth-based telescopes. *(Vilnius Observatory.)*

its energy in the ultraviolet part of the spectrum beyond the limits of terrestrial telescopes. The measured absolute magnitude of such a star, therefore, ignores most of its luminosity, and a large correction to this absolute magnitude must be made to make it a realistic measure of the total brightness of the star. Such a corrected magnitude is called the *bolometric magnitude*; it is usually based on calculations of models of a star's atmosphere that provide an indication of the amount of radiation coming from the star at various wavelengths. Also, space telescopes have allowed a direct observation of total luminosities for bright stars, since they are not limited to the wavelengths for which the earth's atmosphere is transparent. These measurements have provided a further means of obtaining bolometric corrections for stars of different kinds.

In order to convert measured colors to surface temperatures for stars, the relationship between these two quantities must also be established by theoreti-

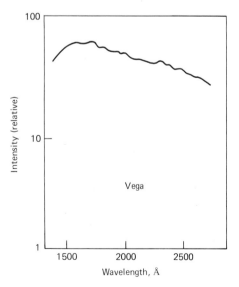

Fig. 10-6. The far ultraviolet part of the spectrum of the hot star Vega was mapped by the Copernicus orbiting telescope. *(From a NASA diagram.)*

cal calculations of the properties for the star's atmosphere. Generally speaking, the temperature is low for stars of redder color, but the exact relationship between the measured colors in a given photometric system and the surface temperatures is difficult to calculate. It is fairly well known now for most kinds of stars; it is found that the relationship depends upon several properties of a star, such as its composition, its radius, and its evolutionary state.

The H-R Diagram

Another plot of data that is frequently used by astronomers is the so-called *H-R diagram,* which is an abbreviation for Hertzsprung-Russell diagram. It was named after two astronomers who first realized the significance of the main sequence. The H-R diagram (Fig. 10-7) plots the magnitude of a star against its spectral type. Normally, the spectral type is given a code name according to the appearance and relative strengths of different lines of different elements in the spectrum of the star. These types are discussed more fully in Chap. 11 and for the purposes of this section it is only necessary to know that the different codes relate to different temperatures. A type O star, for example, is a very high temperature star, while a type M star, on the other hand, is a star of lowest temperature.

10-2 CONDENSATION OF INTERSTELLAR CLOUDS

One of the more difficult problems in understanding the formation of stars is the question of how an interstellar gas and dust cloud initially begins to condense into a discrete small object. A perfectly smoothly distributed interstellar medium, if acted on by no outside force, would remain perfectly smoothly distributed, and no stars would ever form. For stars to begin condensation, some internal inhomogeneity must exist, or some external force must act on the material. There are several possibilities for such effects, and although it is not yet known which of these predominates, it is probably true that all play some part in the formation of individual small gas clouds, the *protostars* (from the Greek word *proto* meaning "the first").

One possible means for triggering the condensation of a star is through the

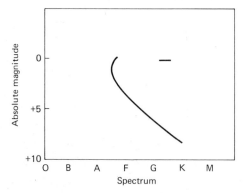

Fig. 10-7. The H-R diagram for the Hyades star cluster shows a main sequence and a few stars off the main sequence. The spectrum types are explained in Chapter 11.

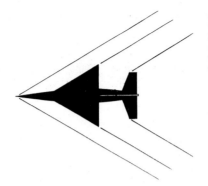

Fig. 10-8. A sonic boom is produced by an object moving through a medium at a speed faster than the speed of sound in that medium.

effects of a shock wave. When an object moves more rapidly through a medium than the natural speed of waves in that medium (for example the speed of sound), then it produces a shock wave, which is very unlike an ordinary wave. Familiar examples of shock waves include those that produce a sonic boom when an aircraft travels at supersonic velocities through the atmosphere. As Fig. 10-8 shows, the sonic boom is the result of the piling up of the disturbance (the noise) that is produced by the aircraft along a cone formed by the expanding spheres of sound waves. The aircraft is at the apex of this cone, and when some portion of the cone intersects the ground a considerable disturbance is felt. Another example is the bow wave created by a boat when it is moving faster through the water than the speed of the natural waves on the water. The disturbance caused by the boat moves off in a triangle, forming a bow wave that is a two-dimensional analog of the three-dimensional shock wave produced by an airplane.

Shock waves are expected in the intergalactic medium in certain areas where disturbances have been produced by one or another different mechanisms. The most likely explanation of the spiral arm structure of our Galaxy (Chap. 19) involves a rapidly moving density fluctuation through the Galaxy, which is expected to produce a shock wave. This shock wave can heat up the medium through which it passes in a manner similar to the piling up of air in the shock wave that produces the sonic boom of an airplane; this small area of heated gas will then cool rapidly by radiation. There will then be a compression exerted on these denser regions after their temperatures reach values lower than the surrounding media. This pressure can cause the cloud so formed to decrease in size further and to become distinct from its surroundings, thus forming a protostar.

Once something triggers the beginning of condensation, hydrogen atoms in the gas can recombine to form hydrogen molecules, thus raising the density by a factor of about 2 because of the greater mass of the particles, and therefore accelerating the condensation process. If dust grains are present in the mixture, these can effectively depress the temperature of the cloud by increasing the rate at which hydrogen molecules can form and by colliding with hydrogen atoms.

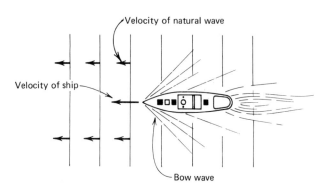

Fig. 10-9. A bow wave is produced by an object moving through a medium at a velocity greater than the natural wave velocity of the medium.

The rate of the resulting decrease in the temperature depends upon the density of the dust, so that a cloud with a high dust density decreases more rapidly in temperature. Another effect is the "shadowing" of grains by each other; this produces a cloud interior that is darker than the surrounding less-dense areas, due to the fact that in the interior less light from stars in the neighborhood reaches the grains there. In addition, gravitational instabilities, due to random fluctuations in the density of the interstellar gas and dust, also can play a role in the eventual condensation of clouds.

10-3
COLLAPSE

Once a gas and dust cloud has separated itself from the general interstellar medium and has reached a sufficiently small size, depending upon its density and mass, it can collapse naturally because of its own gravitational pull. Once the gravitational attraction of the cloud itself on its remote parts is larger than the gravitational attraction of these parts due to the surrounding material in the galaxy, the collapse will proceed unhindered. The only effect that is contrary to the inward pull of the mutual gravitation of the material of the cloud is the heating of the gas due to the resulting collisions. By the various mechanisms mentioned in Sec.10-2, this heating up will be mitigated enough so that condensation will occur; the gravitational collapse will be strong enough to continue in spite of the heating and the resulting increased pressure. It is calculated that a typical protostar—with a diameter of a few light years and a total mass about equal to the mass of the sun—will collapse in about 10 million years. This calculation is simply based on the time that it takes for a particle to "free fall" to the center of an object of that size and mass. It is similar to the calculation of how fast an object would fall under gravity in any situation, as would an apple from a tree or a meteor to the earth.

In the case of gravitational free fall onto the surface of the earth, the fall is impeded slightly by the presence of the earth's atmosphere, which slows the falling body. Eventually the fall is stopped by collision with the earth's surface itself. If the earth were not there, but there were a point at its center with the earth's mass, an object would of course fall all the way to the center, as it is only the blocking of the fall by the hard surface of the earth that stops an object from continuing downward. In the case of star condensation and collapse, the protostar eventually becomes sufficiently small and sufficiently high in density so that the free fall is stopped due to the gas pressure of the hot atoms, which have been heated up through the process of the collapse. When the free fall collapse is so halted, the object becomes a star and then proceeds through the various phases of evolution discussed in the next chapter.

A protostar that begins as a gas cloud a few light years across contains a certain amount of angular momentum, the property of rotation. This is due to the fact that the galaxy of stars out of which it forms is rotating around a central point with a rotation speed that is different at each different distance from that central point. This means that the innermost point in the gas cloud is rotating around the Galaxy with a different velocity from that of the outermost point. When the gas cloud begins to condense, this difference in velocity at different points in the cloud becomes a rotation of the cloud. Typically, the differences in velocity are only about 0.1 kilometer per second at the beginning of collapse, but because angular momentum of an object is conserved (that is, the amount of rotational motion of an object remains the same unless acted upon by an outside force), the velocity of rotation becomes greater and greater as the body becomes smaller and smaller. It is calculated that for a typical protostar, the rotation rate becomes immense when it decreases in size to approximately the size of a star, with calculated surface velocities even greater than the velocity of light. Not only is this very different from what is observed for real stars, which rotate fairly slowly, but it is not even physically possible for an object to rotate faster than the velocity of light. Therefore, astronomers have been led to the conclusion that there must be some process by which the angular momentum of a condensing star can be removed. One way for this to happen is for the angular momentum to be concentrated in smaller objects surrounding the star, such as the planets and comets that might be formed during the

**10-4
ROTATION**

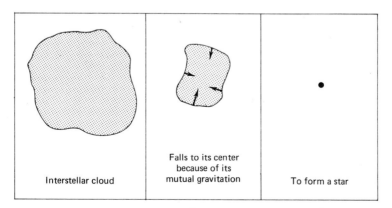

Interstellar cloud   Falls to its center
because of its
mutual gravitation   To form a star

Fig. 10-10. Free fall of a protostar involves gradual collapse of a large cloud of gas and dust.

Fig. 10-11. A protostar acquires angular momentum because its outer parts are moving more slowly in the rotating galaxy than its inner parts.

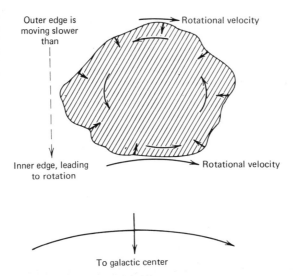

Fig. 10-11. A protostar acquires angular momentum because its outer parts are moving more slowly in the rotating galaxy than its inner parts.

condensation process. In the solar system, in fact, 99 percent of the angular momentum is contained in the planets. A second means of getting rid of angular momentum is for the star to expel material outward either in bursts or continuously, as in the case of the gas of the solar wind (Chap. 6). It is calculated that the solar wind is probably powerful enough during the condensation phase that it can carry away the majority of the angular momentum for most stars, with the remainder primarily going into the formation of a planetary system, if any.

Supermassive Stars Theoretical calculations of star condensation show that exceedingly massive gas clouds, greater than a few hundred times the mass of the sun, cannot form

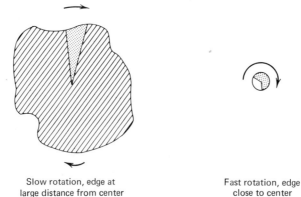

Fig. 10-12. Because of conservation of angular momentum, a small object must rotate faster than a larger one to have the same total angular momentum.

Fig. 10-13. This area is rich in star clusters, dust clouds, and gas clouds, and is an example of a good locus of star formation.

into stable single systems. Instead, they fragment and form multiple stars or star clusters. They might also form what are called *supermassive stars*. For stars of very large mass, more than a million times the mass of the sun, it is found that the gas pressures that exist when it collapses are not sufficient to withstand the

Fig. 10-14. A close binary star. *(U.S. Naval Observatory, Flagstaff station.)*

immense gravitational pull, and the star can collapse to a very small, incredibly dense object called a *black hole*. The remarkable properties of black holes are discussed in Chap. 17.

**10-5
PRE MAIN
SEQUENCE
EVOLUTION**

After the condensation of a gas cloud into a separate entity, the protostar, the force of gravity acts to pull the entity together. The resulting contraction of the protostar proceeds rapidly. In a relatively short time, at least on a cosmic scale, a gas cloud contracts into a small luminous object that eventually becomes a star. Unfortunately, the early stages of this contraction occupy much too small a fraction of a star's lifetime for astronomers to have an opportunity to observe them, unless it is by a lucky accident. Therefore, most of our knowledge of these early stages must come from theoretical investigations. Because of the availability in recent years of highly competent and rapid computers, it is now possible to calculate the properties of contracting stars reliably.

This section is entitled "pre main sequence evolution" because its subject is the stage of evolution through which a star passes before it officially enters the ranks of stars on what is called the *main sequence*. In the temperature-luminosity diagram, it is found that a stable star will occupy a point on a certain line (Fig. 10-15). The line is called the main sequence, and the location of the point is determined by the mass of the star. The position of the main sequence line on the diagram depends somewhat on the chemical abundances in the star, but the main sequences for different abundances lie rather close to each other. Therefore, a grand general main sequence of stars exists in our Galaxy even

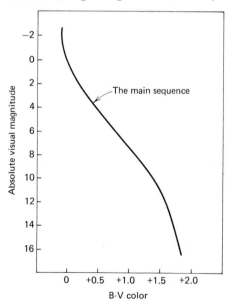

Fig. 10-15. The main sequence is the locus of stars during most of their lifetime.

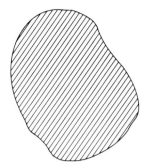

Potential energy is the amount
involved if all the matter

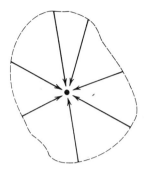

should fall to the center

Fig. 10-16. Potential
energy.

where the chemical abundances are somewhat different from one place to another. By far the great majority of observed stars exist on the main sequence line, and this is due, as described in the next chapter, to the fact that the majority of the time that a star spends as a luminous object is spent in the stable configuration characterized by the main sequence. Stars observed elsewhere in the temperature-luminosity diagram are passing through a fleeting period in their evolution, either on their way to the stable main sequence or on their way away from it toward their eventual death as stars.

**10-6
CONTRACTION**

Before a star can arrive on the main sequence and become a normal, stable, luminous object like the sun, it must contract from the immense size of an interstellar gas cloud down to stellar dimensions. At first, the "velocities of fall" for the atoms mount rapidly. As the protostar becomes smaller, the distances between its different parts become smaller, and therefore the gravitational attraction between them becomes greater. This causes a rapid acceleration of the contraction so that in only about a year or so, the protostar can contract from an immense cloud to a relatively dense sphere of gas (still not a star) about the size of the earth's orbit.

Potential Energy

In the contraction phase, the forms of energy contained in the protostar are converted into other forms. As a large gas cloud, the protostar has a very large amount of what is called *potential* energy. This is the energy that it could potentially realize as motion due to the large distance between its parts and the gravitational attraction of these parts. For example, an apple hanging on a tree branch possesses the same kind of potential energy; if the stem holding the apple to the tree should fail, the apple would fall to the ground because of the earth's gravitational pull on it. Just before hitting the ground, the apple would have a considerable amount of kinetic energy, enough perhaps to split the

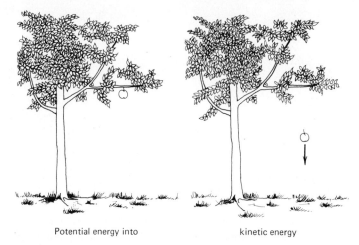

Potential energy into   kinetic energy

apple on contact with the ground (or to awaken the musing Newton, as in the fable). While on the tree branch, that apple is said to have a potential energy equal in amount to the kinetic energy that it would attain if it were released from the tree's grip, and if it could fall all the way to the earth's surface. Similarly a protostar, as it begins its contraction, has a vast amount of potential energy that is equal to the amount of energy of its atoms should they all fall to the center of the protostar under its own gravity.

After it is reduced in size by the contraction, the star has less and less potential energy. Therefore, this form of energy storage is lost continuously. An apple on a tree branch has a large amount of potential energy, but this amount decreases as the apple falls to the ground. Just before it hits the ground, it has the least potential energy. If it were suddenly stopped only an inch or so from the ground and then allowed to drop, the amount of energy it would attain due to that drop would be very much smaller than if it were allowed to fall all the way from the tree. In both the case of the apple and the case of the protostar, therefore, potential energy is lost continuously due to its being converted continuously into another form of energy.

Conversion of Energy

The loss of potential energy in the apple is converted into energy of motion as it falls. Falling slowly at first, the apple rapidly gains velocity as it is hurled toward the ground. Before collision with the ground, it has a large amount of *kinetic energy* (energy of motion) due to its high velocity. In this case, then, the potential energy of the apple has been converted into energy of motion. On striking the ground, the energy of motion will then be converted into some other form of energy—that energy required to cause destruction of a portion of the apple's structure by bruising it or splitting it, and that energy required to produce a dent in the soil at the point of impact.

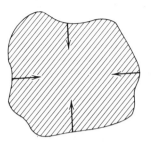 

Fig. 10-18. Conversion of energy—protostar's potential energy is partly converted to kinetic energy when it collapses.

Potential energy into   kinetic energy ($\frac{1}{2}$) and radiation ($\frac{1}{2}$)

In the case of the protostar, the potential energy is converted into two other kinds of energy. At first, before the atoms reach a high enough density in the contracting cloud to collide with each other, the energy is converted into motion, as in the case of the apple. However, as the cloud continues to contract and the density of particles continues to increase, the atoms interact with each other. This causes the material to radiate light, because it is possible for electrons in the atoms to be excited. In some cases the atoms are ionized. When the electrons fall back from higher energy orbits to lower energy orbits, they emit radiation, and as this radiation is a form of energy, it can account for some of the lost potential energy. Approximately one-half of the potential energy during this stage is converted into energy of motion (thermal or kinetic energy), and approximately one-half is lost to the star by radiation, much of which escapes from the star out into space.

Fig. 10-19. A young stellar association like this one is rich in newly formed stars. *(Harvard Observatory.)*

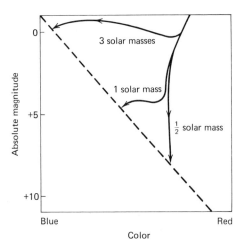

Fig. 10-20. Hayashi tracks in the color-magnitude diagram represent a simplified version of how stars behave as they contract.

3 solar masses

1 solar mass

$\frac{1}{2}$ solar mass

Absolute magnitude

0

+5

+10

Blue

Red

Color

10-7 HAYASHI TRACKS

A comprehensive set of calculations for the properties of contracting stars was carried out in the 1960's by the Japanese astronomer Chushiro Hayashi. These calculations, which were made on a high-speed computing machine, involved an immense amount of mathematics, taking into account each small stage in a contracting star and the detailed physical properties at each point within it. The path that a protostar follows on the temperature-luminosity diagram is called its Hayashi track on the diagram. The star can be plotted on the temperature-luminosity diagram when its contraction has proceeded for a few years and when its surface temperature has reached a value of 2,000 to 3,000° absolute. At this point, the temperature in its center is very high, on the order of 100,000 K. This high temperature results from the conversion of potential energy to energy of motion for the atoms, of which temperature is a measure. More

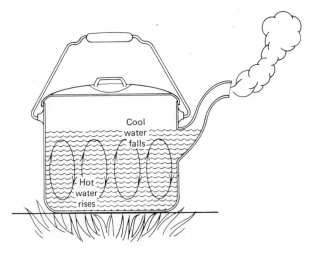

Cool water falls

Hot water rises

Fig. 10-21. Convection in a kettle bears some resemblance to convection in stars.

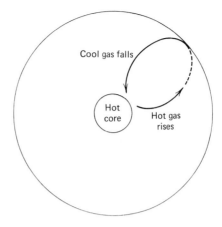

Fig. 10-22. Convection in a protostar mixes the outer layers by bringing hot gas from central regions up to the surface.

recently, R.B. Larson has improved on Hayashi's models of pre main-sequence evolution.

Convection

Because of the great difference in temperature between the central parts of the protostar and its outer parts, the interior of the gas cloud is unstable, and calculations show that convection must occur. The protostar is something like a kettle of boiling water. The water at the bottom of the kettle, the portion that is in contact with the stove, receives energy, thus causing it to rise. For that reason, convection occurs in the form of a rolling boiling motion. The hotter water from the bottom moves to the top and the cooler water from the top is placed closer to the heat source. As this boiling convection continues, the difference in temperatures is kept relatively small (of course, for boiling water there is a further effect due to the evaporation of some of the water to produce steam).

For a contracting protostar, the very hot central temperature leads to convective boiling so that the entire star consists of zones that are circulating outward from the center, bringing the cooler portions into the inner regions where they are heated and then expelled again by the convective motion. The

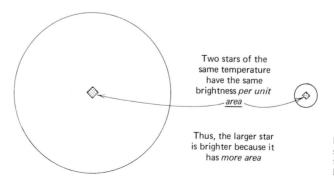

Fig. 10-23. Luminosity vs. surface area for stars of the same temperature but different size.

convection of a protostar is a characteristic of the initial phases on the Hayashi track of a star. Because of it, the temperature of the star at the surface remains fairly constant for a considerable amount of time. In the first few million years, the temperature changes only by 1,000 to 2,000°, going from a surface temperature of about 3,500 K to a surface temperature of about 4,500 K, in a typical case.

During this period, the luminosity decreases because of the continuing contraction of the star. Because the total luminosity of a star depends upon the amount of the star's area that is radiating away energy, and because the surface area of a star depends upon its size, the total luminosity must decrease as the size decreases (unless, of course, internal sources of energy can heat up the star). During these initial phases of nearly even temperature, the luminosity decreases by a factor of about 1,000. For a protostar with a mass about that of the sun, the brightness after only 2 or 3 years is about 500 times the solar luminosity. After about 10 million years of contraction, this brightness has decreased to about half the solar luminosity.

Radiative Stages   After thousands of years (for a high-mass star) or millions of years (for a low-mass star), the configuration of mass within the protostar becomes such that convection gradually stops in the center. Transportation of the energy from the hot center to the outer parts can no longer occur through the circulation of the hot gas outward from the center. Instead, the energy moves outward from the center by means of radiation. Light from the hot central areas is transmitted outward very slowly, because it must move through the process of interaction with each different layer of atoms. By then the star has reached sufficiently high densities that a photon of light has a difficult journey outward through the object. This radiative stage sets in rapidly at a particular time in the contraction phase, and the result is a stabilizing of the luminosity of the star. From then on, its luminosity stays nearly the same, but the temperature of the entire star increases. Thus, an object which for its first million years of life appears as a

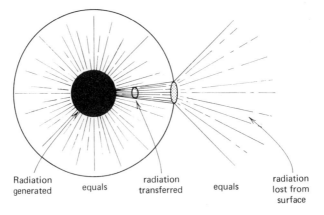

Fig. 10-24. Radiative stars release energy from their hot cores by radiation transmitted through the various layers.

Radiation generated   equals   radiation transferred   equals   radiation lost from surface

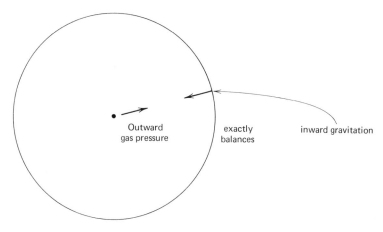

Fig. 10-25. Stability on the main sequence is a balance of two opposing forces.

dull red globe gradually becomes bluer and bluer as it nears the main sequence.

Finally, when the central temperature becomes exceedingly high, on the order of 10 million degrees absolute, there is enough energy in the atoms at its center so that they can begin nuclear reactions. It takes this exceedingly high temperature for atoms to have enough energy of motion to collide with each other to the extent that they can ignite a nuclear reaction; in this case, hydrogen is converted into helium (Chap. 11). As soon as nuclear burning ignites in the center of a protostar, the protostar becomes self-illuminated by means of the nuclear burning at its center, and it then enters the final short step of its journey to the stable main sequence.

Stability of the star on the main sequence is reached when a balance occurs in the interior of the star. Inward pressure, the pull that causes the contraction in the first place, is due to gravitation. Outward pressure, on the other hand, is due to the high gas and radiation pressures caused by the excessive temperatures in the center of the system, which are maintained by nuclear reactions. When the inward pressure equals the outward pressure exactly, contraction stops, and the star becomes stable and constant in size.

Rate of Evolution

Because of the increasing outward pressure, which is the result of the high temperatures in the center of the star as it contracts, the rate of contraction is very much slower in the later phases on the Hayashi tracks than at the beginning. In only 1 or 2 years, the gas cloud contracts from an immense object to something on the order of the size of a planetary orbit. From then on, the contraction becomes slower so that the entire convective phase of contraction occupies millions of years. For a star with the mass of the sun, convection ends after an interval of 10 million years, and the radiative phase of contraction onto the main sequence takes an additional 17 million years. The rate of evolution is much faster for stars of higher mass and slower for stars of lower mass.

10-8
OBSERVATIONS
OF
CONTRACTING
STARS

It is possible to find stars in the Hayashi stages of pre main sequence evolution by looking in areas of the Galaxy where the ingredients for star formation are abundant. For example, in the Orion Nebula, which is a large complex of gas and dust, astronomers have measured the luminosities and temperatures of a large number of stars and have found that some of them lie above the main sequence in the color-magnitude diagram (Fig. 9-12). These stars are among the fainter stars that are associated with the Orion Nebula, and they are clearly in the Hayashi phase of contraction; their position is understood as a consequence of the much longer time that it takes stars of low luminosity and low mass to contract to the main sequence than higher-mass stars. The brighter stars in the Orion Nebula are already on the main sequence because of their shorter contraction times. It is found that most of the stars in the Orion complex have contracted onto the main sequence in the last 1 or 2 million years. Many other areas in the local Galaxy and in other galaxies have stars in these early stages, as shown in Fig. 10-26.

Fig. 10-26. Herbig-Haro objects are groups of young stars mixed together with what appear to be newly forming stars. *(Lick Observatory.)*

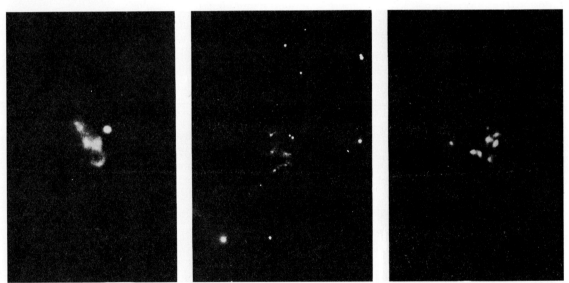

QUESTIONS

1. The sun's absolute magnitude is about +4. What is its color?
2. Plot a color-magnitude diagram and an H-R diagram for the 20 brightest stars. (See Appendix K.)
3.* If a gas cloud 1 parsec in diameter collapses to form a protostar in only 10 millions years, what is the *average* velocity of collapse?
4.* If the velocity of rotation of the outer part of the gas cloud in question 3 is

0.1 kilometers per second, what is this velocity when the cloud is only 100 astronomical units in diameter, if the angular motions are conserved everywhere?

5. What are some common examples of potential energy? In these examples, can the potential energy be converted to some other form of energy?

6. What convective processes in nature can you think of? What are the heat sources and how and where is the heat dissipated?

7.* Calculate the approximate absolute magnitude of the contracting sun when it was only 2 or 3 years old (as measured from the beginning of contraction), and after 10^7 years.

THE MAIN SEQUENCE

This chapter's basic purpose is to explain the remarkable stability of stars during most of their lifetimes. We have a real personal stake in this knowledge; our sun is in its main-sequence phase and its stability or lack of it could mean life or death for us on earth.

A star spends most of its normal lifetime in a situation of relative stability. Stars of various masses existing in this condition make up what is called the *main sequence of stars*. On a color-magnitude diagram, main sequence stars lie on a line extending from very luminous hot blue stars to very faint cool red stars. A random sampling of stars in our Galaxy will contain mostly main sequence stars, because the other phases through which stars pass during their evolution occur much more rapidly, both before and after the main-sequence phase.

**11-1
EQUILIBRIUM**

The reason for the relative stability of the main-sequence phase of stellar evolution is the long period then of the state of equilibrium in the interior of the star. *Equilibrium* occurs when there is a balance of forces; in the case of the main sequence, the balance is between the force of the gas and radiation pressure (caused by the extremely high temperatures in the star's interior) and the pull of gravity.

For an object as massive as a star, which can have a mass many thousands of times greater than that of the earth, the inward gravitational force can be very large. At the surface of the sun, for example, the inward pull of gravity is thirty times greater than that which we experience at the surface of the earth. For other stars, for example for the main-sequence star Sirius, the surface gravity is similar (twenty times that for the earth). This strong inward force, which was the ultimate cause of the formation of the star in the first place, exerts at all levels in the star a strong downward push that must be counteracted by a compensating outward force.

**The Inward Pull
Of Gravitation**

The amount of gravitational force at the surface of a star depends upon both the size of the star and on its total mass. For a star of very large size, but of relatively modest mass, the surface gravity is small. For example, for the star Betelgeuse, which has a radius 400 times greater than the sun's but a mass only about 20 times greater (the mass is actually quite uncertain), the surface gravity is 10^4 times less than that at the solar surface. Very massive stars of small dimensions, on the other hand, can have almost unbelievably large surface gravities (Chap. 15).

To counteract the inward force of gravity, an outward pressure exists due to the gas pressure of the high-temperature stellar materials and to the radiation pressure caused by the tremendous amount of light produced in the hot glowing interior.

**The Outward
Pressure**

The gas pressure is high because of the high temperatures; the amount of pressure that a gas exerts depends directly upon its temperature. This can be visualized in terms of the motions of the atoms of a gas because temperature depends upon the velocities of these atoms. For a high-temperature gas, the atoms are moving rapidly, and if they therefore collide with each other (or with a layer of stellar material above them), they will exert a large pressure. If they were cooler, they would be colliding with less energy and thus the pressure would be smaller. In the interior of a star, the gas pressures are exceedingly large and the temperatures are high. For example, halfway from the surface of the sun to the center, the gas pressure is the equivalent of 10^8 times the atmospheric pressure at the surface of the earth, and the temperature is 2×10^6 K. At the center of the sun, the pressure is calculated to be 10^3 times greater yet, and the temperature is 14×10^6 K.

11-2
NUCLEAR
ENERGY

The gas and radiation pressure from the hot interior of a star cannot by themselves withstand the inward gravitational pressure indefinitely. The star is continuously losing energy by radiation from its surface, and the light that we see represents a drain on the energy of the star that ultimately would lead to its collapse. Depending upon its mass, the collapse time for a star without any heat source in its interior is only a few thousand or a few million years.

When astronomers first calculated this time, it was thought that perhaps the energy from the sun was entirely derived by the loss of the potential energy due to a gradual shrinking of the sun in size. However, calculations showed that such a means of producing sunlight could not possibly have lasted long enough to keep the earth at its present temperature for even a small fraction of geological history. From ample evidence of fossil life, it is clear that the surface temperature of the earth has remained nearly the same for billions, not just millions of years. Therefore, there was for some time a mystery as to how the sun's equilibrium was maintained; clearly some unknown force was responsible for generating the high temperatures and the great amount of energy in the centers of the sun and other stars. Whatever the mechanism, it was readily seen that it must be a very remarkable one to be able to keep stars shining for billions of years. The most reasonable suggestion that people could come up with was that the stars derived their energy somehow from the nucleus of the atom, and in this way astonomers anticipated the discovery and exploitation of nuclear energy long before this immense storehouse of power was demonstrated to exist by laboratory experiments and atomic power plants.

It is now rather well understood how the sun and other stars continue to shine with their observed luminosities. The process is nuclear *fusion*, which involves the fusing together of the nuclei of one element to form the nucleus of another. For stars on the main sequence, the fusion is of four hydrogen atom nuclei (protons) to form one helium atom nucleus. The way this actually occurs is found to depend upon the temperature at the center of a star. If the temperature is less than a few million degrees, fusion cannot occur at all. In the temperature range between 5 and 15 million degrees, fusion occurs by means of what is called the *proton-proton cycle* (see Fig. 11-1). For higher temperatures, a more roundabout process is important, and is called the *carbon cycle* (Fig. 11-2). In both cases, the net change is from four separate protons to one single helium nucleus consisting of two protons and two neutrons tightly bound together. Involved in this transformation are two important changes: first, two of the protons must lose their positive charge and become neutrons; and second, a large amount of energy must be given off. It is the second of these points that is important to the question of energy in the stars.

The conversion of matter into energy was one of the important discoveries of the early part of the twentieth century. Einstein and other contemporary physicists showed the equivalence of matter and energy and derived a mathematical relationship between the two that led to the famous equation

$$E = mc^2$$

where E is the amount of energy released when a mass m is converted from matter into energy. Because c, the velocity of light, is a very large number, and because it is squared in this equation, a very small amount of matter will be converted into a very large amount of energy. That is the principle behind the nuclear energy plants on the earth, and it is the reason why the sun can

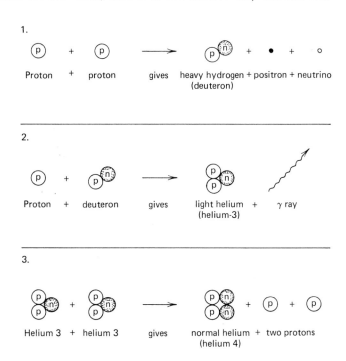

1. Proton + proton gives heavy hydrogen + positron + neutrino (deuteron)

2. Proton + deuteron gives light helium + γ ray (helium-3)

3. Helium 3 + helium 3 gives normal helium + two protons (helium 4)

Fig. 11-1. The proton-proton cycle is the nuclear reaction that provides the sun's energy.

Fig. 11-2. The carbon
cycle is the reaction that
provides the energy of
very massive stars.

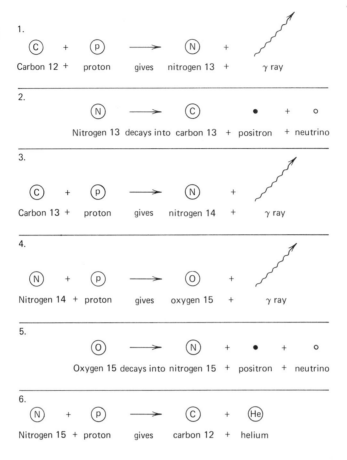

generate large amounts of energy for billions of years without showing a noticeable change (without depletion of a significant amount of its matter).

The Rate of Production of Energy

The amount of energy that is released every time four hydrogen atoms are converted into one helium atom is quite large, considering the very small amount of matter that is lost. The mass of a hydrogen nucleus is 1.673×10^{-24} grams. The combined mass, therefore, of four protons is 6.694×10^{-24} grams. The mass of a helium nucleus, on the other hand, is 6.644×10^{-24} grams. Thus, only 0.050×10^{-24} grams is the mass difference, and it is this amount of matter that is converted into energy by the proton-proton cycle. Using the formula $E = mc^2$, it is found that the energy for each helium nucleus formed is 4.5×10^{-5} ergs. This means that for every gram of helium formed, about 10^{19} ergs of energy is released, which is the equivalent of about 250,000 kilowatt hours of energy. The total rate of energy radiated by the sun is about 4×10^{33} ergs per second. Since it takes 1 gram of helium formation to produce 10^{19} ergs, and approximately one-hundredth of the mass of the helium atom is converted into energy, the rate of mass loss to the sun is 4×10^{12} grams per second, which is about 4 thousand tons per second, or 100 billion tons per year.

Stellar Lifetimes

This process obviously cannot go on forever, but it is simple to see that even at the rate of thousands of tons per second, the mass lost to the sun because of production of nuclear energy can continue for a very long time. The mass of the sun at the present is 2×10^{33} grams. If it is losing mass at the rate of 4×10^{12} grams per second, and if 1 percent or so of its total mass can be converted into energy by this process, the total expected lifetime of the sun must be 5×10^{18} seconds, which is about 1.5×10^{11} (150 billion) years. Thus, clearly even at its very impressive rate of production of energy, the sun can continue to shine for many billions of years.

Actually, calculations of the processes going on in the center of stars show that only about 10 percent of the hydrogen can be converted into helium, because of the fact that nuclear fusion can only occur at very high temperatures which in turn only occur at the center of the star. Therefore, in the outer parts of the star, the temperatures never reach high enough values for this conversion, and only the core of the star is converted into helium by the nuclear processes. Thus, a more realistic estimate of the lifetime of the sun would be about 10^{10} (10 billion) years *(on the main sequence)*.

It is found that the rate of energy production depends upon the central temperature of a star. For a very hot star, the rate of production is very much faster than for a cool star. This is because of the fact that at very high temperatures, the protons in the center of the star have much greater energies and can react with each other to form helium atoms much more readily. The total mass and the total luminosity of a star determine both the rate of energy production and the amount of fuel available in the core of the star. From this fact, astronomers have derived the following formula, which gives the approximate lifetime on the main sequence for a star in terms of these quantities, with the mass M and the luminosity L in units of the sun's mass and luminosity.

$$T(\text{years}) = 10^{10} \, \frac{M}{L}$$

As can be seen by substituting in a value of 1 solar mass for the M and 1 solar luminosity for L, this formula predicts that the solar main sequence lifetime is 10^{10} years.

Another example is the bright star Sirius. Its total mass is 3 times that of the sun, and its total luminosity is approximately 100 times that of the sun.

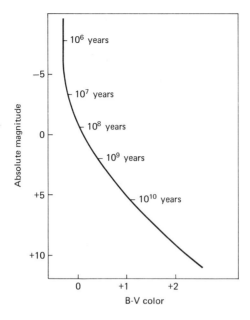

Fig. 11-3. Main-sequence
lifetimes depend upon
the absolute magnitude
of a star.

Therefore the expected lifetime T is 3×10^8 years. This calculation shows that stars that are brighter and hotter than the sun have shorter lifetimes. Sirius, if it had formed at the same time as the sun did 4.5×10^9 years ago, would have long ago disappeared from view, as its life expectancy on the main sequence is only 300 million years.

The brightest stars in our Galaxy have absolute magnitudes of about -10, which is approximately 1 million times the luminosity of the sun. The masses of such stars are about fifty times the mass of the sun, and therefore the expected lifetimes on the main sequence for these most luminous objects are 5×10^5 years. This is a short lifetime compared to that of the sun, and these luminous stars merely produce brief flashes in the history of the Galaxy.

A final example is a faint, cool, red dwarf star, such as we find to be among the most common types of stars in the Galaxy. Consider a star with an absolute magnitude of about $+10$. Its mass is about 0.2 times the mass of the sun, and its luminosity is one-hundredth of the solar luminosity. Therefore, its expected lifetime on the main sequence is 2×10^{11} years, or twenty times the lifetime of the sun. These examples illustrate the way in which stars utilize their natural energy sources and the way in which the rate at which they use them up limits their lifetimes. It is somewhat similar to the situation on earth, where the natural resources that give us energy will be exhausted much sooner the more rapidly we use them. A star's lifetime on the main sequence depends upon the rate at which it uses its nuclear fuel, just as our lifetime as a civilization depends upon the rate at which we use our terrestrial resources.

There are two principal means of measuring the temperatures of stars, and both can be used for stars whether or not the stars lie on the main sequence. In order to determine whether a star is a main sequence star, however, it is necessary to use one of these methods to determine its temperature.

11-3
TEMPERATURES
OF STARS

One method for gauging the temperature of a star is to compare the amount of light it gives off at different wavelengths with the Planck curves for blackbodies of different temperatures (Chap. 3). A precise measurement of the color of a star, which is obtained by determining its magnitude photoelectrically at different wavelengths, can determine the temperature in this way. For most stars, it is sufficient to make measurements in the optical wavelengths available through the earth's atmosphere. For example, a common color system used for this purpose is one in which measurements are made in the ultraviolet, in the blue, and in the yellow part of the spectrum (called "visual" because it corresponds approximately to the colors to which the human eye's vision is most sensitive). Abbreviated as the UBV system, these measures are defined in a particular way according to certain standard stars in the sky to which any other star can be compared by precise photometry.

If a star is found to be much brighter in U than in B and brighter in B than in

Comparison with Blackbody Curves

Fig. 11-4. The brightest stars in a galaxy are very young. *(Kitt Peak Observatory photograph of the galaxy M81.)*

V, then it has more light emitted at the shorter wavelengths that at the longer wavelengths. The amount of these differences can be compared with Planck curves and the temperature derived. For example, the bright star Spica (the brightest star in the constellation of Virgo), is a very hot star with its maximum brightness far in the ultraviolet, beyond even the U wavelength regions measurable from the earth. Its colors can be matched with various blackbody curves, which show that its temperature must be approximately 20,000 K. On the other hand, Alpha Orionis, which was known to the ancients as Betelgeuse, has its maximum brightness in the infrared and is most luminous in the V filter, fainter in the blue filter, and very faint when measured at ultraviolet wavelengths. Its colors are similar to that of a blackbody curve of temperature approximately 3,000 K.

The color of a star is not an infallible means for measuring its temperature. For nearby stars, the measured colors are very nearly the same as the true colors of the stars, but for more distant stars, the colors are affected by the intervening interstellar dust. Therefore, a very red star at a large distance may in fact be a very hot star that appears red because of the reddening due to dust. In order to determine its temperature, some other method must be used.

Spectroscopic Determination of Temperature

If photographic or photoelectric spectra of stars are studied, temperatures can be derived by examination of the absorption lines that appear. Different atoms in the atmosphere of a star absorb in different ways, depending upon the temperature of the atmosphere. The amount of energy in the atoms at the surface of the star determines to what extent the electrons can be excited by collisions and by radiation. Consider, for example, the atoms of hydrogen and helium—the two most abundant atoms in the atmospheres of most stars. It is found by measurements in the laboratory that the electrons of helium atoms are more difficult to excite, i.e., they require more energy than the electrons of hydrogen atoms. A smaller amount of energy is needed for the electron to be thrown out of its lowest energy orbit for a hydrogen atom. Therefore, in order to

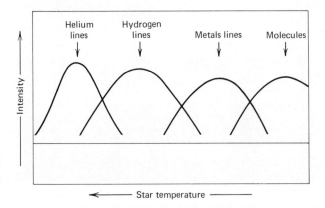

Fig. 11-5. Line strengths of different elements in stellar spectra depend upon the temperature of a star.

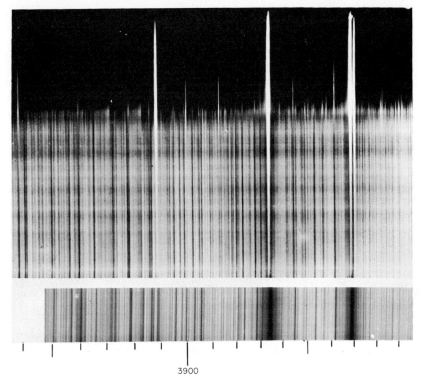

Fig. 11-6. A section of the spectrum of the sun. The upper part shows the spectrum near the edge of the sun, where emission lines are visible. *(Lick Observatory.)*

3900

observe the lines of hydrogen in the spectrum of a star, the amount of energy (and therefore the temperature), can be less than in the case of helium. By examining the strengths of hydrogen lines compared to helium lines, it is possible to determine the temperatures of various stars. Stars with weak helium lines and strong hydrogen lines are cooler than stars with strong helium lines. To carry this example one step further, it should be pointed out that when the temperature of a star is very large, more than 10,000 K or so, much of the

10 Lac	O9
τ Sco	B0
η Ori	B1
γ Ori	B2
η Aur	B3
κ Hya	B5
β Per	B8
α Peg	B9

Fig. 11-7. Spectra of hot stars. *(Copyright University of Chicago Press. Used by permission.)*

Fig. 11-8. Spectra of cool stars. *(Copyright University of Chicago Press. Used by permission.)*

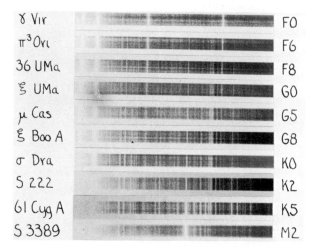

γ Vir	F0
π³ Ori	F6
36 UMa	F8
ξ UMa	G0
μ Cas	G5
ξ Boo A	G8
σ Dra	K0
S 222	K2
61 Cyg A	K5
S 3389	M2

hydrogen in the atmosphere begins to become ionized; the electron of the atom is completely severed from the proton. In this state, hydrogen can no longer absorb light, because there are no electrons to surround the hydrogen nuclei. Therefore, for even hotter stars, the hydrogen lines become fainter as the helium lines become stronger.

At cooler temperatures, other lines show up; for example, lines of ionized calcium and of iron and other metals become apparent. For stars like the sun, the temperature is about 5,800 K, and there are many lines of neutral atoms visible, many more than for higher temperatures where only the lines of helium and hydrogen are conspicuous. For yet cooler stars, these neutral atomic lines are joined by lines of molecules, such as CN(cyanogen) and CH. For the very coolest stars, the molecular lines are very conspicuous, including titanium oxide (TiO) and in some stars molecular carbon (C_2).

Astronomers have worked out a means whereby the ratios of the strengths of certain spectral lines can be used to determine the temperature of a star. A spectral classification system describes these quantitative measures and acts as a simple means of separating stars into classes of different temperatures. It was derived many years ago before the differences in the spectra of stars were completely understood. The various classes and their physical interpretations are described in Figs. 11-7 and 11-8 and in Table 11-1.

TABLE 11-1. CLASSES OF STELLAR SPECTRA

Class	Temperature, K	Diagnostic Lines
O	Over 25,000	Ionized helium
B	11,000–25,000	Neutral helium
A	7,500–11,000	Hydrogen
F	6,000– 7,500	Calcium, iron, hydrogen
G	5,000– 6,000	Ionized calcium, hydrocarbon: CH
K	3,500– 5,000	Neutral metals
M	Under 3,500	Titanium oxide

If you were to look at the spectra of a dozen stars near the sun, stars which had all been chosen because they were of about the same temperature, you would be struck by how similar the spectra would look. Careful study of the lines in each spectrum would show them, in fact, to be virtually identical. This means that the amounts of the different elements making up the spectral lines must be about the same from star to star. Astronomers' surveys show that almost all the stars in our part of our Galaxy have the same chemical composition, as nearly as we can tell from their spectra. Our sun is typical, then, and we speak of the solar abundances of the elements as being "normal." With some exceptions resulting from the process of evolution of the planets and of life, we ourselves share this characteristic; the basic pattern of the abundances of many of the elements in our bodies is just the same as it is in our sun and in most of the stars around us. This doesn't hold for all elements, of course; helium, for instance, is abundant in stars but not in us, as almost all of this very light element was lost to space from the primordial earth (Chap. 3). But with most elements, especially with the large number of trace elements, the pattern of abundance is about the same in people, dogs, mud, the moon, the sun, and the nearby stars. In this sense, the stars that you see when you look out at night are your cosmic cousins.

How can we explain our remarkable chemical relationship to the stars? If they are our cousins, it must be due to our having common ancestors. The pattern of abundance in the elements tells us that our origins are tied closely to the origins of the sun and other stars; all of us must have formed out of the same material. Astronomers now have evidence that our corner of space is largely inhabited by stars that formed out of gas and dust that were already once or twice before part of stars that had previously died, ejecting much of their material back into space. Some of these were supernovae (Chap. 16), which exploded so violently that in their dense, incredibly hot cores fast nuclear reactions built heavy elements out of the more common lighter ones. From calculations based on experiments in nuclear physics, we know which elements are easy and which are hard to form in this way, and it turns out that the observed abundance pattern is just what we would expect. For example, iron forms very easily in such a reaction, and it is abnormally common in stars and on earth, compared to elements of similar weight, such as cobalt, which is

Abundances and Stellar Populations

Fig. 11-9. From the abundances of the elements in a star, it is possible to determine stellar ancestries.

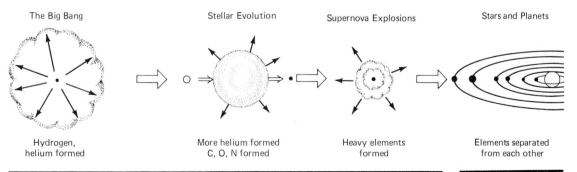

| The Big Bang | Stellar Evolution | Supernova Explosions | Stars and Planets |

| Hydrogen, helium formed | More helium formed C, O, N formed | Heavy elements formed | Elements separated from each other |

more difficult to form in a supernova and which is cosmically a hundred times less abundant. Our common ancestors, then, are the dead stars and supernovae of the distant past.

But not all stars share our cosmic family tree. A few stars near the sun have entirely abnormal-looking spectra. The most common peculiarity is *depletion*; some stars have spectra that have normal-looking hydrogen and helium lines for their temperature, but very faint lines for everything else. Careful measurements of the lines, combined with analysis of the stars' atmospheres, indicates that these stars are depleted of all heavy elements, in some cases by as much as a thousand times compared to the sun. These stars form a different chemical family, with a different family history. They clearly are not made of atoms that were as thoroughly processed by stellar evolution, death, and rebirth as ours.

For ease in talking about these differences, astronomers have roughly divided stars into two families, called *Population I* for stars with abundances like the sun, and *Population II* for stars depleted in heavy elements. There are many gradations in between, but for simplicity most stars are divided into these two classes. The only exceptions are stars that are coming to be called (almost jokingly since we haven't found any) *Population III*, stars with absolutely no heavy elements, just hydrogen and helium. Many experts believe that at the creation of the universe, conditions were such that only these two lightest elements were formed, and that there must have been an early batch of Population III stars in order to produce the heavier elements that we now see in all stars, depleted though they are in Population II.

The two observed populations of stars show other conspicuous differences besides the chemical ones. They inhabit different environments, and they move differently through our Galaxy. Like the sun, the Population I stars move in near-circular orbits around the center of our Galaxy. They are fairly narrowly confined to a flat disk, with all orbits nearly in the same plane. Population II stars, on the other hand, are very old stars that move in more elliptical orbits that are not restricted to the plane of the other stars. Orbits of Population II stars fill up a nearly spherical system centered on the galactic center.

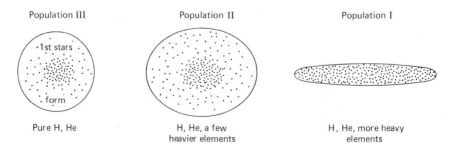

Fig. 11-10. The formation of Population II stars is from nearly pure hydrogen and helium, while Population I stars form out of material richer in heavy elements, because of previous processing in earlier generations of stars.

Population III   Population II   Population I

·1st stars ·

· form ·

Pure H, He   H, He, a few heavier elements   H, He, more heavy elements

There is a striking parallel here with our solar system, the planets of which have nearly circular orbits restricted closely to a common plane while the comets inhabit a larger sphere and have highly elliptical orbits. This is no coincidence—the similarity is due to a roughly similar history. Just as the comets may antedate the collapse of the presolar nebula to the plane in which the planets formed, the old Population II stars apparently antedate the early collapse of our Galaxy to a rotating flat disk, in which the Population I stars formed and are forming. The older stars out of the plane have very little enrichment by heavy elements because they were born before many of the heavy elements could be formed in stars and supernovae, while the disk stars, which are generally younger, have been created from well-enriched material.

There are a few stars whose spectra are so peculiar that they don't fall into any of these normal patterns. Among those being studied with great interest are the *A-peculiar stars*, which have colors and temperatures like normal A stars, but have spectra that are laced with hundreds or thousands of lines of relatively rare elements, such as manganese, chromium, and strontium, and sometimes even such unheard-of elements as yttrium, promethium, holmium, and ytterbium. These elements show abundances that in their complexity have so far defied satisfactory explanation, so several plausible theories have been worked out. One of these theories, which invoke such things as explosions of nearby supernovae or magnetic sweeping of space by a star's strong magnetic field, may someday prove to be true. Some of the A-peculiar stars have incredibly high magnetic fields, which distort the spectral line shapes in a recognizable and measurable way, showing magnetic field strengths tens of thousands of times stronger than the earth's.

11-4 STELLAR RADII

Except for the sun, the radius of a star is difficult to measure directly. The sun is the only star near enough so that its disk can be resolved simply and accurately. Because we know the distance to the sun, a measure of its angular extent, about 30 minutes of arc, tells us that its radius is 7×10^5 kilometers. For other stars, although we know the distances in many cases, it is exceedingly difficult to measure the angular extent because this extent is very much smaller than the resolving power of the telescope. This can be understood if it is considered that the sun, if placed at a distance of 10 parsecs, would have an angular diameter of only 0.001 seconds, while the resolving power of the best telescope under the most perfect conditions is about 0.1 seconds. A star the size of the sun, therefore, if it were 10 parsecs distant, would subtend the same angle as does a penny at a distance of 4,000 kilometers (2,500 miles).

Interferometry

One of the ways of measuring very small angles of extent is by means of interferometry (Appendix E). The first reliable measurements of the radii of stars were made by using a 12-foot long interferometer mounted on the top of the 100-inch Mt. Wilson telescope. The light coming from two mirrors mounted on either end of the beam was fed into the main telescope and analyzed to see if a difference could be detected in the arrival time of waves of light from the two different sides of a distant star. Measurements succeeded for only a very few stars, most of them of very large intrinsic extent (much larger than the sun). A more successful interferometric experiment was carried out at the Narrabrai Observatory in Australia. Two immense telescope mirrors, each 36 feet in diameter and made up of many smaller mirrors, were mounted on railroad cars so that their separation could be altered smoothly as necessary. This experiment is very similar in principle to the way in which radio astronomers perform radio interferometry (Appendix E). The Narrabrai Observatory succeeded in measuring many stellar radii.

Even more recently, mapping of a few very large stars has become possible by means of a technique called *speckle interferometry*. When a bright star is examined through an earth-based telescope, its image is broken up into many tiny separated parts by our turbulent atmosphere. These parts blend together to form a fuzzy image in time, but if the star is photographed or recorded with a very short exposure, it is possible to figure out from individual pieces of the sets

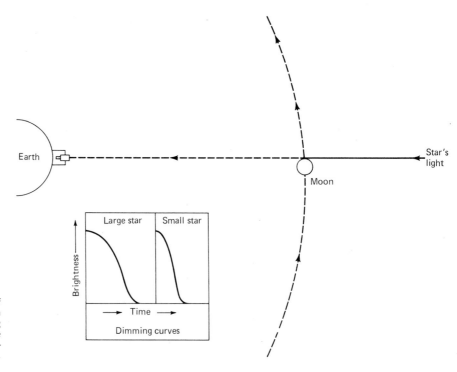

Fig. 11-11. The radius of star can be measured in some cases by recording how long it takes to fade as the moon slowly passes in front of it.

of images (with the help of a computer) what the image of the star looks like, even though it may be extremely small.

Another way to measure stellar sizes is to time how long it takes for a star to dim as it is covered over by the moon. The moon's motion is slow enough so that when it happens to pass in front of a star, an event called an *occultation*, we can see the starlight dim *relatively* slowly for large stars. The dimming curve for the star can be determined by means of rapid electronic recording devices, and from this the apparent angular size of the star can be measured. The moon's motion among the stars across the sky, as a result of its orbit around the earth, averages about $1/2$ second of arc per second of time. Therefore, a star with an apparent size of $1/10$ of an arc second, which is about as big as they come, will take $1/5$ of a second to be completely occulted. Since we can record its brightness every $1/100$ or even $1/1000$ of a second, a good record of the "gradual" eclipse can be made, and its duration immediately tells us the size of the star. For this method to work, of course, we have to be lucky enough to have the moon occult stars of interest.

Lunar Occultations

Another method of measuring the radius of a star is to take advantage of the fact that some stars exist as double systems that revolve around a common center of gravity (as do the earth and the moon). Thousands of such double or *binary* stars are known, and good fortune has oriented some of them in such a way that the plane of their orbits is parallel to our line of sight (Fig.11-12). The result is a continuous series of eclipses of each star by the other. If the distance to the binary is known, and if the eclipse is a complete eclipse (so that one star completely blocks out the other for a period of time), then the duration of the eclipse gives a measure of the diameter of the eclipsing star. Stellar radius data from eclipsing binaries agree well with the information obtained by interferometric and occultation methods.

Eclipsing Binary Stars

Fig. 11-12. The total light of a binary star changes as seen from earth if the orbit is lined up with our line of sight so as to produce periodic eclipses of one star by the other.

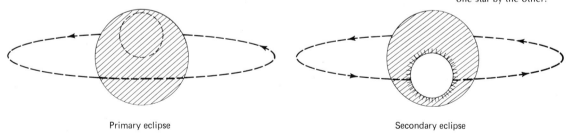

Primary eclipse   Secondary eclipse

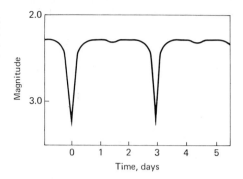

Calculated Radii

A less direct method of determining the radius of a star involves calculating what the radius has to be in order for the star to be able to radiate the measured amount of luminosity for its measured temperature. The laws of blackbody radiation show that a surface of a given temperature will radiate energy at a particular rate, given by the Planck formula (Chap. 3). Consider a square centimeter of surface on the sun; its temperature is measured to be 5,800 K. According to the Planck curve for this temperature, the total amount of energy radiated by this square centimeter will be 6×10^{10} ergs per second. From the measured luminosity of the sun, we know that the total amount of energy for the entire sun is 4×10^{33} ergs per second. A comparison of the amount of energy radiated for every square centimeter of surface of the sun with the total amount of energy radiated by the entire sun tells how many square centimeters of surface the sun must have. From the above data the result is 6×10^{22} square centimeters. The surface area A of a sphere is related to its radius r by the formula $A = 4\pi r^2$. From the blackbody laws and a measure of the temperature and luminosity of a star, it is possible to find A and then to calculate r, the stellar radius. In the above example for the sun, the radius is calculated in this way to be 7×10^{10} centimeters, a value that agrees very well with the radius measured directly.

11-5 STELLAR MASSES

The masses of stars can be determined in either of two ways. The most direct way involves a procedure similar to that used for determining the masses of the planets from the orbits of their satellites. A large fraction (perhaps over half) of the stars in our Galaxy are members of binary systems. For those double stars that are sufficiently close to the sun so that we can measure their orbital motions, it is possible to derive the masses from the sizes of the orbits and from the measured periods.

In order to establish masses of the individual members of a binary system, it is necessary to know the distance to the system and to have an accurate indication of the position of the center of mass of the system. If the two stars

ASTRONOMICAL
COLOR PHOTOGRAPHS

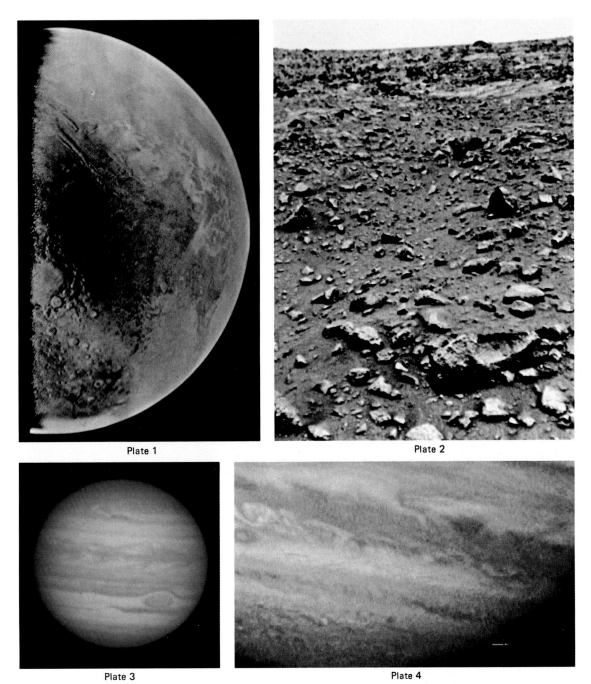

Plate 1

Plate 2

Plate 3

Plate 4

Plate 1. Mars taken from Viking 1 on its approach. The north polar region (top) is shrouded in fog, and the small south polar cap is visible (bottom). The linear feature just above center is the Valles Marineris complex of canyons. Below it, the large circular basin partly in darkness is the Argyre Planita, probably a giant and ancient impact crater. Viking 1 landed in the rather blank-looking area above Valles Marineris, in the Chryse Planita. *(NASA photo.)* **Plate 2.** The surface of Mars as photographed by the camera of the Viking 1 lander. The hollowed-out rock in the foreground was nicknamed the "wooden shoe." Most rocks in this picture show pits or vesicles; some are probably formed by wind-blown pebbles, and others are the remains of volcanic bubbles. **Plate 3.** Jupiter and its Great Red Spot. North is up in this photo. *(Kitt Peak National Observatory photo.)* **Plate 4.** A Pioneer 10 close-up of Jupiter's clouds, showing stormlike patterns in the cloud bands. *(NASA photo.)*

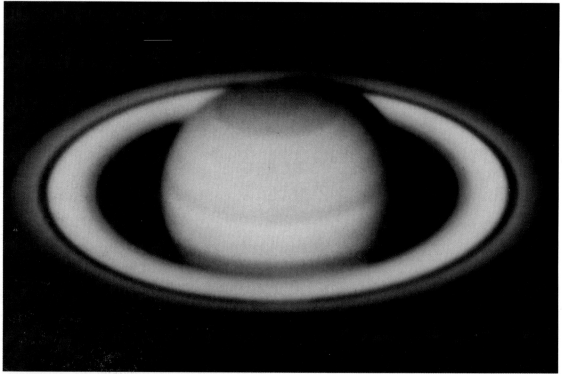

Plate 5

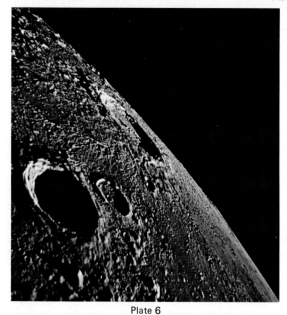

Plate 6

Plate 7

Plate 5. Saturn, photographed with the 60-inch telescope of the Lunar and Planetary Laboratory, University of Arizona. Plate 6. The moon photographed from lunar orbit by the Apollo 12 command module. The large craters are Copernicus (top), Reinhold (center left), and Reinhold B (next to Reinhold). *(NASA photo.)*

Plate 7. An Apollo 15 orbital photo, showing one of the most conspicuous sinuous rilles, known as Schröter's Valley (center). The many mounds in its vicinity are probably volcanoes. The large peaks in the background are Aristarchus (center) and Herodatus (right).

Plate 8

Plate 9

Plate 10

Plate 8. Microphotograph of a thin section of lunar soil, taken with polarized light, giving colors to different mineral types. This is a basaltic rock; the elongated crystals are mostly the mineral plagioclase, which is surrounded by pyroxenes. *(University of Washington photo.)* **Plate 9.** Microphotograph of a thin section of a carbonaceous chondrite meteorite, Allende, showing part of a chondrule (the banded circular object), representative of the features that give this type of meteorite its name. The colors are created by using polarized light and are not the natural color of the meteorite. *(D. Brownlee photograph.)* **Plate 10.** An imaginary landscape on a hypothetical planet located in the Large Magellanic Cloud near a large and relatively old star cluster. *(Painting by Don Dixon.)*

Plate 11

Plate 12

Plate 11. The eta Carina nebula, a giant complex of excited gas and dust lanes located in the southern Milky Way *(© Association of Universities for Research in Astronomy, Inc., Cerro Tololo Interamerican Observatory.)* **Plate 12.** The nebula M 16 in Serpens, a cloud of gas illuminated by a clustering of young stars. Superimposed small dust lanes of complicated structure can be seen. *(© Association of Universities for Research in Astronomy, Inc., Kitt Peak National Observatory.)*

Plate 13

Plate 14

Plate 13. The bright nebula M 20, known as the Trifid Nebula, in Sagittarius. The bright star near the center of intersection of the dust lanes is an 07 star, the main source of excitation of the gas. The fainter blue nebula shines by the reflected light of this bright blue star. (© *Association of Universities for Research in Astronomy, Inc., Kitt Peak National Observatory.*) **Plate 14.** The Orion Nebula, one of the nearest bright HII region complexes, and visible to the naked eye. It is primarily illuminated by four bright, hot stars near the center of this photo but burned out in this exposure. Dust lanes cut through and across the cloud. (© *Association of Universities for Research in Astronomy, Inc., Kitt Peak National Observatory.*)

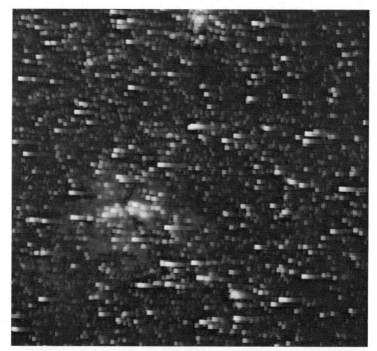

Plate 15

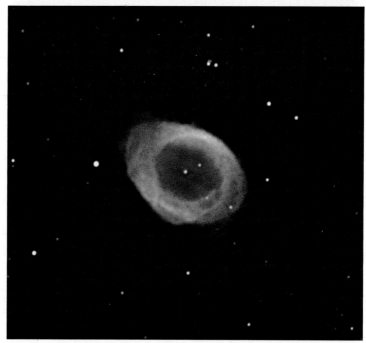

Plate 16

Plate 15. The spectrum of stars in the Milky Way and of the gas cloud eta Carina, taken with a prism placed in front of the telescope objective. (© *Association of Universities for Research in Astronomy, Inc., Cerro Tololo Interamerican Observatory.*)

Plate 16. The planetary nebula M 57, the Ring Nebula in Lyra. The hot blue central evolved star is visible at the center. (© *Association of Universities for Research in Astronomy, Inc., Kitt Peak National Observatory.*)

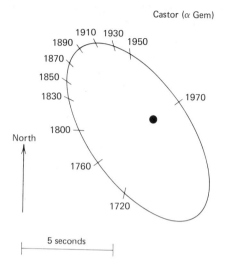

Castor (α Gem)

Fig. 11-14. A binary-star orbit can be determined by repeated observations, if the stars are well enough separated so that both are seen through the telescope.

have equal masses, then the center of mass of the system will be midway between them. On the other hand, if the two stars are very different in mass, then the center of mass of the system will be close to the more massive of the two, as in the case of the earth-moon system. Only a very few of the nearest double stars are sufficiently close so that it is possible to measure individual masses.

As is shown in Fig. 11-15, a plot of the masses of these stars and their luminosities shows a very strong correlation between these two quantities for main-sequence stars. The mass-luminosity relation, which was discovered many years ago observationally, has since been explained in terms of the internal structure of stars and the relationship between the mass, the central temperatures, and the rate of energy generation (Sec.11-1). The more massive stars are the more luminous because their central temperatures are sufficiently high so that the rate of energy generation can be very large.

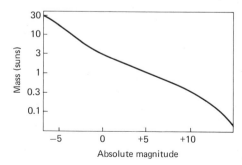

Fig. 11-15. The mass-luminosity relation is determined primarily from orbits of binary stars.

A second method of determining the mass of a star is to resort to the theoretical calculations of stellar interiors. If a star is on the main sequence, then its mass can be found from a measure either of its luminosity or of its temperature. Either of these quantities will locate the star on the line that the main sequence defines in the color-magnitude diagram. Therefore, the mass of the star can be determined by comparison with the theoretical main sequence.

Stars are found to range in mass from as much as 50 times down to as little as one-hundredth the solar mass. It is found that a star much less than one-hundredth the mass of the sun cannot form because the mass is insufficient for there to be a high enough temperature in the central core to produce nuclear reactions. The planet Jupiter, which is a little below this limit, is cold therefore because it is somewhat too small in mass for the proton-proton cycle to begin in its center. If a gas cloud with a larger mass than an upper limit that is believed to be approximately seventy times the solar mass begins to condense, theoretical calculations show that the result will not be a stable star.

11-6
STELLAR
STRUCTURE

When a star is on the main sequence, the events and detailed physical properties in the interior of that star depend upon its total mass. For high-mass, high-luminosity stars, the interior structure is very different from that of the sun; for very cool low-mass stars, the interior is sunlike. Figures 11-16 and 11-17 show the way in which the structure depends on the position of the star on the main sequence.

Radiative
Equilibrium

One of the most important differences between stars of different mass is the location within the star of zones in various states of equilibrium. The two most important means by which energy can be carried up toward the surface of the star from the energy source at the center are by means of radiation and by

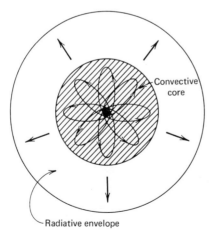

Fig. 11-16. The structure of a high-mass star.

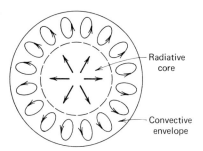

Fig. 11-17. The structure of a low-mass star.

means of convection. In the portion of the interior of a star in which the energy transportation is primarily effected by radiation, astronomers say that the layer is in *radiative equilibrium*. The radiation from the hot nuclear-burning center of the star gradually works its way outward to the surface of the star by its interaction with the matter in the star. Near the center, because of the extremely high temperatures, most of this radiation is in the form of x-rays, which are much more energetic than visible light photons. As the radiation works its way out toward the edge, the wavelengths grow, and by the time the radiation is emitted from the surface of a star, it is primarily in the wavelength range of ordinary light. Because of the great *opacity* (how opaque the material is to radiation) and very high densities of matter in the interiors of stars, it takes a very long time for the energy of a photon emitted from the center of a star to

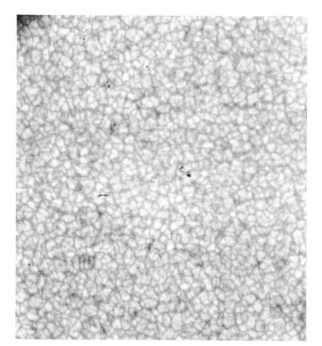

Fig. 11-18. Stratoscope photographs of solar granulation show the tops of convective cells. *(National Science Foundation and Princeton University Observatory.)*

make its way to the surface. In the case of the sun, the average time is on the order of 10^7 years.

Convective Equilibrium

Under certain circumstances, the change in temperature outward in a star is so great that radiation cannot maintain the energy transport fast enough for the star to remain in a stable configuration. In such layers of a star, convective currents are set up, and the principal means of transporting energy from the inner part of such a layer to the outer part is by means of convection. In a convective current, the hot matter at the bottom of the layer is carried up to the top of the layer, pushing the cooler matter at the top down to the bottom to be reheated. The outer layers of the sun (just below the surface) are in convective equilibrium, and it is possible to see the convection cells on the solar surface. In the case of the sun, it is found that the "turnaround" time for a convective cell is about 5 minutes; that is, a pocket of hot air from the bottom of the convection zone takes about 5 minutes to rise to the top, where it cools, and then to return again to the bottom.

11-7 STELLAR ATMOSPHERES

The *atmosphere* of a star is defined as that portion of the star that is transparent to radiation. The *surface* of a star is a fuzzy area, rather than a hard surface as in the case of the earth, and it is defined as the layer of the star below which the material is opaque. The opacity of stellar material gradually builds up below the atmosphere until finally virtually no light can pass through without interacting with the gas. In that sense, the surface is a layer rather than a plane. It is overlaid by a very much thinner and cooler layer, the stellar atmosphere. Stellar atmospheres are important because they are the portions of stars in which the spectroscopic absorption lines that are so easily observed and studied are formed.

Solar Activity

The only stellar surface that can be studied in detail is the surface of the sun. Since the sun is a main-sequence star, the study of its various surface and atmospheric phenomena has told astronomers a great deal about main-sequence star activity.

Sun Spots

The sun's luminosity and temperature have remained virtually unchanged as long as they have been measured, and geologists find evidence that they have probably remained unchanged for billions of years. However, short-term

Fig. 11-19. The sun-spot cycle has been recorded for over a century.

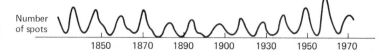

Number of spots

1850   1870   1890   1900   1930   1950   1970

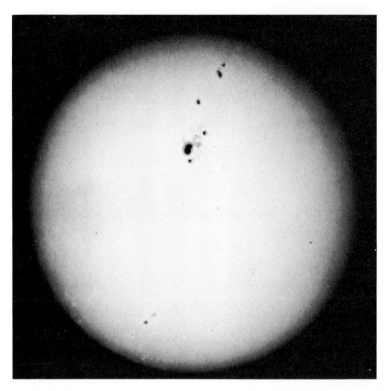

Fig. 11-20. Sun spots. *(Harvard Observatory.)*

changes of details on the surface of the sun do occur, though they are in all cases too small to affect the overall properties of the sun noticeably. Among the conspicuous features of the solar surface and atmosphere are the *sunspots*. These represent areas of disturbance (something like terrestrial storm systems) that are darker than the surrounding solar disk because of somewhat lower temperatures. The sunspots are clearly connected with magnetic field disturbances in the sun and show a regular pattern in their distribution across the disk of the sun, as well as variation in their numbers over an 11-year cycle. They often appear together as pairs of spots, with one spot having a north and the other a south magnetic polarity. The arrangement of these pairs is reversed in the two hemispheres of the sun. There is good evidence that the number of spots, and thus the intensity of solar activity, is sometimes abnormally low. This apparently results in long-term change in the earth's weather.

Other conspicuous solar features are the *filaments*, dark sinuous features that extend across portions of the solar disk. When observed at the edge of the sun, the filaments are seen to be high wisps of material that extend to great distances above the surface of the sun—as high as hundreds of thousands of kilometers. They appear dark when seen against the sun because of their

Filaments and Prominences

Fig. 11-21. Filaments (Black narrow features) and sun-spot groups show up in this hydrogen photograph. *(Sacramento Peak Observatory.)*

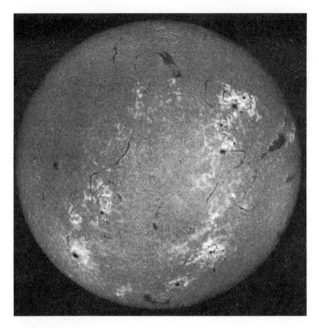

temperature, which is lower than that of the solar surface. However, when seen at the rim of the sun (either during a solar eclipse or through a special telescope that blocks out the bright disk of the sun), they appear bright against the dark, sunless background. When seen at the limb, they are called *solar prominences.*

Prominences usually involve motions (which can be measured by movie cameras) that include both outward motions from the solar surface and, in some cases, an apparent raining downward to the surface. The prominences are often associated with sunspots and sunspot groups, and are understood to

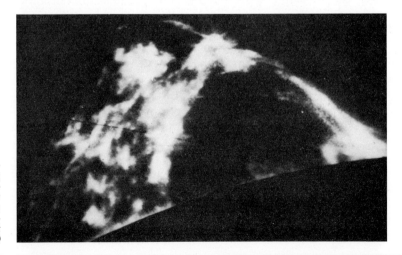

Fig. 11-22. A solar prominence is a huge moving cloud of gas ejected upward above the solar disk. The gas then rains back down to the surface. *(Sacramento Peak Observatory.)*

be related to the magnetic-storm-like disturbances that produce the spots. Most of the material that is thrown above the solar surface to form the prominences and filaments eventually falls back down also.

Solar Flares

A solar flare is a rare and energetic disturbance on the solar surface. Flares occur primarily during sunspot maximum, and they usually are found to be associated with a highly disturbed portion of the sun's surface (marked by a sunspot group). Flares last only a few minutes and involve exceedingly bright flashes of light and large amounts of emission of energy. They are caused when a portion of the much hotter gas below the surface of the sun is allowed to escape briefly to the surface, due to the magnetic disturbance marked by the spots. Solar flares are often the cause of highly energetic particles that are flung out from the sun into the solar system, occasionally encountering the earth and causing *terrestrial magnetic disturbances* (storms in the upper atmosphere and auroae). (See Chap. 6.)

The Corona

The outer parts of the atmosphere are highly unusual in their properties. At heights of about 10,000 kilometers above the surface of the sun, the temperature is found suddenly to rise to very high values, on the order of 10 million degrees. This high rise in temperature occurs at distances from the solar surface where the gas density is exceedingly small, and therefore the high-temperature region is difficult to see. Called the *corona*, this region is viewed most easily

Fig. 11-23. A solar flare is conspicuous in the lower right part of this hydrogen photograph of the sun. *(Sacramento Peak Observatory.)*

Fig. 11-24. The solar corona is the faint outer atmosphere of the sun, and is conspicuously visible during solar eclipses. *(Sacramento Peak Observatory.)*

during a solar eclipse, though there are certain specially made telescopes called *coronographs,* that can detect it. The high energy of the corona is contributed by the convective layer of the sun just under its surface. The convective currents can cause shock waves to pass out through the solar atmosphere, and these shock waves carry energy with them that is then temporarily trapped in the corona. A small amount of the matter that is carried out to the solar corona in that way is eventually lost to the sun in the form of the solar wind (Chap. 6).

QUESTIONS

1. If a star is measured to be red, can you say anything about its temperature?
2. A very red star is observed spectroscopically and is found to have conspicuous helium lines, weak hydrogen lines, and no other visible lines. If it is a normal main-sequence star, what can you say about its temperature?
3. What is the radius of a star if its temperature is the same as the sun's, but its luminosity is twice the sun's?
4.* What is the sum of the masses of the components of a binary star separated by 100 astronomical units and having a period of 100 years?
5. If it were possible to stand on the sun's surface, how much would a 150-pound person weigh (in kilograms)? Why could a person not stand there?
6. What is the surface gravity of a giant star with the sun's mass but 50 times larger than the sun? (Give the answer in terms of the earth's surface gravity).
7.* Calculate in ergs the amount of energy that your weight represents, according to the principle of the equivalence of matter and energy.
8.* How much energy is generated in the solar interior by the creation of a million metric tons of helium?

1. Using a small telescope, project an image of the sun onto a sheet of white paper and examine any sunspots that may be visible. If this is done several days in succession, it is possible to observe the rotation of the sun and to follow the development of sun spots and sun spot groups. Never look toward the sun through the telescope; even a brief glimpse can cause permanent eye damage or blindness.

2. Some college observatories have or can get solar hydrogen alpha filters, which transmit only the light of hydrogen gas. If you have access to one of these, it can be used to observe prominences beyond the limb (visible edge) of the sun and hydrogen structure on the solar disk.

3. Binary stars are exceedingly numerous, and some are widely enough separated so that they can be seen with a small telescope. The following are good examples:

Fall objects	Separation, 1970	Period, years
1. σ Cas	3 seconds	—
2. η Cas	11 seconds	480
3. τ Cyg	1 seconds	50
4. β Cyg	35 seconds	—
Winter objects		
1. γ And	10 seconds	—
2. α Psc	2 seconds	720
3. α Gem	2 seconds	420
Spring objects		
1. γ Leo	4 seconds	620
2. ξ Ursa Maj	3 seconds	60
3. γ Vir	5 seconds	170
4. ζ Ursa Maj	15 seconds	—
Summer objects		
1. ϵ^1 Lyr ⎫	3 seconds	1200
2. ϵ^2 Lyr ⎬ a quadruple star	2 seconds	600
3. ϵ^1-ϵ^2 Lyr ⎭	208 seconds	—
4. α Her	5 seconds	—

These, as well as the stars listed below, may be found by reference to the charts at the back of the book.

4. Eclipsing binary stars are sometimes bright enough to be observed without a telescope. To record their variation, estimate their brightness with respect to nearby comparison stars, as shown on a star map.

Fall object βPer (Algol), varies from magnitude 2.1 to 3.3 with a period of 2.87 days.

Winter object λTau, varies from magnitude 3.3 to 4.2 with a period of 3.9 days.

Spring object δLib, varies from magnitude 4.8 to 6.2 with a period of 2.33 days.

Summer object βLyrae, varies from magnitude 3.4 to 4.3 with a period of 12.93 days.

**OBSERVATIONS
AND
EXPERIMENTS**

12

THE INSTABILITY STRIP

What happens to a star when it uses up the nuclear fuel in its central furnace? This chapter provides a description of the events that we can see taking place in a star when its central fuel is gone.

When a star has exhausted the hydrogen fuel in its core, it can no longer remain on the main sequence. With no hydrogen atoms to react together, there is no source of heating in the core, and the star therefore collapses. When the core falls in on itself in collapse, the outer parts of the star grow in size and become cooler and redder. After a long, stable, and uneventful period of time on the main sequence, the star begins to undergo rapid changes. Seen from a distance, its temperature decreases and its luminosity increases. The reason for these changes would be clear if the star could be watched from up close, because both changes are connected with the increase in the radius of the star following the collapse of its core. The star turns off the main sequence, rising above it, and to the right in the color-magnitude diagram.

Shortly after the star has turned off the main sequence, it passes through a portion of the color magnitude diagram called the *instability strip*. This long, narrow section of the diagram represents a set of conditions in the interior of the star that are unstable. Any star with a luminosity and temperature falling within the confines of the area defined by this strip cannot remain long with those conditions, because they are inherently unstable. From study of clusters it is found that stars tend to be absent from this part of the color-magnitude diagram. The only way that a star can avoid it is by evolving rapidly enough through it so that the star quickly obtains a stable configuration. The interior of the star rearranges itself in order to become stable again. An analogy might be thought of in terms of a person riding on an elephant, which is, relatively speaking, a stable configuration. If in the process of a long journey the elephant should become exhausted, it might propose a new configuration, with the elephant riding on the human. Such a configuration is inherently unstable, and eventually it will rearrange itself whether or not the elephant or the human desires it. In the case of the star leaving the main sequence, this instability

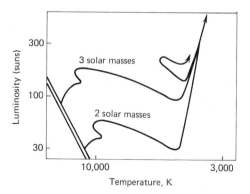

Fig. 12-1. The paths of stars that are leaving the main sequence carry them into the realm of high luminosity and cool temperature.

Fig. 12-2. Instability and stability.

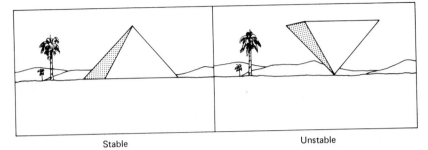

Stable   Unstable

similarly causes a star to go through a period of readjustment so that eventually a more stable, evolved arrangement of matter in the star can take over.

In the case of the human and the elephant, the unstable configuration of elephant riding on human would undoubtedly lead to a considerable amount of wobble. In the case of the star, when instability of this sort sets in, the star similarly tends to go through a stage of contraction and expansion. The star first falls in toward its center, and then when it reaches a sufficiently high density and temperature so that the outward pressure becomes too great, the star then expands outward again. Because of the instability, this expansion is not checked until it is too great for the amount of gravity; therefore, the star once again collapses inward. The succeeding contractions and expansions continue in a regular way, leading to a pulsation of the star that shows up from the earth as variations of brightness and temperature. These pulsations continue until the instability is stopped by rearrangement of the material of the star, at which time the star leaves the instability strip and once again becomes a fairly stable object. Depending upon the mass and the evolutionary stage of a star, the pulsational properties are distinguished from each other and given different names. Some of the more important are described below.

12-2 CEPHEID VARIABLES

Probably the most important pulsating stars, the *Cepheid variables*, show large and conspicuous variability. Named after the star Delta Cephei, the Cepheid variables constitute a class of objects that occupy the color-magnitude diagram in the instability strip above absolute magnitudes of about −1. Most if not all stars found in that portion of the diagram are Cepheid variables. Any star, therefore, that is massive enough to be in that bright part of the color-magnitude diagram will pass through the instability strip and become a Cepheid for at least one short period of time after its main-sequence lifetime.

Light Curves

A plot of the brightness of a Cepheid through its full cycle of variability is called its *light curve*. Fig. 12-3 shows examples of light curves for Cepheids of various periods. During the thousands of years that a star remains in the Cepheid phase, the light curve is repeated over and over again, and is almost exactly the same

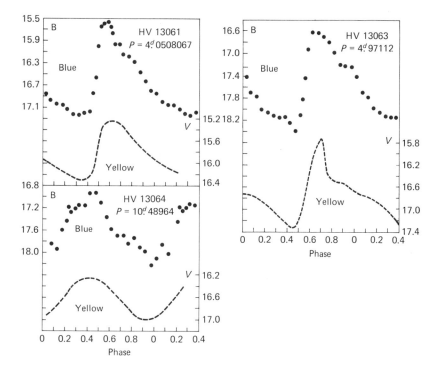

Fig. 12-3. Light curves of Cepheid variable stars. *(After F. W. Wright.)*

Fig. 12-4. A Cepheid at maximum and at minimum light. Compare its brightness in each case with one of the other stars of the same luminosity.

for each cycle. The light curve usually shows a rapid increase of brightness to maximum, and then a more gradual decrease in brightness to minimum, although a few Cepheids do show light curves that are almost perfectly symmetrical.

Velocity Curves

At the same time that the Cepheid is changing brightness, the radial velocity is also changing over the cycle with exactly the same period. Fig. 12-5 shows the way in which the velocity of the star changes at different times in its light curve. The velocity is at a maximum when the size of the star is about average, and it reaches a value of 0 (with respect to the center of the star, which may have a constant velocity relative to the sun) when the star is either at maximum or at minimum size. This can be understood in terms of the way in which a star pulsates during its cycle. When the star is shrinking, we see a large red shift velocity, indicating that most of the surface of the star that we can see from the earth is moving away from us. When the star reaches its minimum size, we see no velocity shift because at that point the atmosphere of the star has stopped moving with respect to its center. Then the star is near minimum brightness because of its minimum size. The star begins to expand, showing a blue shift, and continues to expand until its gravitation finally stops it at its maximum size, which precedes somewhat its maximum brightness. At this point, the velocity again becomes 0 with respect to the center of the star, and we see no radial velocity shift. As the star contracts again, we again see a red shift and decrease in brightness, and the cycle repeats itself.

Periods and Luminosities

The periods of Cepheid variables are found to lie within the range of about 1 day to about 150 days. The very long-period Cepheids are quite rare. Luminosities of the Cepheids also show a wide range, from absolute magnitudes of −7 or −8 for the brightest to −1 for the faintest. During their cycle of variation, the luminosities of Cepheids vary by factors of as much as 2 or 3; that is, the Cepheid can be 100 percent brighter at maximum than minimum, or even more. Some Cepheids show an amplitude of brightness of only 20 percent or so, and the most common values are on the order of 50 percent. The magnitude difference, called the *amplitude*, can range from about 0.2 to 2.0

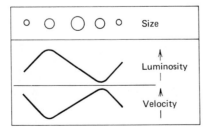

Fig. 12-5. A light curve, a velocity curve, and the pulsation that causes them.

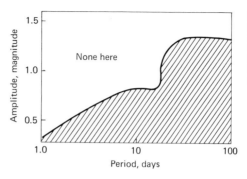

Fig. 12-6. The period-amplitude relation for Cepheids shows that most short-period stars change very little in brightness as they pulsate.

magnitudes. (It is useful to keep in mind that a difference of 0.75 magnitudes corresponds to a factor of 2 multiplication of brightness).

In 1914 the astronomer Henrietta Leavitt, who was working at the Harvard College Observatory, made a remarkable discovery about Cepheid variables. She had been examining photographic plates of the Magellanic Clouds, two nearby galaxies observable from the southern hemisphere. Miss Leavitt had discovered many hundreds of variable stars in these galaxies and was using a large number of plates taken at different times to determine their properties—their light curves, their periods, and their magnitudes. In the process of this examination, she discovered a strong correlation between the periods and the luminosities of the Cepheid variables in her sample. The Cepheids of short period were all faint, while the Cepheids of long period were comparatively bright. When she plotted up the data she found something similar to what is shown in Fig. 12-8. This correlation, called the *period-luminosity relationship,* had not been discovered before because of the difficulty of determining the

The Period-Luminosity Relationship

Fig. 12-7. Henrietta Leavitt was a pioneer in the study of Cepheids. *(Harvard Observatory.)*

Fig. 12-8. Miss Leavitt's period-luminosity relation for the Magellanic Clouds as she plotted it in 1914.

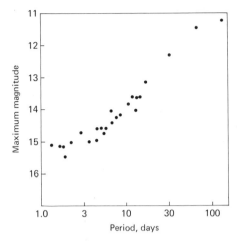

exact distances to Cepheids in our own Galaxy. No Cepheid is close enough for an accurate parallax to be obtained, and the other methods for getting distances had not yet been sufficiently developed (Chap. 8). Therefore, it remained for Miss Leavitt to make the discovery. It was possible for her to do so because of the fact that all of the Cepheid variables in the Magellanic Clouds are at about the same distance from us. It was not even necessary for her to know the distance to discover the correlation between the two features, because the uniformity of the distance made it quite apparent even when plotting only the *apparent* magnitudes and periods.

The period-luminosity relationship has become one of the most important means of obtaining distances in the stellar universe. It is a fundamental calibration for the distances to nearby galaxies and forms the basis for the calibration of distance criteria for more distant objects in space. It can be used

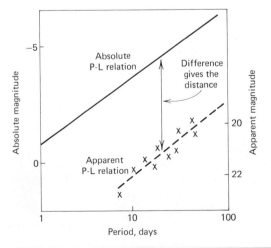

Fig. 12-9. Using the period-luminosity relation to measure distance involves comparing the apparent magnitudes of Cepheids of the same period.

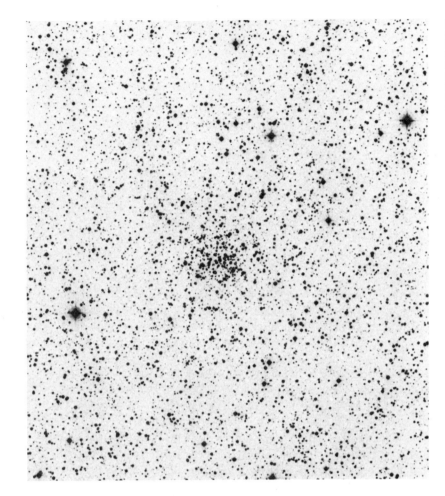

Fig. 12-10. An open star cluster such as this can have its distance accurately measured by study of its main sequence, or, if it contains Cepheids, from the period-luminosity relation. *(Lick Observatory.)*

anywhere that Cepheid variables can be detected, because the only thing that needs to be measured is the period of the Cepheid and its apparent magnitude, once the period-luminosity relationship has been calibrated in terms of absolute magnitudes.

The calibration of the period-luminosity relationship was very difficult at first and was carried out by various indirect means. In 1952, it was found that the calibration had been made incorrectly and it was redone in the 1950s with a more accurate method which then became available—main-sequence fitting (Chap. 8). Cepheid variables in open clusters in the nearby portion of our Galaxy are used to establish the absolute magnitude-period relationship, because their distances can be determined accurately by means of fitting the main sequence of the cluster.

By such calibration, the period-luminosity relationship is found to be that

Fig. 12-11. The absolute period-luminosity relation. *(After Sandage and Tammann.)*

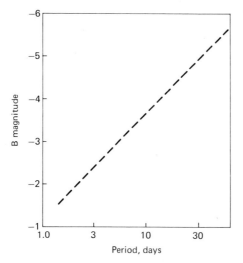

given in Fig. 12-11 where the period and the absolute magnitude are plotted. If a Cepheid variable found anywhere in the universe can have its period determined and its average apparent magnitude measured, then the distance to that Cepheid can be directly read off from Fig. 12-11. One merely reads off for the period measured what the absolute magnitude of the Cepheid is and then makes a comparison between the measured apparent magnitude and the absolute magnitude to get the distance.

For example, a Cepheid variable in another galaxy is found to have a period of 10 days and an apparent blue magnitude of 21.5. We know that the absolute magnitude of this Cepheid must be −3.5 by examining Fig. 12-11. This means that there is a difference of (21.5 + 3.5 = 25) magnitudes between the apparent and the absolute magnitude of the Cepheid. A factor of 25 magnitudes is a factor of 10 billion (10^{10}) in luminosity. Remembering that the absolute magnitude of an object is defined as the apparent magnitude that it would have if it were at a distance of 10 parsecs, it can be said that this Cepheid therefore must be 10 billion times fainter than if it were 10 parsecs distant. The brightness of an object varies inversely as the square of its disance, so its distance must be $\sqrt{10^{10}} = 10^{5}$ times greater than 10 parsecs, or 10^{6} parsecs. This is 1 million parsecs, and it corresponds approximately to the largest distance that Cepheids can be studied well with the world's largest telescopes.

12-3 RR LYRAE VARIABLES

A type of variable star even more common than the Cepheid variables is the fainter, shorter-period object called the *RR Lyrae variable*, named after the star RR in the constellation Lyra, the Lyre. These objects have light curves and velocity curves similar to those of Cepheid variables, but with periods only a

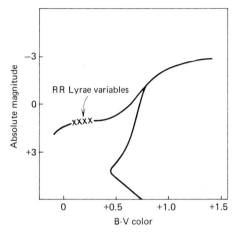

Fig. 12-12. RR Lyrae variables in the color-magnitude diagram of an old population of stars.

fraction of a day in length, ranging from about 0.3 to 0.9 days. While Cepheid variables are usually found in areas where there is an abundance of young stars with ages of only a few million to a few hundred million years, the RR Lyrae variables occur only in areas where there are stars that are several billion years old. They are all of approximately the same brightness, with an absolute magnitude of about +0.5, regardless of period. Fig. 12-12 shows that the RR Lyrae variables fall in the instability strip and that they are surrounded by other nonvariable stars on either side of the strip, which suggests that stars evolve through the RR Lyrae phase along this horizontal sequence, called the *horizontal branch* of a color-magnitude diagram. Calculations of evolutionary models of stars indicate that the RR Lyrae variables occur much later in the evolutionary pattern of a star's life than do the Cepheid variables. A star becomes an RR Lyrae variable only if its mass is in the appropriate range, which

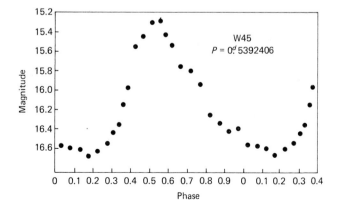

Fig. 12-13. Light curve of an RR Lyrae variable with a half-day period. *(After F. W. Wright.)*

would be somewhat smaller than the mass of the least massive Cepheids. It goes through the RR Lyrae phase after it has already become a red giant star (Chap. 13).

Most RR Lyrae variables are stars of low heavy-element abundance, as determined by the absorption lines in the spectra. They occur frequently in globular clusters (Figs. 13-27 and 13-29) and in the halo of our Galaxy (Fig. 19-10), both of which locations are populated by very old stars, most of which also show a much smaller abundance of heavy elements than the sun possesses. The reasons for this are tied to the evolution of our Galaxy and the history of the creation of the elements, as discussed in Chap. 19.

12-4 CEPHEIDLIKE VARIABLES

In the 1950's it was found that some of the variable stars that had been called Cepheids were actually quite different objects. They had been recognized as having certain obvious differences from the normal Cepheids; for example, they possess somewhat unusual light curves and a different distribution of periods. But it was not until 1952 that it was discovered that they are intrinsically very different; i.e., they belong to an old star population instead of to a young population, and have a markedly different period-luminosity relationship. These stars have been called *W Virginis variables*, after a bright example in the constellation Virgo, and *RV Tauri variables*, after an example in Taurus. The W Virginis classification is used for the shorter period examples, and RV Tauri for the longer period.

W Virginis Variables

The W Virginis variables are found to range in period from about 1 to 30 days. Their light curves (Fig. 12-14) are different from those of Cepheids of the same

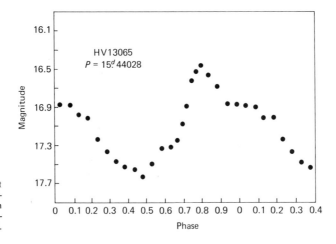

Fig. 12-14. The light curve of a W Virginis variable is rather similar in shape to that of a Cepheid.

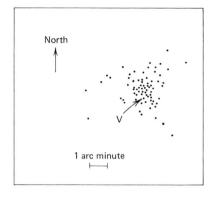 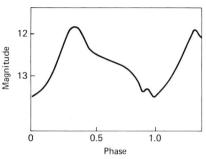

Fig. 12-15. A W Virginis variable in the globular cluster M 3. *(After F. W. Wright.)*

period and are much more irregular; they do not repeat themselves exactly from period to period. Their spectra are also somewhat different from the spectra of Cepheid variables, because they frequently show emission lines and give velocity curves that differ from those of normal Cepheids.

Unlike the Cepheid variables, which are strongly concentrated to the plane of our Galaxy and confined to positions in the flat arrangement of young stars of the Milky Way, W Virginis variables are commonly found out in the stellar halo of the Galaxy. Many are found in the globular star clusters, which are also largely in the halo of the Galaxy. In location in space, therefore, the W Virginis stars are clearly associated with old star populations. The globular star clusters, with ages of 10 to 15 billion years, are among the oldest objects in the Galaxy. The stars found in the halo of the Galaxy are also exclusively old; therefore, W Virginis stars must also be old.

The W Virginis variables occupy the instability portion of the color-magnitude diagram, and they pulsate like normal Cepheids. However, because of their great age, it is believed that they are not at all similar to normal Cepheid variables in mass or in interior construction. They are only about as massive as the sun and pulsate over a much larger range of size than Cepheids.

From the fact that globular star clusters have distances that are readily measured either by measuring the apparent luminosities of the RR Lyrae variables or by obtaining the color-magnitude diagram for the main-sequence members, it is possible to obtain the period-luminosity relationship of W Virginis variables. This relationship is found to be very much less precisely defined than that for the normal Cepheid variables. The reason for this is now believed to be the relatively high sensitivity of the luminosity of such an object to age and chemical composition. A small chemical compositon difference shows up as only a very slight change for the normal Cepheid period-luminosity relationship, while for the W Virginis stars, it can cause a large difference (Fig. 12-16).

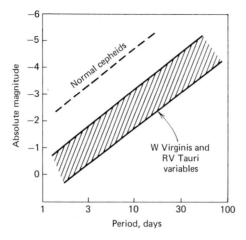

Fig. 12-16. The period-luminosity relation for W Virginis and R V Tauri variables.

RV Tauri Variables

The longer period cepheidlike variables, with periods between 40 and 100 days, are termed RV Tauri variables. They show a wide period-luminosity relationship and are, therefore, not very useful for determining distances. Their light curves, as in the case of W Virginis stars, are liable to change somewhat from period to period, and their spectra show both changes and anomalous features, such as emission lines.

The exact role that the cepheidlike variables have in the evolutionary pattern of stars is not yet known precisely. From the circumstantial evidence that we have, it is clear that they are very old objects and probably not very massive, in both senses very different from normal Cepheids. The small number of them compared to the numbers of other kinds of stars, for example in globular star clusters, indicates that they represent a phase of evolution that is brief. Figure 12-17 suggests certain possible patterns of evolution that include these variables.

**12-5
SEMIREGULAR
VARIABLES**

Another kind of variable star that is associated with old star populations and that is relatively common in globular star clusters is the *semiregular variable*. These stars are not in the instability strip, but represent the manifestation of instabilites that can occur in the giant phase of stellar evolution (Chap 13). Semiregular variables are redder than Cepheids or cephedlike variables and, unlike both, they are nonrepetitive; they change brightness in an irregular and fairly erratic way. They are called "semiregular" because of their tendency to take several days to increase in brightness and decrease in brightness, with a roughly equal interval of time between maximum or minimum.

Fig. 12-18 shows a light curve of a semiregular variable. They are luminous objects, always more luminous than RR Lyrae variables, with absolute magnitudes averaging about −3. Apparently, they are stages in the later portion of the life history of a star. It is not yet known for certain the exact role that

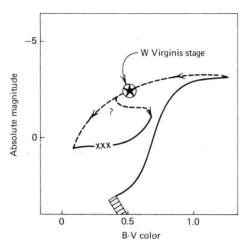

Fig. 12-17. Possible patterns of evolution for a globular cluster with W Virginis stars.

semiregularity of this sort plays in the development of a star. Normally, semiregular variables occupy a part of the giant branch of a globular star cluster (for example) that is also occupied by many apparently stable stars.

Long-period variables are objects that represent the onset of instability at very much cooler temperatures than those that define the instability strip in the color-magnitude diagram. Long-period variables have average temperatures of about 3,000 K, whereas Cepheid variables have temperatures in the range of 8,000 to 9,000 K. Long-period variables are stars of large size, and they undergo much larger changes of size and luminosity than do Cepheids. Whereas Cepheids during a cycle change their size by about 10 percent, long-period variables can be several times as large at maximum than at minimum. Light variations for long-period variables are also often much larger than in the case of Cepheids, with ranges commonly as large as 3 or 4 magnitudes. The periods of long-period variables are longer than those of Cepheids, and lie in the range from about 100 to 800 days. These variables do not repeat their light curves exactly from cycle to cycle, but tend to show differences both in the shape of the light curve and in the brightness achieved at maximum and minimum. At minimum light, the long-period variables are among the cooler stars so far observed, with such low temperatures that many molecules can be seen in their spectra (including the water molecule, indicating that the atmospheres of longer-period variables contain "steam.")

From statistics of their location in our Galaxy and in other galaxies, it is concluded that long-period variables exist both in old star populations and in relatively young star populations. The variability is apparently related to their extreme coolness rather than to their mass or mass distribution. There is, however, a noticeable difference in the periods for the very old long-period

12-6 LONG-PERIOD VARIABLES

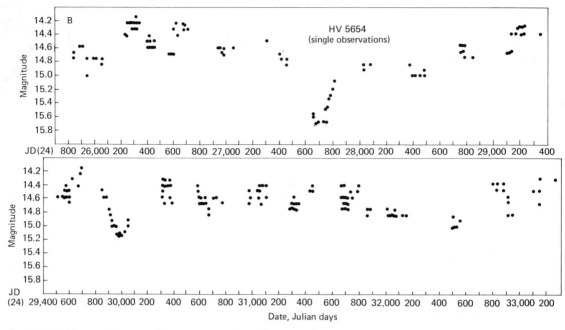

Fig. 12-18. Light curve of a semiregular variable. *(After F. W. Wright.)*

variables compared to the ones found in young star groups. The shorter period long-period variables are associated with the old stars, and the longer period examples occur in young star environments. This is most likely a result of the relatively larger mass of the long-period variables of young age.

**12-7
OTHER
VARIABLE
STARS**

Another type of variable is the *Delta Scuti class*, which consists of stars of periods of only a few hours and very small light changes, only a few hundredths of a magnitude. These objects may be the equivalent of heavy-element-rich RR Lyrae variables.

Also important are the so-called *flare stars*, which are probably the most common type of star with variable luminosity. Flare stars are faint, red main-sequence stars that irregularly flare up in brightness for a few seconds or minutes, then gradually return to a normal brightness. Most flare stars are of spectral types M5 or later, and thus are among the coolest stars on the main sequence. The flares, which amount sometimes to as much as a tenfold increase in visible brightness, are probably hot spot disturbances on the star's

Fig. 12-19. Light curve of a long period variable in the constellation Cassiopeia.

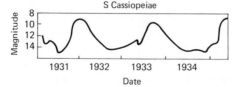

surface, like solar flares except on a bigger scale. Sometimes they are hot enough and big enough to be detected with radio telescopes.

A rather rare kind of variable is the bright, blue pulsating type called *Beta Canis Majoris* stars. These are B-type stars with temperatures of about 20,000 K and absolute magnitudes between −2 and −4. They are young stars, still near the main sequence, in which instabilities have caused oscillations of the surface with periods between 2 and 7 hours. The pulsations have small amplitudes and are difficult to detect, causing only a 1 or 2 percent change in the luminosity of the star. Some have two different periods occurring at the same time, giving rise to a "beat" phenomenon. The two periods are only slightly different, so that sometimes they reinforce each other and the star varies a lot, and at other times they cancel each other out and the star hardly varies at all.

QUESTIONS

1. If the sun were a Cepheid variable, would we be able to notice its variability? What might the effects be on the earth? Will the sun ever become a Cepheid variable?
2. What is the absolute magnitude of a Cepheid variable with a period of 50 days? How much more luminous is it on the average than a Cepheid with a 5-day period?
3. Which is likely to be younger, a Cepheid or a RR Lyrae variable?
4. How much more distant is a certain RR Lyrae variable if it is 1.5 magnitude fainter than another (both are equally reddened)? Must you know the periods of these stars to answer this question?
5. Two new variables are discovered; one has a bluish color and a period of 12 hours, the other has a deep red color and a period of 400 days. What kinds of variables are they likely to be?

OBSERVATIONS

Variable stars are observable in many cases without a telescope. Examples of some that can be observed with the naked eye, binoculars, or a small telescope are tabulated below and shown on the charts at the back of the book.

Season	Star	Type	Range	Period, days
Fall	R Aql	long-period	6.5–12.0	290
	δ Cep	Cepheid	4.1– 5.2	5.37
	η Aql	Cepheid	3.7– 4.5	7.18
Winter	o Cet	long-period	3.4– 9.6	332
	R Cas	long-period	5.5–12	432
Spring	R Leo	long-period	5.5–10.5	313
	ζ Gem	Cepheid	4.4– 5.2	10.15
	α Her	semiregular	3.0– 4.0	—
Summer	RR Lyr	RR Lyr	6.9– 8.0	0.57
	R Sct	RV Tau	6.3– 8.6	144
	χ Cyg	long-period	4.2–13.7	407
	W Sgr	Cepheid	4.8– 5.8	7.59

13

RED GIANTS

This chapter explains how and why stars, near the end of their lifetimes, puff up into huge, distended, cool giants.

After spending most of its life on the main sequence, a star has a final burst of glory as a *red giant*. The term is derived from the fact that in these advanced stages, the star is much cooler than it was on the main sequence (and therefore much redder), and from the fact that its size is considerably larger than its stable main-sequence size. Although the way by which the star passes through the red-giant phase depends on both its mass and its chemical composition, the general pattern for all is approximately the same. In the first section of this chapter, an account will be given of the calculated evolutionary history of the sun through the red-giant phase. Most stars of solar mass in our Galaxy have chemical abundances sufficiently similar to that of the sun so that they will pass through the same phases in the same way as outlined here for the sun. For a star that is more massive than the sun, the events occur considerably more rapidly, while stars less massive than the sun pass through the various phases of the red-giant stage more slowly.

The sun is now approximately 4.5 billion years old. It lies slightly above the main sequence, having spent about half of its available time as a main-sequence star. As calculated in Chap. 11, the expected main-sequence lifetime for the sun is about 10^{10} years. During the latter half of this period, the sun will be increasing in brightness slowly and imperceptibly, moving slightly above the main sequence as it ages. Although observations of the sun's brightness and temperature show that the sun is indeed slightly above the main sequence now, all the other aspects of the sun's future development are based on detailed and very complicated calculations of events that cannot be observed for the sun at present. However, as later sections of this chapter show, it is possible to confirm these calculations by comparing them with stars in star clusters.

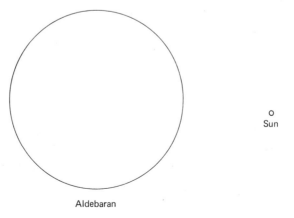

Aldebaran

Fig. 13-1. A red-giant star is huge compared to the sun.

Fig. 13-2. The next 4 1/2 billion years in the sun's development involve a slow brightening.

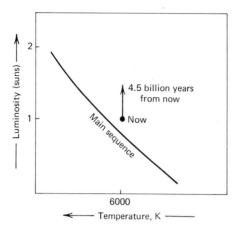

Fig. 13-2. The next 4 1/2 billion years in the sun's development involve a slow brightening.

The Helium Core

During the next 4 billion years, the sun will continue to brighten, and it will grow slightly larger in size. The brightening compensates for the increase in size, so that the temperature will remain approximately the same as now. In about 4.5 billion years, the luminosity of the sun will be about 50 percent greater than at present, and its diameter will be about 25 percent greater than its present size. At this time, the center of the sun will have exhausted all its hydrogen; all of it will have been converted by nuclear burning into helium. At the center there will be what is called a *helium core*, i.e., a small dense region consisting almost exclusively of helium. At this point the helium core will become a very important feature of the sun, and its influence on the development of the sun will cause a rapid series of events to occur.

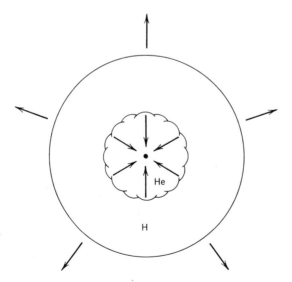

Fig. 13-3. In 1 billion more years, collapse of the helium core of the sun occurs.

During the following billion years of the solar life, the sun will grow in size, rapidly becoming two and one-half times bigger than its present size when its age is 10.3 billion years. The temperature during this growth will decrease because of the fact that the luminosity remains approximately the same; there will be no new source of energy production. At this point, the helium core is calculated to be approximately 40,000 kilometers across, only one one-hundredth of the then size of the sun. The helium core contracts, eventually becoming a size on the order of the earth. The remarkable fact about this core, however, is that it contains almost one-quarter of the mass of the entire sun at this stage. It is almost 1 million times more massive than the earth when its diameter is only twice the earth's diameter. The density of the gas in the core becomes about 50,000 times greater than the density of iron, which means that a thimble full of this material would weigh approximately a ton.

The Hydrogen-burning Shell

The gas surrounding the helium core still contains large amounts of hydrogen. This hydrogen was too far from the core of the sun on the main sequence to have temperatures high enough to produce fusion. As the core contracts, the temperatures in the surrounding part of the sun will become high enough so that a hydrogen burning shell will form. This shell develops rapidly, and as the rate of increase of hydrogen burning proceeds, the luminosity of the sun will increase rapidly also. In only about 100 million years, the sun will increase in luminosity so rapidly that its brightness will become 1,000 times brighter than its present brightness. At the same time, the sun will grow rapidly in size; the outer envelopes will reach a diameter as great as 100 times the sun's present size. Then the sun will be a true *red giant* with a temperature at the surface of only about 3,500 K. Because one-quarter of the mass of the sun will be compressed into a tiny central helium core (tiny by comparison with the immense size of the sun at that period), the rest of the material in the sun will be

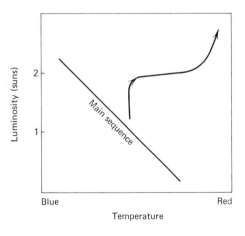

Fig. 13-4. The sun eventually enters a red-giant phase.

Fig. 13-5. The red-giant sun's structure.

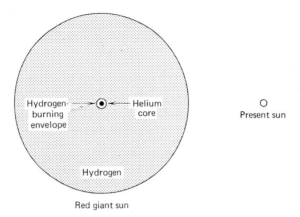

Red giant sun

spread out exceedingly thinly in an enormous globe of hydrogen. The density of this distended envelope will be very small, the equivalent of what is considered a very good vacuum in a terrestrial physics laboratory.

The large size and cool temperature of the sun would make it appear very strange to us if it could then be viewed from the earth. The sun will subtend an angle of almost 60° from the earth and will be a dull red color. Seen at noon, its diameter would extend over almost one-third the extent of the sky.

Helium Burning

As the helium core of the sun continues to contract, the temperature will rise further. Finally, the temperature of the core will reach a certain temperature (about 100 billion K) that is sufficiently high so that helium nuclei can combine together to form heavier elements. The most likely reaction is the combination of two helium nuclei to form a berylium nucleus. The berylium that is formed,

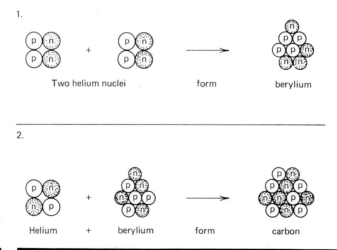

Fig. 13-6. Carbon formation in a red-giant star occurs by means of this process.

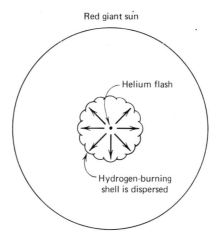

Red giant sun

Helium flash

Hydrogen-burning
shell is dispersed

6 billion years from now

Fig. 13-7. The helium
flash for a star disperses
the hydrogen-burning
shell.

however, is an unstable form of berylium that decays back into two helium
nuclei in only a very short period of time (10^{-12} seconds). If there are many
helium nuclei around (as in the case of the sun's future helium core), another
helium nucleus can combine with this unstable berylium before it decays to
produce a stable form of carbon. By these two steps, therefore, the helium core
in a star like the sun can eventually convert its helium into the heavier element
carbon. In the process, large amounts of energy are released in exactly the
same way that hydrogen fusion produces energy when it forms helium

In the case of calculations for the future structure of the sun, it is found that
the helium burning does not occur in a gradual way, but rather is marked by an
explosive event, called the *helium flash*. This explosion only occurs for stars of
about the mass of the sun or smaller. It is the result of the fact that for such stars
the core has reached such an extemely high density by the time the
temperature is high enough for helium burning to occur that it acts more like a
solid rigid body than like a gas. The electrons (and the helium nuclei) are
packed together so tightly that the electrons form a nearly incompressible solid
medium. Because it is more like a sphere of solid metal than a sphere of gas, the
onset of helium burning causes the temperature to rise rapidly, but the size of
the core increases only slightly. We know from our own experience in the
laboratory that when heated, a gas will expand considerably, while a solid
when heated will expand only very slightly. In this way, therefore, the solar
core will not expand greatly, but the temperature will rise to very high values,
since there is no compensating expansion as there would be if the core were
behaving like a free gas. This higher temperature increases the rate at which the
helium is converted into carbon, and this in turn increases the temperature still
more. The core becomes so exceedingly hot that it acts like an uncontrolled
nuclear fusion bomb. It only takes about a day for the sun to pass from the stage
of beginning helium burning to the explosion of its helium core.

The Second "Main- Sequence" Phase

The helium core of the sun will be small in size when it explodes about 6 billion years from now, and the effects will not be seen at the sun's surface immediately. The interior of the sun, however, will change rapidly and markedly. Following the core's explosive growth in the interior, the hydrogen shell surrounding the core, which has until the explosion been the source of the sun's heat, will be dispersed. The temperature of the core will drop quickly because of its increased size, and similarly the shell hydrogen burning will drop in intensity. Therefore, the helium flash, instead of causing a brilliant outburst of light, will cause the sun to decrease in brightness.

A certain amount of mixing of the interior materials of the sun will occur. The radius of the sun will decrease along with the decreasing brightness, and an increase in temperature will be noticeable. The sun will move along the color-magnitude diagram in the direction toward the main sequence, arriving close to but not on the main sequence after only a few million years. It will remain in this quasi-main-sequence stage for a comparatively long period of time, as the hydrogen-burning envelope around the helium core is aided by a small helium-burning area at the center of the star. The sun's energy thus will come both from the release of energy by a small kernel in the core of the star (where helium is being converted into carbon) and by a more distant shell where helium is being formed from hydrogen.

It is calculated that the sun will remain in this second "main-sequence" phase for several million years. At the end of this period, the core will again contract because of the exhaustion of fuel, and the sun will undergo a second increase in brightness, increase in radius, and decrease in temperature, to become an immense red giant for a second time. As yet, the details of the second red-giant phase for the sun are largely unknown because of the great complications that arise in the structure of a star under such circumstances. The approximate calculations that have been made show that, following the second

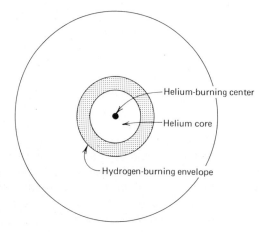

Helium-burning center

Helium core

Hydrogen-burning envelope

Fig. 13-8. The very brief "second" main-sequence period involves this complex structure of a typical star.

main-sequence phase, the sun will rapidly move through the red-giant phase again and then promptly collapse to become a "white dwarf" (Chap. 15).

The Future of the Earth

It is interesting and somewhat alarming to calculate what the properties of the earth will be while the sun is passing through these advanced stages of evolution. The temperature of the earth has remained relatively stable for billions of years during the time that the sun's luminosity has been continuously the same (on the main sequence). During the next 4 billion years, this situation will change by only a small amount as the sun very slowly increases in brightness. Then, when the sun rapidly changes from a normal star into a red giant, the earth's temperature will rise drastically. During the 5 billion years that it takes the sun to increase to 100 times its present radius and 1,000 times its present luminosity, the earth will rapidly become uninhabitable. Even though the sun's temperature will be decreasing, the earth's temperature will increase because of the increasing luminosity of the sun. The earth's warmth is produced by the energy received from the sun, and as the sun becomes a red giant, its size will be sufficiently large to radiate far more energy to the earth than at present. At that time, the oceans of the earth will boil away into the atmosphere, leaving a dry, scorched surface beneath humid skies. Because of its high temperature, part of the terrestrial atmosphere will be lost to space.

Following this intense period of heat on the earth, a brief period of almost normal conditions will occur. During the sun's stay on the second (quasi) main sequence, the surface temperature of the earth will return almost to normal, the oceans will condense out and fall back to the basins that they had left, and what remains of the atmosphere will return to moderate temperatures so that

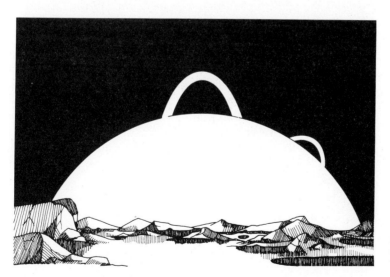

Fig. 13-9. The red-giant sun as it may appear from the scorched earth.

Fig. 13-10. The white-dwarf sun as it may appear from the frozen earth.

Fig. 13-10. The white-dwarf sun as it may appear from the frozen earth.

the earth can briefly harbor life again. This respite is short lived, however, as in only a few million more years the oceans will again be evaporated into the atmosphere and the sun will again become a red giant. Following this second red-giant stage, and the succeeding collapse of the sun into a white dwarf, the earth will again become what we might consider normal in its condition briefly, but it will then change instantly into a frozen wasteland, with not only the oceans frozen but with the atmosphere frozen out as well, covering the land with ices.

These prospects for the future of the earth have lead to some speculation regarding the future of civilization on the earth. Life on the earth is expected to be destroyed completely when the sun becomes a red giant. Therefore, if humanity or its successors in the civilization chain hope to persist, it will be necessary in the next 4 billion years to develop an effective means of interstellar transport. What is left of civilization in that part of the future will have to seek out another star, younger or less massive than the sun, with a planet that will have a longer expected lifetime. It is tempting to anticipate in such an eventuality a return to the earth for a nostalgic visit during those two brief periods when conditions here return to normal. However, it is unlikely that anything in the way of relics of our civilization here will be able to withstand the experience that the surface of the earth will go through during these stages.

13-3
MASSIVE
RED GIANTS

For stars of large mass, the pattern of evolution into the red-giant phase is similar but not identical to that calculated for the sun. Such stars have no helium flash episode because their cores are sufficiently large so that helium burning begins gradually in the core. The stars then move, as for the sun, into a quasistable, second main-sequence phase, becoming Cepheids for a second time if they intrude into the instability strip. The evolution for such objects is much more rapid than in the case of the sun, and they spend only a few moments, cosmically speaking, at each stage. Figure 13-11 shows four different evolutionary patterns for stars of different masses larger than the sun.

As in the case of solar evolution, there is still uncertainty in the exact pattern of development that follows the second giant stage. There is some possibility that the stars will experience carbon burning in their cores, producing even heavier elements and giving the star a third high-temperature phase, with a possible third trip across the color-magnitude diagram to the main sequence. Evidence from stellar interior calculations for this is still highly uncertain, and astronomers must rely principally on observations of young stars that are possibly in these phases for clues as to the events occurring in these far advanced phases.

Because calculations of stellar evolutionary patterns are exceedingly difficult and involve many good but not always entirely certain approximations, astronomers have checked them against observations of star clusters. It is found that a star cluster forms with virtually all of its stars condensing at the same time, and therefore all the stars in a given cluster have very nearly the same age. This makes it possible to study stellar evolutionary patterns by examining the temperatures and luminosities of all the stars in a given star cluster; the cluster will have stars of a variety of masses that all have the same age and therefore will show stars in each of the different evolutionary phases (provided there are enough stars present).

Among the star clusters recognized in the local Galaxy are two types: the youngest are called *open* or *galactic* clusters (and are discussed below), the older, very large, populous ones are called *globular* star clusters (discussed in Sec. 13-5). The galactic clusters are primarily limited to the flat plane of our

**13-4
YOUNG STAR
CLUSTERS**

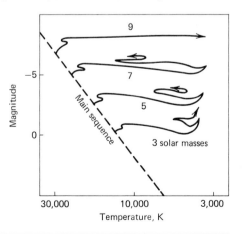

Fig. 13-11. Evolutionary patterns for massive stars. *(After Iben.)*

Fig. 13-12. NGC 2264—a young star cluster involved in nebulosity. *(Cerro Tololo Inter-american Observatory.)*

local Galaxy. There are about a thousand of them known in our Galaxy, but it is estimated that many thousands more exist that we do not see because of the intervening heavy obscuration of dust. Galactic clusters range in population from only a few dozen stars to several hundred. They range in measured age, determined from their main sequences, from less than a million years to five or more billion years.

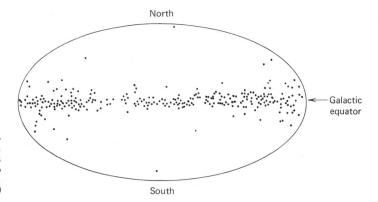

Fig. 13-13. The distribution of galactic open clusters in the sky shows a strong concentration to the Milky Way plane. *(After H. S. Hogg.)*

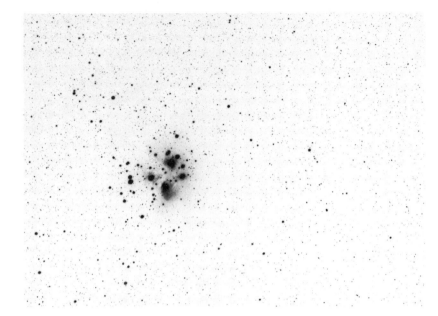

Fig. 13-14. The Pleiades star cluster. *(Lick Observatory.)*

Fig. 13-15. A close-up of the center of the Pleiades cluster. Light reflecting from dust near the bright stars is visible. *(Kitt Peak National Observatory.)*

Fig. 13-16. The color-magnitude diagram of the Pleiades. *(After Johnson and Mitchell.)*

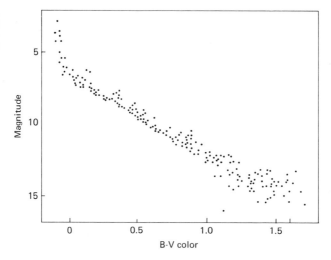

The Pleiades An example of an easily observed galactic cluster is a small group of stars in the constellation Taurus called the Pleiades, sometimes referred to as *The Seven Sisters*. Normally, six of the brightest of these stars are visible without a telescope, but the use of field glasses will show dozens of stars. A study of the individual motions of the stars in the area shows that more than a hundred stars are members of the Pleiades. The color-magnitude diagram for the Pleiades (Fig. 13-16) shows that this cluster has a main sequence that reaches absolute magnitudes of approximately +2.0. The stars brighter than this magnitude lie above the main sequence and are beginning their journey into the red-giant stage. The Pleiades, however, is not a rich enough cluster to show any stars in

Fig. 13-17. The Hyades star cluster *(Lick Observatory.)*

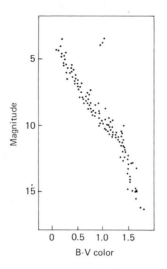

Magnitude

B-V color

Fig. 13-18. The color-magnitude diagram of the Hyades. *(After Johnson.)*

the giant stages at this time. The evolution of stars more massive than those leaving the main sequence of the Pleiades is so rapid that it is all over in just a few million years, and we would have to be fortunate to happen to be observing this cluster at a time when one of the stars is going through this rapid evolution.

The Hyades

The Hyades is another galactic cluster, not far in the sky from the Pleiades, but much nearer to the sun and therefore spread out across the sky in such a way as to make it less obviously a cluster. The Hyades is an exceedingly important cluster for its ability to provide us with an independent means of measuring the stellar distance scale (Chap. 11). It is sufficiently close so that its distance can be determined from the motions of the individual stars within it, all of which are moving together in space and which, therefore, give a measure of space motion that could not otherwise be obtained. The Hyades is also of interest because of the fact that it does contain a few stars in the giant region that confirm the stellar evolutionary calculations as far as can be done with such a small sampling.

Populous Clusters

In order to make a really reliable comparison between a star cluster and calculated stellar evolutionary patterns, it is necessary to have star clusters that contain much larger numbers of stars in the giant phase than the typical open clusters near the sun. For that reason, astronomers were pleased to discover that one of the nearest galaxies, the Large Magellanic Cloud (about 50,000 parsecs from our local Galaxy) contains an abundance of such star clusters. This galaxy is rich in young stars, gas, and dust, and therefore provides an excellent hunting ground for clues as to the formation and evolution of young massive stars. Among the star clusters that have been studied in this galaxy are

Fig. 13-19. The open cluster NGC 457 *(Kitt Peak National Observatory.)*

Fig. 13-20. NGC 2164, a massive young cluster. *(Cerro Tololo Interamerican Observatory.)*

Fig. 13-21. NGC 1866, an intermediate-age rich star cluster in the Large Magellanic Cloud. *(Boyden Observatory.)*

about thirty-five clusters that are unusually populous, having many more stars in them than any of the young clusters near the sun that have been studied in our Galaxy. Examples are shown in Figs. 13-20 and 13-21.

One of the most populous and brightest clusters in the Large Magellanic Cloud is NGC 1866 (Fig. 13-21). This cluster has an estimated 10,000 to 20,000 stars and a reasonably large number of these (about 50) are in the red-giant stages of evolution. In addition, it has several Cepheid variables that are members, due to the fact that the second excursion toward the main

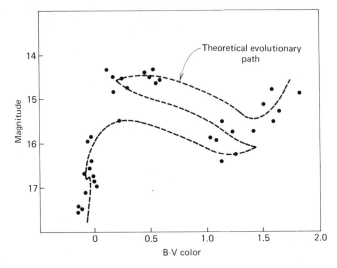

sequence that shows up in the color-magnitude diagram falls in the instability strip.

13-5 OLD STAR CLUSTERS

When the ages of star clusters near the sun in our Galaxy are measured, a surprising result is obtained. Although there are some clusters as old as several billion years in this volume of space, the number of very old star clusters is surprisingly small. The clusters that we observe near us seem to be mostly rather young objects less than a billion years old, with an apparent preponderance of clusters that are only a few million years of age. A first look at the distribution of ages almost suggests that clusters are a recent phenomenon in our Galaxy and that in the old days, back before about 1 billion B.C., clusters

Fig. 13-23. The star cluster Praesepe. *(Lick Observatory.)*

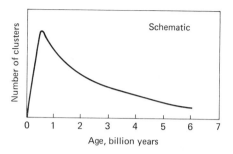

Fig. 13-24. The distribution of cluster ages for the sun's neighborhood (schematic.)

were very rare. By contrast, the distribution of stars of different masses and luminosities (shown by the luminosity function, Chap. 8) indicates that in those ancient eras there must have been very large numbers of stars formed. It even looks probable from the statistics that the rate of formation of stars has been steadily decreasing since the formation of the Galaxy (10 billion years ago).

If the rate of star formation has been decreasing since the beginning, why should the rate of cluster formation be increasing? The answer is that it probably is not, but only appears to be because of the instability of small clusters. Unless a cluster of stars is very massive, strong forces of gravity external to it act on it over the lifetime of the galaxy, gradually disintegrating it.

Fig. 13-25. The old star cluster M 67. *(Kitt Peak National Observatory.)*

Calculations of the effect of such forces on a cluster like the Pleiades or the Hyades show that they can only last as a cluster a few hundred million years or so, after which they will be completely broken up into individual stars going their own way. Therefore, it is likely that cluster formation has proceeded apace with star formation during the history of our Galaxy, but the individual clusters (with a few exceptions) have only a comparatively short life expectancy as clusters.

Old Galactic Clusters

There are a few very old clusters near the sun, and most of these are objects that contain several thousand stars at least. An example is the cluster called M 67 (the 67th object catalogued by the eighteenth-century French astronomer Charles Messier). Located in the constellation of Cancer, M 67 is an inconspicuous object, invisible without a telescope and unspectacular with a telescope. The reason for its inconspicuous appearance is its old age; in its 5 billion years, all its bright luminous stars have evolved away and virtually disappeared from view, leaving only the fainter stellar members. The brightest stars in the main sequence of M 67 are only slightly more luminous than the sun.

Figure 13-26 shows a color-magnitude diagram of M 67. It compares well with theoretical evolutionary tracks for stars 5 billion years old. The excellent agreement gives confidence to the interpretation of theoretical calculations of the red-giant stage of stellar evolution.

Globular Star Clusters

The oldest objects in our Galaxy are immense, luminous, very populous clusters called globular star clusters. About 120 of them are catalogued, and most are distributed in a nearly spherical halo centered on the center of our Galaxy. Unlike the galactic clusters, they are not limited to the flat plane, but extend up to high distances above the plane. There is also a strong concentration toward the center of the Galaxy.

Because the globular clusters are so populous, they have many members in

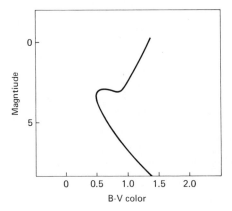

Fig. 13-26. A color-magnitude diagram for M 67. *(After Sandage and Johnson.)*

Fig. 13-27. The globular
cluster 47 Tucanae.
(Cerro Tololo Interameri-
can Observatory.)

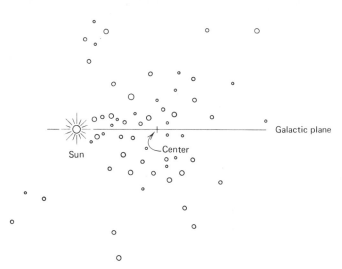

Galactic plane

Sun

Center

Fig. 13-28. The distribu-
tion of globular clusters
in the Galaxy (schemat-
ic.)

the giant phases of stellar evolution. Therefore, they are exceedingly useful objects for the study of the evolutionary pattern for old stars. From them, we can obtain information about the red-giant phase for stars of relatively small mass. It should be mentioned that there is no way of determing observationally the evolutionary pattern for stars with a mass much smaller than that of the sun, because of the fact that it would take longer than the estimated age of the universe for such a star to evolve off the main sequence.

Figure 13-30 shows a color-magnitude diagram for globular star cluster M 3. It shows a richly populated main sequence, above which the stars are seen to rise in brightness and decrease in temperature following a line similar to that derived from theoretical calculations. The heavily populated region at cool temperatures is called the *giant branch* of the globular star cluster color-magnitude diagram, and it represents the initial red-giant phase of evolution.

The level accumulation of stars at absolute magnitude about 0 is called the *horizontal branch* of the color-magnitude diagram. It represents the path of stars that have already passed through the giant branch, have experienced a helium flash, and are now burning helium in the core and hydrogen in a shell around it. For some globular clusters, the horizontal branch intersects the instability strip; pulsation in the form of RR Lyrae variables results. M 3, for example, contains 201 RR Lyrae variables; it is, however, unusual in this respect, as most globular clusters contain only a few.

Fig. 13-29. The globular cluster M 3. *(Harvard Observatory photograph by Martin Schwarzschild.)*

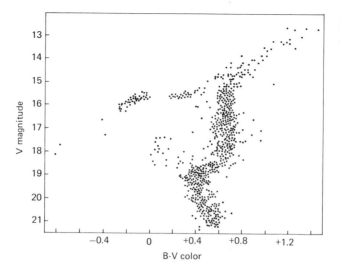

Fig. 13-30. A color-magnitude diagram for M 3. *(After Sandage.)*

Especially for old star clusters, the effects of the abundances of elements of large atomic weight are very important to the luminosities of stars in the giant branch. It is found, for example, that the cluster M 3 (a globular star cluster in the halo of the Galaxy) and the cluster NGC 188 (a galactic cluster nearly in the galactic plane) are both about the same age. However, their color-magnitude diagrams are conspicuously different (Fig. 13-31). This difference shows up most conspicuously in the luminosity of the giant branch, which is about 4 magnitudes brighter for M 3 than for NGC 188. Calculations of stellar evolutionary tracks with different assumed heavy-element abundances show

The Effect of Heavy Element Abundances

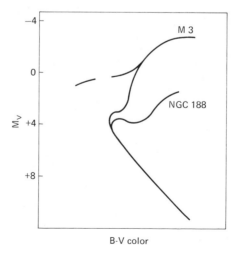

Fig. 13-31. Color-magnitude diagrams for M 3 and NGC 188 compared, showing the difference in the brightness of the giant stars that is caused by the difference in the compositions of the stars in these two clusters.

that this can be explained as a result of a smaller percentage of heavy elements in M 3 (and in fact in almost all globular star clusters) than is typical for the sun and other stars in the plane of our Galaxy. Individual spectroscopic measurements of the elements in the atmospheres of the stars in these star clusters confirm this difference. For example, for M 3, the heavy elements are depleted by a factor of about 100 compared to those in the sun. For other globular star clusters, a similar depletion (sometimes even greater) is observed.

The explanation for this difference in the brightness of the giants has to do with the fact that the electrons in the atmospheres of these stars are one of the important sources of the absorption of light. In stars such as those in NGC 188 (or the sun), electrons are quite abundant because these stars contain comparatively large amounts of heavy elements. A heavy atom, like iron, has many electrons, while hydrogen has only one. Thus, the heavy elements can contribute comparatively large numbers of electrons to the stellar matter, and the opacity of the materials in these stars is larger than in the case of stars that have fewer heavy elements, and therefore fewer electrons. The increased opacity leads to a smaller luminosity for the giants in clusters that are heavy-element rich than for those that have almost no heavy elements.

13-6 LUMINOSITY CLASSES

Giant stars can be distinguished from main-sequence stars by close examination of their spectra. Because of their extended size, the gravity at their surfaces is low, and the amount of pressure broadening (see Sec. 15-2) is therefore less than for a main-sequence star. A giant star will thus have much narrower lines than a main-sequence star of the same temperature. Astronomers indicate this by adding a roman numeral to the star's spectral type. A main-sequence star is signified by a V, while giants are given luminosity classes ranging from I to IV, depending upon their luminosity. The most luminous stars of a given temperature are luminosity class I supergiants. For the rare stars found below the main sequence, the roman numeral VI is assigned.

QUESTIONS

1. Determine the distance to the Large Magellanic Cloud using the period-luminosity relation. Does this distance agree with that given in the text? What might explain any differences?
2. Why are there no known star clusters with giant branches at absolute magnitudes of $M_V = +5$?
3. What might explain the difference in the luminosities of the giant-branch stars in NGC 6356, whose brightest giants have $M_V = -1$ and in M 3, whose brightest giants have $M_V = -3$? Both clusters are globular clusters of about the same age.

1. Star clusters are the most important objects available for studying the observational effects of advanced stellar evolution. The following are galactic clusters that can be observed with a small telescope, binoculars, or in a few cases, the naked eye. Many have giant stars as members, detectable by their yellow or reddish color. These, and clusters in list 2 below, can be found on the charts in the back of the book.

OBSERVATIONS

Cluster name	Constellation	Distance, parsecs	Age, million years	Notes
Fall objects:				
h and X Persei	Perseus	2,300	5	Double cluster
Pleiades	Taurus	125	100	"Seven Sisters"
Hyades	Taurus	40	500	Very spread out
Winter objects:				
M 35	Gemini	870	50	—
M 41	Canis Major	670	20	—
Spring objects:				
M 44	Cancer	160	250	Praesepe, "The Beehive"
M 67	Cancer	830	8,000	Very old cluster
Summer objects:				
M 7	Scorpio	550	250	—
M 25	Sagittarius	600	100	Contains a Cepheid
M 11	Scutum	1,720	250	Rich

2. Globular clusters are all too distant for a small telescope to resolve them into stars, but many are large and bright enough so that they can be seen as faint, unresolved patches of light, many even through binoculars. Some good ones to observe are:

Cluster name	Constellation	Total magnitude
Fall and Winter Objects:		
M 55	Sagittarius	7
M 15	Pegasus	7
M 2	Aquarius	7
Spring and Summer Objects:		
M 3	Canes Venatici	7
M 13	Hercules	6
M 92	Hercules	7

14

ERUPTION, COLLAPSE, AND DEGENERACY

After its brief period of glory as a red giant, a star dies, either in a quiet collapse or in an explosive flash. This chapter explains how and why.

The final stage of stellar life is reached in different ways, depending upon the mass of the star. A very massive star apparently dies in the aftermath of a brilliant explosion that scatters much of its matter into space. Less massive stars the size of the sun or even smaller achieve their final state more quietly. For the more massive stars, it is possible that several (at the very least two) trips are made back and forth between the hot, pseudo-main-sequence stage and the giant stage before death occurs. For less massive stars, probably only one giant stage is reached before the star expires. In both cases, the period on the giant branch is relatively short.

One of the important changes that comes over a star while it is a red giant is a decrease in its total mass. Up until this stage in its life, the mass of the star has remained almost exactly constant. It is true that a small amount of material has left the star, because of stellar winds analogous to the solar wind, and it is also true that a small amount of matter has been gained by the star because dust has fallen into it. A small amount has also been converted into energy and radiated away. But all these effects do not affect the star appreciably.

A red-giant star is much more susceptible to mass changes, because at its surface the force of gravity is so much weaker than that for a main-sequence star. A red giant is so exceedingly large that gravity at the surface is not sufficiently strong to hold the gases there tightly enough to prevent "leaking" of mass into interstellar space at a fairly high rate. Some red giants, for example, show absorption lines in their spectra that are shifted toward the blue as compared to the other lines in the spectrum. The fact that the lines are absorption (dark) lines tell us we are seeing a cloud of cool gas in front of the star, and their blueward shift tells us that this gas is moving toward us with respect to the star. This is interpreted as indicating the slow but steady leaking of mass away from the tenuous outer envelope of the red giants.

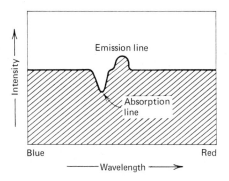

Fig. 14-1. Complex spectrum of a star with a hot shell.

Fig. 14-2. We see from blue-shifted lines that Alpha Orionis is losing mass from its surface.

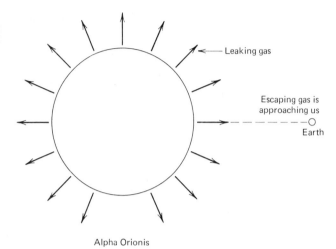

Leaking gas

Escaping gas is approaching us

Earth

Alpha Orionis

Binaries with Emission Shells

Another indication that red giants lose mass comes from the study of some of those that are members of binary star systems. The brightest star in the constellation Hercules, Alpha Herculis, is a binary star consisting of a red giant and a main-sequence star that are revolving around their center of mass. These facts are learned not because we can see the individual stars moving. Other binaries are so far away that even with the most powerful telescopes their angular separation is too small to be resolved. We see only a single image when we look visually at or take a photograph of such a binary. However, when its spectrum is examined, it is found to consist of a composite made up of two spectra, e.g., of a red giant together with the spectrum of a normal main-sequence star. It is difficult to disentangle such a composite spectrum from one spectrogram, but by examining the changes that the spectrum undergoes while the two stars are revolving around each other, it is possible to see the two sets of spectral lines move back and forth with respect to each other. By measuring the wavelengths of any spectral feature and by plotting its motion over the period of the binary star, it is possible to disentangle the lines and to discover the nature of the stars that produce them.

Fig. 14-3. The composite spectrum of a spectroscopic binary shows most lines doubled because of different Doppler shifts of the two stars. *(Courtesy G. Wallerstein.)*

For Alpha Herculis, the main-sequence star shows something unusual that is

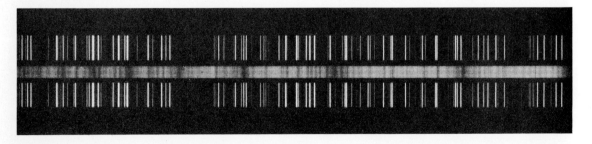

not found for single main-sequence stars that are not members of such binary systems, i.e., a tenuous gaseous shell surrounding it (detected by its absorption spectrum). It is likely that this shell of gas was captured by the main-sequence star from gas that was expelled by the slow leaking process from its red giant companion. Some of the gas could not escape from the system as a whole and is likely to have fallen into the area around the main-sequence star after expulsion. Several other binary stars of a similar nature have been found to show similar effects.

The reason that the material can be lost to a giant star is that the outer parts of such a star are continuously in motion due to turbulence of the gas and due to acoustic (sound) waves that are propagated through the star. These gas motions and waves are produced by the effects of convection and rotation. They give the outer parts of the giant star sufficient motion so that, since the force of gravity there is very low, some of the matter will move away from the star (unchecked by this weak force of gravity) to be lost in space.

14-2 PLANETARY NEBULAE

It is not yet known exactly how important the mass loss mechanisms described above are in reducing the mass of a red-giant star. It is calculated that a star like Alpha Orionis (Betelgeuse), an immense red giant in the constellation Orion (the Hunter), may lose a very substantial amount of its mass if it continues to leak matter at the rate that we observe. However, there is a further evolutionary stage for stars that carries away a measurable amount of mass, and this stage is now recognized as likely to be inevitable for all stars of low mass.

Final Evolution of Red Giants

A red-giant star ends with a very dense carbon core, made up of carbon produced by the fusion of helium during the red-giant stage, surrounded by a thin shell of helium that is burning. This helium burning produces most of the energy that we receive as light from the star. Surrounding the shell is an extensive helium-rich envelope. As the star evolves and the helium burning continues, the carbon core becomes increasingly smaller and hotter. This change is gradual in stars of moderate mass, and it results in an increase of the temperature of the surrounding helium-burning shell, and therefore an increase in the rate at which the helium fuses to form more carbon. This effect makes the star become larger and cooler at its surface. Finally, the red-giant star becomes so cool at its surface that the gas in the surface layers begins to become made up of neutral atoms again. This is because the temperature is sufficiently low so that the nuclei of the atoms can capture and hold the electrons. The temperature is so low that collisions between the particles making up the gas in the outer parts of these stars are so infrequent and so slow that they do not dislodge electrons from the nucleus of atoms continuously, as has been the case throughout the lifetime of the star until this stage.

Fig. 14-4. Forming a neutral shell.

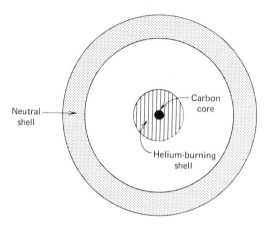

The formation of neutral atoms in the envelope of a red-giant star marks the beginning of the end for this stage. When formed, neutral atoms emit light waves because of the change in the energy of the system when the electron is captured by the nucleus of the atom. A nucleus, together with electrons forming a neutral atom, has less total energy than that possessed by a nucleus and electrons separately. This is because there is a certain amount of "binding" energy that holds the electrons to the nucleus in its neutral state. Therefore, whenever a nucleus captures an electron, it must give up some energy, and this energy is emitted in the form of a photon of light (see Fig. 14-5). When the envelope of a red-giant star begins to become neutral, many photons are thus emitted by the forming atoms. These photons are subsequently absorbed by surrounding material, thus heating up the envelope slightly. The heating in turn causes the envelope to expand to a larger size, and the expansion in turn leads

The Formation of a Shell

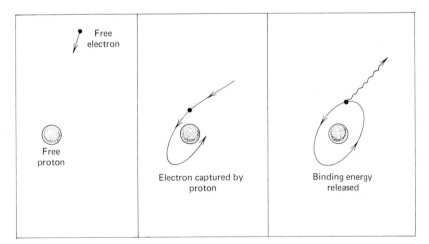

Fig. 14-5. Release of binding energy by the capture of an electron by a proton.

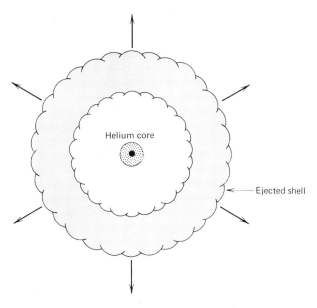

Fig. 14-6. Ejection of the shell.

to a cooling of the envelope that allows more neutral atoms to form and more photons to be emitted. This process continues in a cycle, with the envelope getting progressively larger and cooler. Finally, the envelope reaches such a size (expanding sufficiently rapidly) that it completely leaves the star and becomes a separate entity, a ring or shell surrounding the star. Calculations made theoretically to explore the amount of material that might be emitted in this way suggest that perhaps 20 percent or so of the mass of a star like the sun could be ejected gently into space by this process. What is left is the bare core of the star, consisting of the carbon center and the thin helium-burning envelope. Because the envelope is very hot, the remaining part of the star will appear to be very blue.

Many examples of these expanding stellar envelopes can be found in space near the sun; in our Galaxy, about 1,000 such objects are now known. They are called *planetary nebulae*, because early discoverers, who saw small bright ones, noted their resemblance through the telescope to the faint distant planets, such as Uranus.

Observation of Planetary Nebulae

The planetary nebulae are made up of thin, tenuous gaseous envelopes surrounding hot blue stars. The various panetary nebulae observed range in size from those just barely bigger than the resolving power of the best telescopes to some that are very large, many minutes of arc in diameter. The larger planetaries are faint and thinly spread out, while the small planetaries are often exceptionally bright. From this fact, it is understood that the small bright

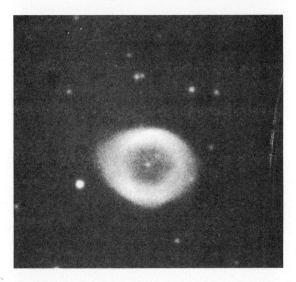

Fig. 14-7. A planetary nebula, the Ring Nebula in Lyra. *(Manastash Observatory.)*

planetaries are the most recently formed. The larger ones are old, having formed long ago and since expanded out into large volumes of space, where they cooled off and spread out to become fainter. Measurements, both of the radial velocities of the gas in planetary nebulae and of changes in their

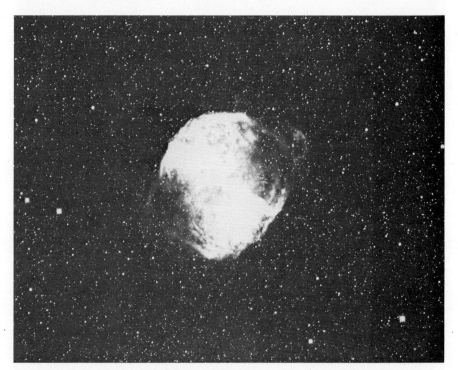

Fig. 14-8. A large, bright planetary nebula, the Dumbbell Nebula in Vulpecula. *(Kitt Peak National Observatory.)*

Fig. 14-9. A small, compact, young planetary nebula, NGC 6828. *(Kitt Peak National Observatory.)*

diameters over many years, indicate that the expansion is indeed continuing for these objects, typically with expansion rates on the order of 20 kilometers per second. The largest planetaries have diameters of about 1 parsec, and both the expansion rates and theoretical considerations agree that the implied age of such objects is about 5×10^4 years. This is an indication that the lifetime of the envelope is on that order, and that after that long an interval the gas is so tenuous and scattered out in space that we can no longer detect the shell.

After the outer parts of the red-giant star have been blown away to form a planetary nebula, the core star continues without any noticeable change caused by the precipitous loss of its outer mantle of gas. The carbon core, unable to produce energy by the nuclear reaction itself, slowly continues to contract while the surrounding helium shell burns. This contraction finally stops when the material in the core is sufficiently dense so that the electrons once again form a component that acts like a solid body, which is similar to its behavior when the helium core of the star had collapsed earlier in its lifetime. At this point, the star can no longer contract. The matter in this stage is called *degenerate*, and the star becomes a white dwarf (Chap. 15).

Evolution of the Core Star

**14-3
SURFACE
EXPLOSIONS—
NOVAE**

On the average of once or twice each year, astronomers discover a star that suddenly brightens to tens of thousands of times its normal brightness in a period of just a few hours. Such objects, called *novae* from the Latin word for "new," have been seen by astronomers for hundreds of years. They were given the name novae because at first they were thought to be new stars that had not existed before. It was subsequently learned that in fact each nova phenomenon is an example of an already existing star's greatly increased brightness, as a result of an explosive event's occurring on its surface.

**Number and Light
Curves**

It is estimated from surveys of our Galaxy that novae occur in our local stellar system at the rate of approximately 20 each year. The phenomenon shows a great increase in luminosity in a very short time (Fig. 14-10), followed by a gradual decline that takes years before the brightness returns to the preoutburst level. Individual novae show a variety of different shapes of light curves. Study of novae in the Andromeda galaxy by Halton Arp many years ago showed that the novae which reach the greatest intrinsic brightness decline in brightness most rapidly, while those that reach the least anomalous luminosity decline more slowly.

Spectra of Novae

The spectrum of a nova shows bright emission lines from gas that must be moving away from the star with velocities on the order of 1,000 kilometers per

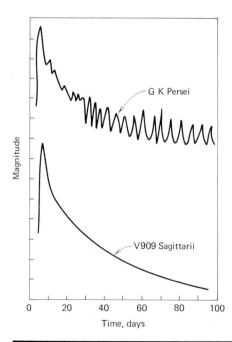

Fig. 14-10. Light curves of two novae of very different short-term character.

second. Figure 14-11 shows an example of a typical nova spectrum. It shows the absorption lines due to the star itself, and the broad emission lines that occur first only on the blue side of the normal position of the line, and then later form on both the blue and the red sides. This can be interpreted as resulting from an ejected shell of gas, which first we see only silhouetted against the star on the side facing us, where the gas is moving toward us, and then later we see as gas also moving in other directions (including away from the sun) as the shell expands. It is calculated from study of the spectrum and the intensities of the emission lines that the total mass of material ejected in the shell is only about one ten-thousandth of the mass of the sun, a very small percentage of the star's mass.

Fig. 14-11. A nova spectrum, showing complex emission and absorption features. *(G. Wallerstein.)*

There are many different theories for the nova explosion, but modern research has indicated that there is one model of the outburst that seems best able to explain the many different observations of these objects. This model depends upon the recently established fact that many novae seem to be binary stars. In most cases, the binary nature is determined by the measured wavelength shifts (Doppler shifts) that show up when the spectrum of the nova is examined with very high dispersion.

Explanation of the Novae

Since all novae appear likely to be binaries, the most attractive theory of their nature is one which depends upon the fact that they are binaries. It can be calculated that if two stars of different mass are members of a close binary pair, one will evolve more rapidly than the other and will be the first to become a red giant. The red-giant star will eject its envelope, possibly as a planetary nebula, and then will collapse to form a white dwarf with a very hot surface (Chap. 15). Then, in due time, the less massive member of the binary star will become a red giant also. When it becomes so large that it begins to lose mass, some of its ejected material will fall into its companion's sphere of gravitational influence. When this matter falls to the very hot surface of the white dwarf, it will produce

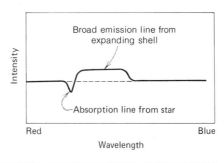

Fig. 14-12. Interpreting a nova's spectrum.

a sudden rise in the temperature and gas pressure of the surface material because of the high surface gravity of the white dwarf. Conditions calculated for this gas are found to be sufficiently high in temperature and pressure so that the proton-proton cycle (Chap. 10) can be ignited suddenly, and a nuclear explosion results. This bright flare is what we observe when we see a nova outburst, and the explosion causes ejection of matter away from the system into space, which we can see later as the ejected shell of gas that shows up in the spectrum of the nova.

Although most novae have been seen to explode only once over the history of astronomy, some do in fact reoccur every few years. Apparently the recurrent novae represent cases of continually expanding red giants, whose envelopes are continuously spilling over onto the white dwarf's surface, thus giving rise to a series of nuclear explosive events.

QUESTIONS

1.* How long would it take a red-giant star of 5 solar masses to lose all but 1.4 solar masses of its material, if its gas "leaks" out from its surface at a rate of a billion tons per second?

2.* How many grams of matter are present in a typical planetary nebula?

3. What evidence suggests that novae are binary stars?

4. Does the explanation of novae given in the text as a possible model require that all observed novae must be binaries?

5. Is it likely that the sun will become a nova after it becomes a red giant?

OBSERVATIONS

1. Planetary nebulae are difficult objects to observe with binoculars, but can be interesting objects when seen through small telescopes. The following include some of the most easily observed. (See the charts at the back of the book.)

Planetary Nebulae

Name	Constellation	Magnitude	Diameter, seconds of arc
Fall objects:			
NGC 7009	Aquarius	8	30
NGC 7662	Andromeda	9	20
Winter objects:			
NGC 2392	Gemini	8	20
M 97 ("Owl")	Ursa Major	11	200
Spring objects:			
NGC 6210	Hercules	10	20
Summer objects:			
M 57 ("Ring")	Lyra	9	80
M 27 ("Dumb-bell")	Vulpecula	8	360

15

WHITE DWARFS

15-0
GOALS

This chapter's purpose is to explain and describe the remarkable properties of "dead" stars—the white dwarfs.

15-1
PROPERTIES OF
WHITE DWARFS

The white-dwarf stars, which are fairly common in the Galaxy, are the final stage in the evolution of most stars. While at least some stars of large mass may become more spectacular and bizarre objects when they die (Chap. 16), the stars with masses like that of the sun or smaller probably all become white dwarfs. This term is used to describe the stage of evolution in which the star has finally stopped contracting after being a red giant. The stellar material in this stage is compacted into a faint object of very small size, on the order of the size of the earth. The star then cools off slowly since it no longer has any source of new energy.

Distances

White-dwarf stars are common enough in our Galaxy so that there are enough close to the sun for astronomers to be able to measure their distances accurately by means of trigonometric parallaxes (Chap. 8). Within 20 parsecs of the sun, about twenty five white dwarfs are known, and accurate distances

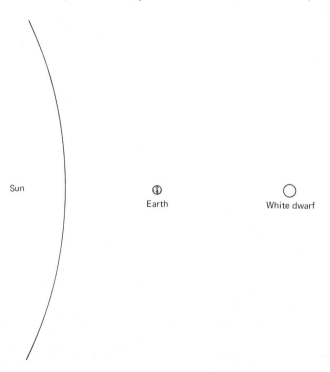

Sun

Earth

White dwarf

Fig. 15-1. The sun, a white dwarf, and the earth—sizes compared.

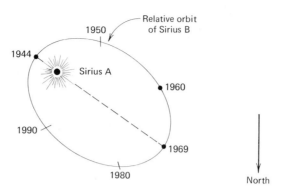

have been determined for these. This is still, however, a small percentage of the more than 1,000 white dwarf stars that have been discovered. For the rest, distances must be obtained by more approximate methods.

One of these methods works for white-dwarf stars that are members of binary systems, for which it is possible to measure the properties of the companion. When this companion is a normal star, its distance can be estimated fairly reliably from its spectrum and apparent magnitude. The distances to about fifty white dwarfs have been estimated in this way. Similarly, when white dwarfs have been discovered in star clusters, their distances can be taken as the same as the distance of the cluster, which is determined by fitting the main sequence (Chap. 8). A final approximate method is a statistical one in which the apparent motions of white dwarfs are examined and approximate distances are assigned according to whether the apparent motion is large or small. Statistically, the amount of apparent motion for a star from year to year might be expected to be large for the nearer stars and small for the more distant stars, as explained in Chap. 8. Astronomers have used their motions to estimate statistical distances for hundreds of white-dwarf stars.

From distances and from measured apparent brightnesses, it is possible to calculate the luminosities of white dwarfs, and it is this property that makes them remarkable. All are exceedingly faint in absolute brightness, most of them more than 10 magnitudes fainter intrinsically than the sun. An example is the system of Sirius and its white-dwarf companion. The star Sirius is the brightest star seen in the sky in apparent magnitude and is a normal star on the main sequence. Its white-dwarf companion, on the other hand, is about 10 magnitudes fainter and is impossible to see through a telescope except under exceptionally excellent conditions. This is because of the fact that light from Sirius, which is about 10,000 times more intense, blocks it out. That white dwarf is fairly typical of the class, but there is a wide range in luminosities from absolute magnitude of about +10 to about +20.

Luminosities

Space Densities   Astronomers' surveys have by no means discovered all the white dwarfs that must exist in the immediate neighborhood of the sun. Especially difficult to detect are those of very low intrinsic luminosity. The most common means used to detect them is the painstaking one of comparing plates taken over many years in an attempt to locate stars of very faint apparent brightness, but large apparent motion. Estimates from the surveys presently in existence suggest that about 100 white dwarf stars may exist in the volume of space within 10 parsecs of the sun. This large number makes them among the more common objects in our neighborhood in space.

From these statistics, we can obtain a rough idea of the total number of white dwarfs in our Galaxy. If we assume that the local density of these objects is about average for our Galaxy, then we can find the total number by comparing the volume of our Galaxy with the above sample. Our Galaxy has roughly the shape of a pancake, with a thickness of about 1,000 parsecs and a radius of about 15,000 parsecs (Chap. 19). The volume of a cylinder is given by the formula

$$V = \pi r^2 h$$

where r is the radius and h is the height or thickness of the cylinder. Thus, the volume of the Galaxy is roughly 7×10^{11} cubic parsecs. On the other hand, the volume of space within a radius of 10 parsecs from the sun, in which we find 100 white dwarfs, is $^4/_3 \pi r^3$, which approximately equals 4×10^3 cubic parsecs. Our estimate of the total number of white dwarfs, then, is approximately 200 million for the entire Galaxy. The true value is probably larger than this estimate, because the sun lies in an outer part of the Galaxy and the density of white dwarfs around us is no doubt smaller than the average.

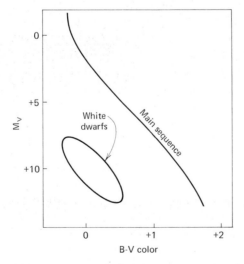

Fig. 15-3. White dwarfs in the color-magnitude diagram.

Fig. 15-4. The spectrum of the white dwarf van Maanen 2 shows very few lines. The two darkest lines are of calcium and the fainter lines to the left are of iron. (*Courtesy K.H. Bohm.*)

Colors

Colors of almost 500 white-dwarf stars have been measured, and they are found to range from very blue to yellowish. The range of temperatures represented by this range in color extends from about 100,000K for the bluest down to about 5,000K for the reddest objects that are considered white dwarfs. Actually, it is found that the colors and temperatures of objects like white dwarfs could be extended farther to cover objects that are much cooler than this, but which are so red that they cannot be included in the somewhat arbitrarily set-up definition of a white-dwarf star. As shown below, the range in colors can be explained in terms of the cooling of these degenerate, collapsed stars, and this cooling eventually allows them to have temperatures that can be very low.

Spectra

The spectra of white dwarf stars are unlike spectra of normal stars. The majority show very few spectral lines. The only lines that are conspicuous are the lines of hydrogen. Other white dwarfs, however, show somewhat more unusual spectral features; for example, there are some that show no lines of any element whatever. There are some white dwarfs that only show helium lines, with no hydrogen lines being visible, and there are some that show only the lines of heavier elements, such as calcium and iron. Many attempts have been made to understand the differences between these different spectroscopic phenomena. They probably must be understood as resulting from differences in the amount of material that is left in the atmosphere of the collapsed star, the core of which is mostly or entirely carbon, according to present, still tentative conclusions. Those white dwarfs that show no hydrogen lines are apparently stars for which the hydrogen in the outer envelope was completely lost while the star was a giant, and for which the hydrogen in the core was completely exhausted by nuclear burning.

Masses

Fortunately, some of the white dwarfs that have been discovered are members of binary systems, and so it is possible to determine reliable masses for them. One of the most famous white dwarfs, Sirius B, is a binary to which we have already referred. Another of the brightest stars in the sky, Procyon, also has a white-dwarf companion, called Procyon B, for which we can get an accurate mass. A measurement of the orbits of these stars shows Sirius B to have a mass of 1.05 times the mass of the sun, and Procyon B to have a mass of 0.63 times the mass of the sun. These are probably typical values for white-dwarf stars,

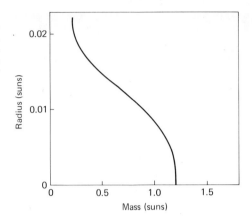

Fig. 15-5. The radius-mass relationship for white dwarfs as calculated first by Chandrasekhar. The more massive the star, the smaller its radius.

and there is good reason to believe that no white-dwarf star is very much more massive than Sirius B. Calculations of degenerate star configurations show, in fact, that no white dwarf can exist with a mass much larger than the mass of the sun. This limit is connected to the fact that the size of a white dwarf is related to its mass in a rather strange way, as shown by calculations of the properties of a "degenerate" gas, in which all the electrons have been packed into all possible available space and all possible velocities. It is found that the more massive the white dwarf, the smaller its size. The calculations furthermore show that if a star is more massive than about 1.2 solar masses, it cannot form a stable system because it would have a negative radius, which is of course absurd. More massive stars must die in some other way than by directly becoming a white dwarf. Chapter 16 describes some of the other means available to them.

Densities White dwarfs are very small objects with radii all found to be very close to about one one-hundredth the radius of the sun, i.e., about the size of the earth. Since the masses of white dwarfs are about the same as those of the sun, this means that the density must be about 100^3, or about 1 million times the density

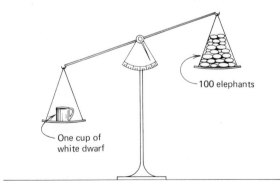

Fig. 15-6. The density of white dwarfs is almost inconceivably greater than that of anything on earth.

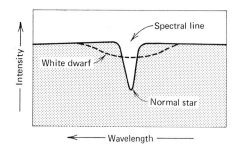

Fig. 15-7. Pressure broadening of spectrum lines allows the detection of high-surface gravity.

of the sun. The average for typical white dwarfs is about 100,000 times the density of iron, so that a cupful of white-dwarf material would weigh about 100 tons. The amount of gravity at the surface of a white dwarf is also almost incredibly great, about 100 billion times the surface gravity at the surface of the earth. Any object that might attempt to land on and explore the surface of a cool white dwarf would immediately be crushed by the incredible gravitational pull.

Calculations on the nature of the kind of degenerate gas that makes up a white-dwarf star show that this material acts more like a solid than it does like a gas, and that the star is held together rigidly. There is no change in the white dwarf's size or structure as its temperature cools. Its interior is held steady by the electrons, which form a crystal lattice, somewhat like that of a rock or a piece of iron.

**15-2
STRUCTURE
AND
EVOLUTION**

The remarkable spectra of white dwarfs show that the atmospheres of these stars vary considerably from one to another. Certainly it is expected on the basis of the very large surface gravity that the structure of the atmosphere of a typical white dwarf will be quite unusual. There must be tremendous pressure and a very thin size, with a typical height of only about 100 meters (300 feet), which is less than "paper thin" by comparison with a white dwarf's typical radius of 10,000 kilometers (10 *million* meters). For some white dwarfs, the spectral lines are found to be broad compared to the lines of the sun, and this can be understood in terms of the broadening of these lines by the very high pressure due to the high surface gravity. For a gas that is undergoing these immense pressures, the wavelengths emitted by the electrons can be significantly altered, thus giving a variety of different emitted wavelengths that in turn melt together in the spectrum as a broadened line.

**Atmospheres of
White Dwarfs**

Because of the large gravity, a further complication arises in the spectra of white dwarfs. Einstein showed when he introduced the concept of general relativity that light can be affected by a large gravitating mass. For example, one

The Einstein Shift

Fig. 15-8. The bending of light seen at solar eclipse and resulting from general relativistic effects.

Earth   Moon   Sun   Distant star

of the best confirmations of Einstein's theory of general relativity is the bending of light rays around the sun, which is observable during a solar eclipse. This effect on light is most noticeable when the force of gravity is exceptionally high. In the case of white-dwarf stars, the force of gravity is so high that it acts on the light emitted by the atmosphere, "pulling it back" somewhat, so that when we observe the light it is shifted in wavelength toward the red. This is called the *gravitational redshift* and is one of the more conspicuous examples of the importance of the discovery of general relativity.

Cooling   The theory for degenerate white dwarfs shows that the star's internal temperature will decrease, but its radius will virtually never change. The radiation that we detect, feeble compared to that from normal stars, represents a loss of energy for the system that can never be regained and that cannot be made up by any interior nuclear processes. It is the energy of motion of the atomic nuclei existing in the interior of the white dwarf, and as this energy leaks away from the surface of the star, these motions quiet down. After an extended interval of time, calculations show that the nuclei of the atoms become degenerate in the same way that the electrons have been degenerate throughout the white-dwarf stage. When that happens, the nuclei can no longer move, and they also form a crystal lattice structure, so that from then on both the electrons and atomic nuclei act like a solid.

The cooling of a white dwarf is very slow. It is calculated that it takes an average white-dwarf star about 10^{10} years (10 billion years, about the age of the universe) to cool down to a temperature of 3,000K. This suggests that we shouldn't find any cooler white dwarfs in our Galaxy, and indeed we do not,

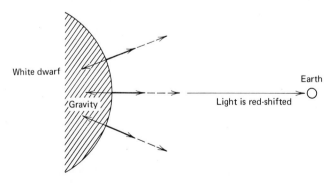

White dwarf

Gravity

Earth

Light is red-shifted

Fig. 15-9. The Einstein red shift for high-density stars.

although there are many white dwarfs in the solar neighborhood that have apparent temperatures only slightly higher than this value. These cool objects, not white in any sense, should perhaps be called "red degenerate stars;" they have not been adequately explored as yet.

By the time a degenerate star has reached such a cool temperature as 3,000K, its luminosity is calculated to be very low, and it almost could be referred to as a "black dwarf." A very long-term prognosis about stellar evolution suggests that eventually almost all of the matter in our Galaxy will be in the form of black dwarfs. Because of the continual expulsion of part of the mass of stars into interstellar space, where it then forms again into other stars, new stars will continuously form for a long time, even as matter continues to become locked up in white and black dwarfs. Nevertheless, it appears likely that eventually almost all the mass of our Galaxy will be lost from view in this way, the rest of it being either dispersed as a thin interstellar gas, ejected from the system by explosive events, or caught up in neutron stars or black holes (Chaps. 16 and 17). It may take a hundred billion years or a million billion years, but it looks inevitable that our Galaxy will eventually die and become an invisible mass of cool stellar cinders. Only some unknown and unsuspected phenomenon could prevent such a dark, quiet end for our stellar system and, in fact, for the universe as a whole.

QUESTIONS

1. What is the source of energy that allows a white-dwarf star to shine?
2. What is the ultimate future of a white-dwarf star?
3. In 100 billion years or so, what kind of stars will predominate in our neighborhood? What will our Galaxy look like?
4. If the sun were a white dwarf, about how big would it appear to be in the sky? (Express your answer as a fraction of the sun's present apparent diameter.) What would its luminosity be, in terms of its present luminosity? Assuming that the earth's temperature is about proportional to the sun's luminosity, what would the earth's temperature be? Use the Kelvin (absolute) temperature scale. Would the earth be colder, even if the sun, as a white dwarf, were hotter?

16

SUPERNOVAE, PULSARS, AND NEUTRON STARS

Sometimes stars explode, destroying themselves and becoming supernovae. This chapter's purpose is to explain what we have learned about supernovae so far and to describe the strange objects that remain behind after the explosion.

**16-0
GOALS**

While moderate and low mass stars quietly pass into oblivion from their final stages as red giants, stars that are much more massive than the sun apparently can die in a spectacular way, becoming objects with almost unbelievable properties. The explosive destruction of a star can produce what is called a *supernova* (an object very much brighter than a nova), and it can leave the dregs of the star in the form of a remarkable object called a pulsar or a neutron star, or perhaps a black hole.

**16-1
SUPERNOVAE**

When a supernova goes off, the result is a very much more spectacular increase in brightness for the star than in the case of a nova. Whereas a nova can at maximum reach a luminosity that makes it one of the brightest stars in a galaxy, a supernova can reach a brightness equal to the sum of all the brightnesses of the stars in a galaxy. The brightest supernovae observed in other galaxies have occasionally been several times brighter than the entire galaxy itself. The total luminosity of a supernova can reach values of about a billion (10^9) times the brightness of the sun.

Luminosities of Supernovae

Fig. 16-1. A supernova in the galaxy NGC 4303. The galaxy in this photograph is underexposed to show how brilliant the supernova is by comparison. *(Lick Observatory.)*

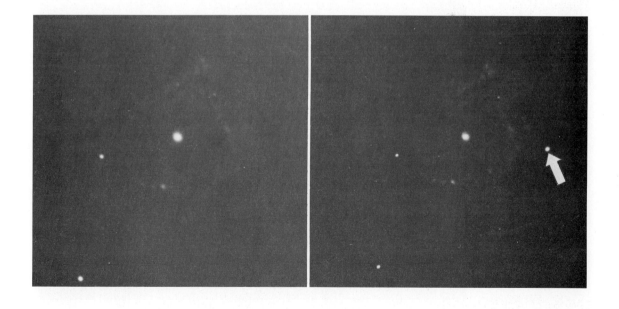

Light Curves and Spectra

Only a very few supernovae have been observed close at hand, but hundreds have been photographed in other galaxies in various parts of the universe, and we have obtained from these observations a knowledge of their various properties. When a supernova explodes, its brightness increases to maximum in a day or so. At maximum, the spectrum of a supernova is very complex.

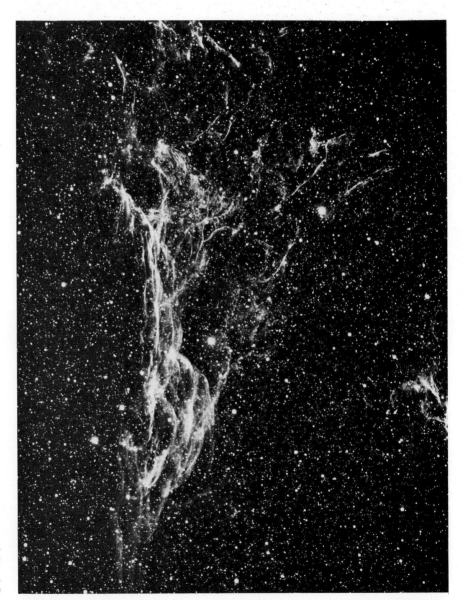

Fig. 16-2. The Veil Nebula in Cygnus is a remnant of an ancient supernova. *(Kitt Peak National Observatory.)*

There are at least two different classes of supernova spectra, and both are sufficiently complicated that astronomers have not yet completely pieced together the spectral evidence in terms of the physical properties of the exploding body. Following maximum, the spectrum changes and the luminosity decreases. For each type of supernova, the pattern of luminosity decrease is different. Typically, the brightness decreases slowly so that not until several months have passed will a supernova in a nearby galaxy disappear from view.

Modern astronomical instrumentation has never been used on a supernova in our Galaxy, and so the details of the process of supernova outburst have not been observed up close. That is why we have no information on the pre-explosion luminosities of supernovae, as those near enough for observation occurred too long ago to be observed with telescopes.

The first well-documented supernova outburst in our local Galaxy occurred in the year 1054. Chinese, Japanese, and American Indian records all seem to represent that object, which reached a sufficiently bright luminosity that it was visible for a while during the daytime. The position of the object in the sky corresponds with a strange diffuse object, called the Crab Nebula, that has been found to be an exceedingly remarkable gas cloud radiating energy strongly over the entire electromagnetic spectrum from radio waves to x-rays and gamma rays. The next well-recorded supernova in our Galaxy is the one called Tycho's Nova, which occurred in 1572 and was studied extensively by the great astronomer Tycho Brahe. This object was also sufficiently bright to be visible during the daytime. In 1604, the astronomer Kepler was treated to the privilege of observing a third supernova in our Galaxy, one which was somewhat fainter than Tycho's but brighter than any other stellar object in the sky. This object is referred to as Kepler's Supernova.

Galactic Supernovae

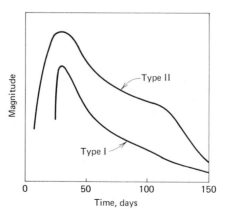

Fig. 16-3. Supernova light curves can be separated into at least two general classes.

Fig. 16-4. Indian pictograph of the 1054 supernova drawn in relation to the crescent moon and the star Aldebaran. *(Courtesy J.C. Brandt.)*

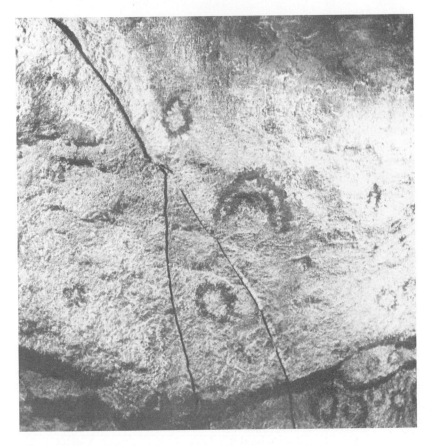

The Rate of Supernovae

By studying the statistics of supernovae in other galaxies, it has been found that a typical galaxy will have a supernova outburst on the average of every 50 years or so. The fact that we have not detected a supernova in our local Galaxy since 1604 is probably mostly due to statistics and is suggestive that we may be able to see one of these remarkable objects up close in the near future. However, it is also possible that supernovae have occurred elsewhere in our Galaxy since 1604, but on the other side of the Galaxy, too much obscured by intervening dust to have been detected.

Supernova Remnants

Supernovae produce visible remnants of material that are usually most conspicuous at radio wavelengths. The spectrum of the radio radiation has the same shape as that of radiation produced in large "atom smashers" that physicists use to study the properties of elementary particles. These machines are called *synchrotrons*, and the radiation that is emitted by a beam of particles in a synchrotron is called *synchrotron radiation*. This light is produced in ways

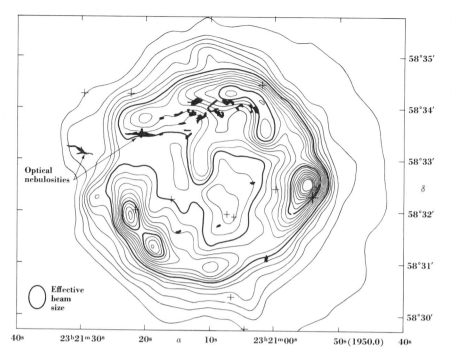

58°35'

58°34'

58°33'

δ

58°32'

58°31'

58°30'

Optical
nebulosities

Effective
beam
size

40s   23^{h}21^{m}30^s 20^s α 10^s 23^{h}21^{m}00^s 50^s(1950.0)   40^s

Fig. 16-5 A radio map of
a supernova remnant in
Cassiopeia. *(From Kraus,
Radio Astronomy,
McGraw-Hill Book Co.
By permission of the
publishers.)*

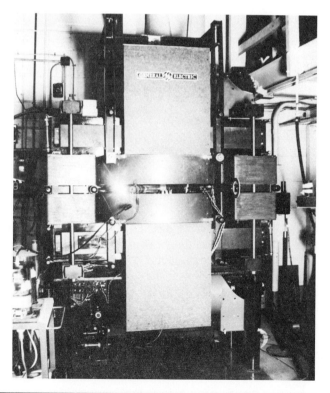

Fig. 16-6. A nuclear particle accelerator, called a
synchrotron, showing
synchrotron radiation
near the center of the
photograph. *(Courtesy
General Electric.)*

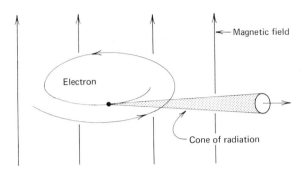

quite different from the light that we normally see emitted by stellar objects. Rather than energy generated by electrons moving from one orbit to another around an atomic nucleus, the light is produced by electrons that are spiraling exceedingly rapidly through a magnetic field. The velocities of these electrons must be almost as great as the velocity of light before very much synchrotron radiation will be emitted, and so it is clear that the very bright synchrotron emissions from supernovae remnants indicate that exceedingly violent events must have produced them. The radio properties of supernova remnants in our Galaxy have enabled astronomers to identify dozens of such objects, found by their radio radiation, which can be distinguished from other sources of radio waves in our Galaxy by the characteristic synchrotron spectral shape. Most of the supernova remnants are seen only by means of their radio radiation, because of intervening dust that in many cases obscures any optical parts. For a few, including the Crab Nebula and the supernova of 1572 (Tycho's nova), optical features have been discovered.

Theoretical Models of Supernovae

Calculations of what happens to a red-giant star of mass considerably greater than the sun's throw some light on the cause of supernova explosions. It is found that the carbon core slowly collapses toward the end of the red-giant phase, eventually reaching exceedingly high temperatures. For less massive stars, these very high temperatures are never reached, but in the case of a massive star, temperatures as large as 600 million degrees are attained. Calculations and experiments indicate that when such temperatures occur, the carbon of the core will begin to fuse as did the helium and hydrogen before it, producing even heavier elements such as silicon and magnesium. This fusion then heats up the core still further, and the pressure produced by the energy generated temporarily stops the contraction of the core. After a short period, however, the carbon of the core will be used up, and the core will once again begin to contract because there is no longer any source of outward pressure. Once again, when the core has contracted further and heated up to an even higher temperature, other nuclear reactions can begin, such as silicon burning. This series of steps continues until many heavy elements have been produced in the core. The process occurs relatively rapidly, and in only a few thousand years or less, depending upon the mass of the star, these steps finally come to a natural halt.

The reason for the eventual stoppage of element building is the very peculiar nature of the element iron. Unlike the previous, lighter elements that have been formed, iron does not give up energy when it engages in a nuclear reaction, but rather absorbs it. Therefore, when iron is formed, it acts to drain the core of energy instead of supplying further energy to it. Thus, iron is the final element, and it provides the final step of the core collapse.

Without further energy sources, the iron core of the star has nothing to stop it from contracting further. It falls violently in on itself, so rapidly that in only a few minutes it reaches a size of only 10 to 50 kilometers. At this point, the density is so high and temperatures so excessive that even heavier elements than iron may be produced, but only in very small amounts. In fact, it is observed in nature that elements heavier than iron are very much rarer than the elements that are lighter than iron, probably for this reason. The collapse of the core at this time is so violent that there follows an equally violent bounce back of the material, which explodes out into space with a tremendous amount of energy. This explosion is observed as a supernova outburst, and it is the material that is thus dispersed into space that eventually forms what is observed as a supernova remnant.

A reasonably large fraction of the total mass of the star, perhaps half of it, is lost forever to the star in this way. This material eventually is dispersed into the general interstellar medium and mixed with the hydrogen gas that is prevalent there. From such evidence, astronomers now believe that most elements heavier than hydrogen and helium were formed in supernova eruptions. The sun and earth, which contain considerable amounts of such heavy elements, obtained them from supernova explosions that occurred in a period in the history of our Galaxy before the sun itself formed out of the interstellar material. Thus, many of the atoms that make up this book and its reader were formed in a series of violent events, culminating in supernovae that exploded more than 5 billion years ago.

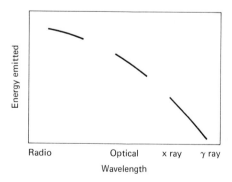

Fig. 16-8. The spectrum, from radio to γ rays, of the Crab Nebula. This spectrum shape is typical for synchrotron radiation.

The Crab Nebula   It is mentioned above that the early Chinese, Japanese, and American Indian records tell us about a supernova that occurred in our Galaxy in the year 1054. We know now that the remnant of that supernova is an object in the constellation Taurus known as the Crab Nebula. Originally thought to be a planetary nebula, the Crab is now well established as one of the most thoroughly studied and thoroughly understood of the supernova remnants. It is a remarkable object, both in its observed properties and in the wide range of kinds of radiation that we receive from it and that give us information. Most of this radiation is found to be produced by the synchrotron process mentioned above. Because synchrotron electrons give off almost equal amounts of radiation at all wavelengths, being in that way very unlike blackbodies in their radiating properties, we receive considerable amounts of energy from the Crab Nebula at wavelengths ranging from very long-wavelength radio radiation to exceedingly short-wavelength x-rays. There is even a likely identification of gamma radiation, which has wavelengths even shorter than those of x-rays.

The Crab is found to be expanding with a velocity of about 10,000 kilometers per second. It is expected that in the next few thousand years, it will gradually become dimmer at all wavelengths, and eventually it will disappear as it diffuses into the general interstellar medium. In the meantime, it is one of the strongest sources of radio radiation and x-rays in the sky, and it may also be one of the agencies responsible for the large numbers of "cosmic rays" that impinge upon the earth. Cosmic rays are exceedingly high-velocity particles (protons, helium nuclei, and heavier elements) that have sufficient energy to penetrate down through the atmosphere of the earth to its surface, and in some cases even to pass through the earth entirely.

One of the most remarkable features of the Crab Nebula is a central object that is now identified as the remains of the star that produced the explosive supernova. This object is not at all a normal star, but rather is luminous only for a few thousandths of a second, flashing on and off with a period of about 0.03 seconds. It also emits pulses of radio waves and x-rays and is called a *pulsar*.

16-2
PULSARS

Discovery   An accidental discovery by a group of astronomers at Cambridge University in England in 1967 led to the first clear understanding of what happens to a

Fig. 16-9. Cosmic rays made these tracks in a nuclear emulsion when they collided with a proton in the emulsion. Part of the energy of the collision went into creating a shower of particles called *mesons*.

massive star when it collapses. The Cambridge astronomers had just put a new and very remarkable radio telescope into operation in the English countryside. They were in the process of making the final tests of the instrument, which involved mapping all the interfering sources of local noise that might confuse their measurements of cosmic signals. Their radio receivers were so sensitive that a wide variety of different kinds of local interference could be detected, including nearby power lines, cars on roads nearby, and other sources. When they were nearly complete, they found there was one source that could not be identified. It signalled pulses that were repeated with a period of a little more than 1 second, with the pulses themselves very much briefer in extent.

Identification

A graduate student at Cambridge, Jocelyn Bell, studied this source long enough to notice that it reappeared regularly every 23 hours and 56 minutes. This immediately told the astronomers that it must be a cosmic source, since that is exactly the period of the rotation of the earth with respect to the distant stars. A radio source in space had never before been found to pulse in this way. Repeated study showed that the burst of radiation lasted about thirty-thousandths of a second, and that the period between bursts was exceedingly regular at exactly 1.3372275 seconds.

Shortly after this discovery, radio astronomers in various parts of the world found similar pulsing objects. About 150 *pulsars* are now known, and their periods have been found to range from values as short as 0.03 seconds to as long as 4 seconds.

The Crab Pulsar

For almost two years, the pulsars could be studied only by means of radio astronomy, because despite efforts to detect the objects optically, no positive identification could be made. Then in 1968, astronomers at the National Radio Astronomical Observatory in West Virginia discovered a pulsar called NP0532

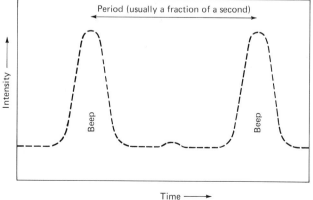

Fig. 16-10. Pulses of a pulsar.

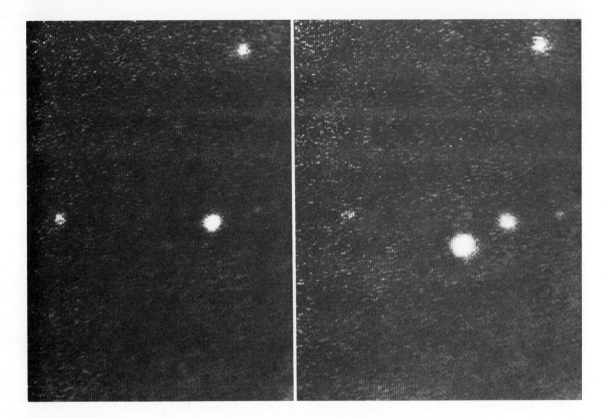

Fig. 16-11. The Crab pulsar shown when "off" in the left photo and when "on" in the right. *(University of Arizona.)*

that appeared to be in the direction of the Crab Nebula. Three months later, three astronomers at the Steward Observatory of the University of Arizona discovered optically that one of the stars near the center of the Crab Nebula could be identified with the pulsar, because the star pulsed optically with the same period and in the same manner as the radio pulses. By using a stroboscopic technique, they showed that the optical star disappeared and reappeared, blinking with a period of 0.03 seconds. The importance of the discovery of the pulsar in the Crab Nebula is that the nebula itself is known to be a supernova remnant (Sec. 16-1), and therefore the pulsar must be understood as the remains of the star that produced the supernova outburst.

It is now believed that the pulsars are probably all collapsed cores of stars that have exploded as supernovae. The Crab Nebula pulsar is one of the youngest pulsars, a fact which not only comes from the known age of the nebula itself, but also from the rate at which the period of the pulsar is increasing. Measurements indicate that the slow-down rate is approximately 36 millionths of a second per day. It is found furthermore that the slowing down is not perfectly continuous, but occasionally shows a sudden change. These

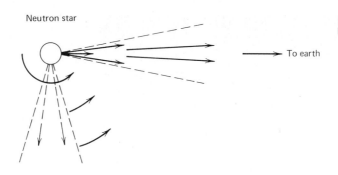

Neutron star

To earth

Fig. 16-12. The pulsar beacon mechanism.

sudden changes are understood as being due to events that are analogous to terrestrial earthquakes.

Neutron Stars

Before the discovery of the pulsars, astronomers and physicists had been studying the properties of purely hypothetical kinds of objects called neutron stars. These were calculated to have very small sizes, on the order of 10 kilometers, but masses greater than the mass of the sun. It was calculated that if such massive objects as the cores of exploded supernovae should collapse following the supernova explosion, they might form such dense configurations of neutrons, in which no chemical elements would ever exist, as they all would be broken down to their basic building blocks (mostly neutrons) by the incredibly high pressures in the star. Presently, the pulsars are believed most likely to be such neutron stars, and the pulses from them are thought to be the result of "hot spots" on the surfaces that pulse on and off (as seen by us) because of the rapid rotation that such stars are expected to have.

X-ray Stars

X-ray sources have recently been discovered by means of various special detectors flown in satellites orbiting the earth. Of the nearly 100 x-ray sources discovered in the first years of such searches, the most common seem to be neutron stars like the pulsars, but most do not show pulses. Probably only those with the necessary "hot spots" and the necessary orientation with respect to the solar direction can be detected as pulsars, while all neutron stars may emit x-rays. X-ray astronomy is a rapidly developing field that has benefited greatly by the interaction of satellite techniques, new radio astronomical techniques, and new developments in theoretical astrophysics. Some x-ray sources are believed to be binary stars in which one member is a neutron star that has collapsed from a red giant, and the other member is a normal star. A light curve of one such x-ray source, Hercules X1, is shown in Fig. 16-15. A star called HZ Herculis was known for many years as a peculiar variable star before it was discovered in 1972 to be an x-ray source. It is now known that the reason for

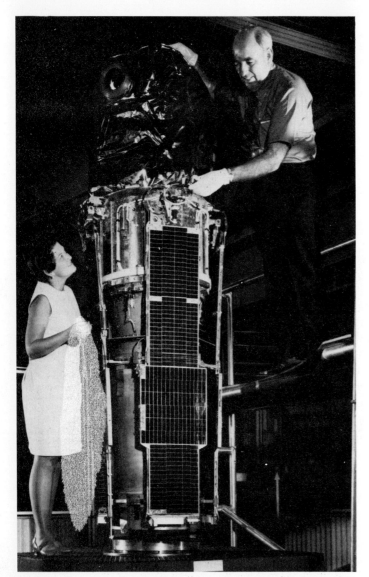

Fig. 16-13. An x-ray satellite (Explorer 42) being checked by NASA scientists *(NASA.)*

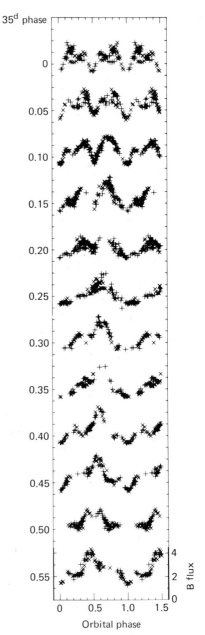

Fig. 16-14. Light curves of the Hercules x-ray source, HZ Herculis, showing how its behavior varies over a 35-day period. *(Courtesy D. Gerend and P. Boynton.)*

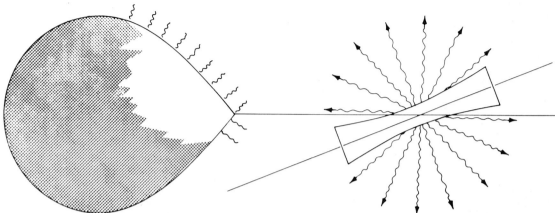

Fig. 16-15. A proposed model of HZ Herculis involves a hot disk surrounding the neutron star that rotates around the companion star every 1.7 days. The normal star is distorted in shape, is slowly losing mass to the hot disk, and is brighter on one side than the other because of the disk.

the optical variability is the eclipse every 1.7 days of both the visual light and the x-rays. The neutron star is believed to affect the temperature of the companion star, thus causing it to have a hot and a cold side. There is evidence that material is being lost from the "normal" star to the x-ray source (the neutron star), causing further, more erratic changes in the brightness of the combined source of light.

QUESTIONS

1. According to present ideas, is it likely that the sun will become a supernova? What do you think would happen to the earth if it did?
2. If Sirius should become a supernova, how bright would it appear from the earth?
3. If supernovae have always occured in our Galaxy at an average rate of one every 50 years, about how much mass in total has been ejected by supernova explosions during the Galaxy's lifetime? What fraction is this of the total mass in the Galaxy?

OBSERVATIONS

• Visual observations of supernovae are rarely possible; occasionally, one in a nearby galaxy becomes bright enough to be seen with a small telescope. The Crab Nebula, a supernova remnant, is visible as a faint patch of light in a small telescope. The Cygnus Loop Nebula, a possible supernova remnant in the form of a large filamentary loop, can also be seen with a small telescope. (See the charts at the back of the book.)

Name	Constellation	Magnitude	Diameter
Crab Nebula (M 1)	Taurus	11	6'
Cygnus Loop	Cygnus	—	60'

17

BLACK HOLES

This chapter introduces you to a brand new problem in astronomy—the study of objects that have disappeared entirely. The purpose of the chapter is to make clear the almost incredible properties of black holes and related phenomena. The chapter begins by first relating some of the implications of Einstein's theory of relativity.

17-0 GOALS

To understand many of the aspects of black holes, it is necessary to have some familiarity with the nature of relativity. Our concepts of how the world works were altered dramatically in the early twentieth century when Albert Einstein (1879–1955) found a new and better way of understanding the physical world. Several odd results had been turning up in physics labs, especially results dealing with the properties of light. Einstein realized that everything made better sense if the nature of light were different from what had previously been assumed by physicists.

17-1 THE SPECIAL THEORY OF RELATIVITY

A century earlier, light had been found to be understandable as a wave phenomenon involving exceedingly rapid oscillations. Experiments that showed a connection between magnetism and electrical charge were used to argue for the existence of *electromagnetic radiation,* which involves oscillating electrical and magnetic fields, and which we perceive as light, including visible light, radio radiation, x-rays, and so forth (Appendix D). But if these are waves, they must have a *medium,* just as waves on a lake are carried in water and sound waves are transmitted through air. Thus physicists had come to think that in space there must be an all-pervading substance that provides a medium for light waves. They called this substance the *ether,* and its nature remained both mysterious and somewhat dissatisfying. Did the earth really plow through the ether in its race with the other planets around the sun? Did the sun have to push its way through the ether in its rapid flight among the stars? Apparently so, the physicists argued, for otherwise light could not be transmitted through space and we would exist in eternal darkness.

Fig. 17-1. The velocity of light is an ultimate barrier. The motion of the source toward or away from the observer has no effect on the velocity of the light observed.

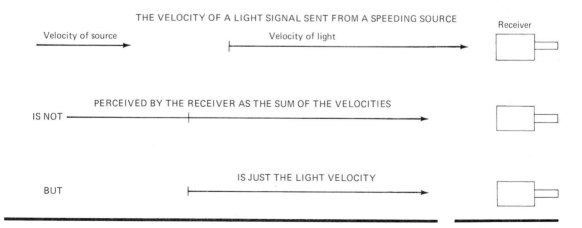

THE VELOCITY OF A LIGHT SIGNAL SENT FROM A SPEEDING SOURCE

Velocity of source

Velocity of light

Receiver

PERCEIVED BY THE RECEIVER AS THE SUM OF THE VELOCITIES

IS NOT

IS JUST THE LIGHT VELOCITY

BUT

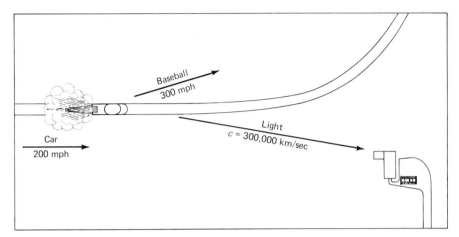

Einstein seems to have gotten his clue to the heart of relativity from an experiment that had been set up to detect the ether. Just before the turn of the century, Michelson and Morley, two American physicists, used a complicated optical apparatus to measure the effects of the earth's motion through the ether, and they found to their surprise that not only was there no effect as a result of the ether, but there was no effect at all even from the motion of the earth itself. Apparently, contrary to what they thought was common sense, the velocity of light was measured to be exactly the same no matter whether the light was emitted along the direction of the earth's motion or perpendicular to it. It was as astounding as it would be to discover that an airplane's ground speed would always be the same no matter how fast or in what direction the wind blew. To use another analogy, imagine a car traveling 200 miles per hour on the Bonneville salt flats. The driver, a pitcher for the major leagues, stands up and throws a baseball straight ahead at a velocity of 100 miles per hour with respect to the car. As seen by an observer sitting at rest on the ground, the baseball would be moving forward at a velocity of 200 + 100 = 300 miles per hour. However, if the driver should shine a beam of light forward instead of a baseball, the velocity of the light would be c (the symbol used in astronomy for this velocity, which is 186,000 miles per second or 300,000 kilometers per second), *not* c + 200 mph! Similarly, if the light is shined in the opposite direction, its velocity with respect to the ground would be c, not c − 200 mph. Furthermore, its velocity with respect to the moving car is also c!

Einstein saw the Michelson-Morely experiment as evidence that light is not like an airplane or a baseball in a medium or like a wave on a pond, but it travels through empty space without the need of a medium. Furthermore, light has certain fundamental and unusual properties that are tied up with the basic nature of matter, space, and time. He saw that the velocity of light is a fundamental quantity. This seemed to come directly from the experiment, and

Einstein saw how the consequences of this hypothesis all fit together to provide a new understanding of nature.

The first inkling of an understanding of relativity can come only when it is realized that "common sense" cannot be relied on completely in this connection. The constancy of the velocity of light, which is a proven fact, does not make common sense—it seems contrary to our usual experience. But this is a clear warning that at the extremes of physics we must rely on experiments and careful mathematics, rather than preconceptions based on common experience.

Einstein showed that the constancy of light, combined with careful formulations of how we define and measure such basic things as length, time, and mass, leads to amazing conclusions when we consider motions that approach the speed of light. We will not use the mathematics here to demonstrate how he derived the special theory of relativity, because this book is intended to be nonmathematical, but we do recommend that an interested reader read Einstein's brief book, written in simple terms, entitled *Relativity, The Special and General Theory: A Popular Exposition* (3rd edition), and published in 1921.

The basic results of Einstein's discovery are as follows. As an object speeds up and reaches a velocity near the speed of light, time as measured by an observer at rest begins to differ from time as measured by an observer on the object. For example, if the baseball pitcher driving a car at high speed on the Bonneville salt flats should accelerate up to, say, 90 percent of the velocity of light, an observer at rest would think that the clock on the speeding car was faulty, as it would appear to be slowing down. It would not be just the clock, of course, but time itself that would be slowing down, and since the driver would be experiencing that time, he or she would not perceive anything unusual in the clock. However, when the car finally stopped back at the starting point

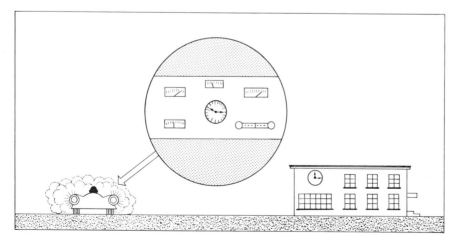

Fig. 17-3. Time as perceived inside a relativistic race car slows down when compared with time as measured by the clock on the stationary building.

Fig. 17-4. A relativistic race car is foreshortened in the direction of its motion as its velocity nears the velocity of light.

again, the driver would be a few minutes younger (depending upon how long the trip was) than he or she would be if the trip had not been undertaken. This may sound unbelievable, but it is true that this sort of thing happens, as we have measured the slowing of clocks when they are put into rapid flight in earth orbit. The slowing of time is just as Einstein predicted back in 1905.

A second surprising effect involves the length of something moving near the velocity of light. To continue the racing car example, if the driver were speeding along at 90 percent of the velocity of light, the stationary observer would be surprised to see that the car would have shrunk to only about half its normal length. It would still look as wide, but the length, the size parallel to the direction of motion, would be remarkably reduced. The driver, of course, who would be moving along with the car, would not perceive anything unusual about it. Again, physicists have checked the theory in the laboratory and confirmed this relativistic contraction of length.

The third effect has to do with mass. Einstein showed that the mass of a moving object is also perceived by a stationary witness to depend upon velocity. If the racing car at rest, for example, weighs 500 kilograms (1,100 pounds), when it speeds up to 90 percent of the velocity of light, if the people at

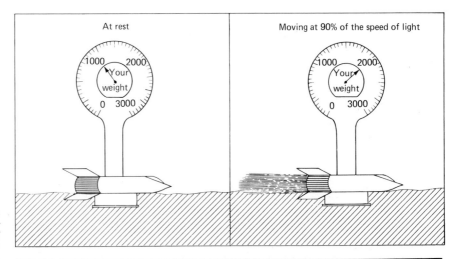

Fig. 17-5. The mass of a rocket increases as its velocity increases to nearly the velocity of light.

rest could somehow measure its weight, they would have determined that it had doubled to 1,000 kilograms (2,200 pounds).

The three main predictions of the special theory of relativity, then, all tell of remarkable effects that occur when an observer looks at an object that is traveling near the speed of light: time seems to slow down, length in the direction of motion shrinks, and mass increases. These effects, according to the equations and to numerous corroborating experiments, increase drastically right near the speed of light, so that when an object has almost reached that velocity, time almost stops, everything is squashed flat like a pancake, and mass is almost infinite (as perceived by an external observer).

After developing the theory of special relativity, Einstein set about the task of revising Newton's theory of gravity to make it conform to the relativistic principles. Newton had supposed that the force of gravity acted instantaneously; the pull of the sun on the earth, he thought, exerted itself in no time, and the gravitational interactions occurred in an instant, even over the immense distances between stars. Einstein realized that this must be wrong, as no interactions (as well as no objects) can propagate faster than the velocity of light. Therefore he set about developing a new theory of gravitation, one that spreads out at the velocity of light but that otherwise explains the motions of the planets as well as Newton's theory.

The task was difficult, both mathematically and conceptually. Einstein was forced to abandon several preconceptions about the nature of space, time, and motion before he finally put together a complete picture of gravitation. It was

**17-2
GENERAL
RELATIVITY**

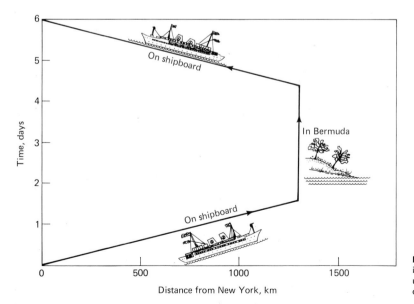

Fig. 17-6. A vacation trip in a two-dimensional universe, with one of the dimensions being time.

necessary for him to give up our comfortable view of space as being simple, flat, and three-dimensional. Instead, Einstein showed that a gravatational field in space, resulting, for instance, from the presence of the sun, amounts to a distortion of space, a curvature depending in degree upon the mass of the gravitating body. Furthermore, he found that thinking only in terms of a three-dimensional world was no longer convenient for its understanding, and he therefore used time as a fourth dimension in his equations. We normally think of an event as taking place only in space, and we think that we can describe it with just three numbers. For instance, an earthquake can be described as having occurred with its center of latitude 33° 30'N, longitude 118° 20'W, which gives two of the dimensions of its location. The earthquake occurs because of movement below the surface, say at a depth of 1000 feet, so that is the third dimension. With these three numbers, we can pinpoint where it occurred, but this still does not completely describe the event, because it doesn't say when it took place. If we add time as a fourth dimension, e.g. 8:01 A.M. PST on February 29, 1984 A.D., then the event is described usefully.

Einstein used space-time as his description of events, but was even forced to go a step further and concede that the four dimensions of space-time could be *curved*. This problem, which is difficult to grasp at first, is more fully described in Chap. 24, where determining the general curvature of space-time is shown to be a key problem in cosmology.

In 1916 the theory of general relativity finally emerged in its full mathematical development, which consisted of equations that describe the curvature of space-time around a gravitating body. For convenience in visualizing it, think for a moment of space-time as being two-dimensional instead of four-

Fig. 17-7. A two-dimensional empty universe (left) compared with one containing a massive object that warps the otherwise flat space.

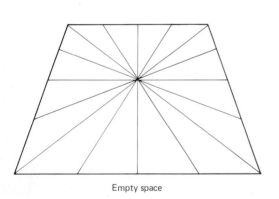

Empty space

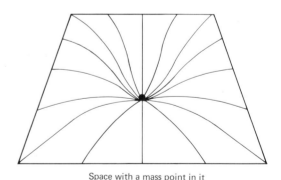

Space with a mass point in it

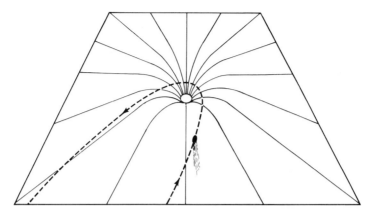

Fig. 17-8. A comet approaches the sun, falls into the dent in space produced by the sun's mass, accelerating as it falls, and then swings out onto the nearly flat, more distant space, slowing down as it moves away again.

dimensional (just ignore two dimensions for a moment, as we did above at first when we gave only the latitude and longitude of the earthquake). We could represent this two-dimensionally as a flat plane (Fig. 17-7), if no object with mass were nearby. But in the presence of an object, such as the sun, the flat plane will be distorted, and a dent will appear in it to represent the curvature of space-time resulting from the gravitational field of the object. The bigger the mass of the object, the larger the dent it causes in space-time. Any other object coming close enough to the dent will have its path altered. If it is going too slowly and heads too close to the center of the depression, it falls into the hole and comes to rest at the bottom (e.g., a spaceship that is pointed toward the sun will fall into it, accelerating as it nears its destruction). If it has enough speed as it zips almost toward the sun, it will make it by, but it will be deflected from a straight-line path. While traversing the dent it follows the straightest possible path, which is called a *geodesic*, but of course this cannot be a straight line since the surface is curved. Thus a body, such as a comet, as it speeds past the sun, has its path altered by the sun's gravity so that it proceeds in an entirely

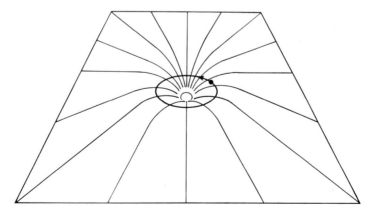

Fig. 17-9. A planet moves around the sun nearly uniformly partway down in the warp in space that is caused by the sun's mass.

different direction after the passage. Similarly, a planet like Earth, in spite of its high speed through space, is captured in the sun's dent in space-time and revolves around and around, part-way down the slopes of the dent, forever.

There have been numerous experiments carried out since 1916 to check Einstein's theory of general relativity, and so far all have been in close agreement. The scientific approach to any such theory is to ask: What does this new theory predict about things that have not been looked at and/or things that are not predicted by other theories? Then the predictions are put to test by experiment, and the theory stands or falls with the experimental results. One of the predictions of general relativity, to give a particularly nice example, is that light should be deflected as it passes near a massive body. For example, consider light from a distant star that passes near the sun. Since light must travel along a geodesic, it will fall part-way into the dent in space-time that is caused by the mass of the sun. When it comes up over the other side and moves on, toward the earth for example, its path has been altered slightly in direction (Fig. 17-10; also see Fig. 15-8).

By arranging the moon so that occasionally it totally eclipses the sun, nature provided scientists with a marvelous opportunity to check this prediction of Einstein (Chap. 4). At such times it is possible to see stars near the sun and to measure their apparent positions, thereby checking to see if their light, when passing close to the sun, has been shifted by the amount predicted (Fig. 17-11). According to classical physics, of course, the apparent positions should be exactly the same as those determined by measurements made at other times of the year, when the sun is not in front of them.

Three years after Einstein published the general theory, a total solar eclipse was observed by groups of English astronomers in Africa and Brazil, and when

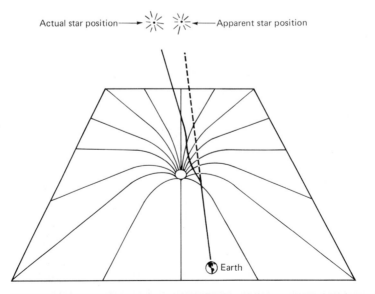

Actual star position⟶   ⟵Apparent star position

🜨 Earth

Fig. 17-10. Light from a distant star has its path altered as it dips down into the curved space near the sun.

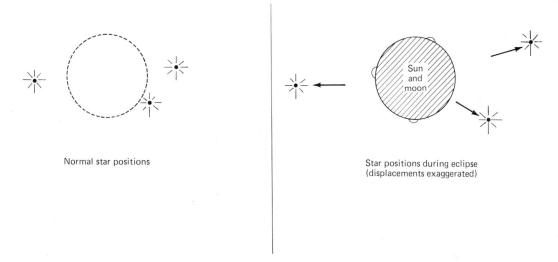

Normal star positions

Star positions during eclipse
(displacements exaggerated)

they had measured the star positions on their photographs they announced triumphantly that Einstein's predictions were confirmed. All measured stars were dispalced in positon away from the sun by just the amount that Einstein predicted.

The general theory of relativity still stands as our best and most tested understanding of space, time, and gravitation. Various physicists have tried since 1916 to find alternative explanations for the way nature works, but so far none of them has supplanted Einstein's solution. Einstein himself tried for the last 40 years of his life to improve upon the theory by searching for some more general principles or formulation that would encompass all of physics, including the ways in which subatomic particles act, but he did not succeed. If an elegant, simple concept that encompasses all of physical reality exists, it will probably not be found until someone comes along, as Einstein did, and perceives a new, surprising way of interpreting what we see and measure around us.

Fig. 17-11. The bending of starlight by the effects of the sun's mass can be measured during a solar eclipse.

Shortly after Einstein published the general theory, the German astronomer Karl Schwarzschild looked at the equations and found a solution to them that described what happens close to a very massive compact object, which would have a very strong gravitational field. (Remember that the force of gravity depends not only upon mass but also upon distance, so that the force (e.g., surface gravity) is largest for a massive object of small size for which the mass is highly concentrated.) Schwarzschild found that if the mass of an object is so compressed as to be virtually all at the center, then at a certain distance (now called the *Schwarzschild radius*) from that center strange things happen because of the space-time geometry. A spaceship, a particle, or even light

**17-3
SIMPLE
BLACK HOLES**

Fig. 17-12. A black hole swallows anything, including light, that comes too close.

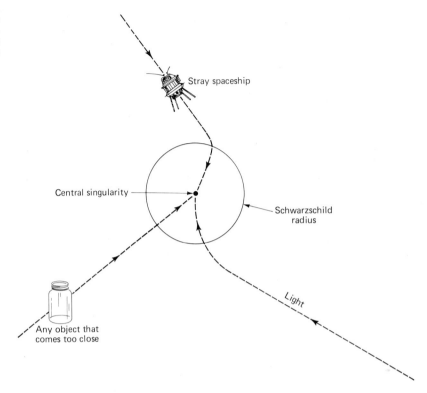

would be unable to escape from the massive object past the Schwarzschild radius, no matter how it might try. The field would be so strong that even light rays would be bent so much that any light emitted would fall right back onto the central mass.

From the outside this massive object would be invisible. No light could escape from it and outsiders could only detect it because of its gravitational field. A spaceship foolish enough to get too close would find itself uncontrollably drawn into the hole in space-time and would plummet through the Schwarzschild radius, at which time its ability to radio for help or sympathy would vanish forever.

Because of relativity, what happens near the Schwarzschild radius is quite different, depending upon whether the observer is external and at rest or is on the spaceship heading in toward the central massive object, which is called a *black hole*. From outside we would see our friends being pulled toward an invisible point, as if by a magnet, but in spite of their headlong progress, it would seem to take forever for them to penetrate the Schwarzschild radius toward the center. The victims, on the other hand, because of the slowing down of their time, would notice no such infinity of time, but rather would be pulled through the barrier in a finite time, probably all too short a time for them.

The universe would fade and quickly disappear above them as they crossed the Schwarzschild limit, and then they would fall inexorably to the center.

Figure 17-13 shows a simplified two-dimension representation of a black hole. The dent in space-time now has a hole at the bottom, with the size of the hole representing the Schwarzschild radius. It is a bottomless hole; anything that falls into it can never climb out. It is gone from our visible universe forever.

What is at the center of a black hole, where all this mass has fallen? Physicists at first preferred not to think too much about this problem, because the mathematics gave a rather unsatisfactory answer. It predicts that everything that is caught by a black hole, including its own mass, is drawn rapidly to a point, called a *singularity*, right at the center. No known physical forces are strong enough to counteract the immensely strong gravity of a black hole, so nothing can hold it up. Atoms crush atoms, nuclei crush nuclei, neutrons crush neutrons, until finally it is all in the same place, an infinitesimal point. Or is it?

Physicists and astronomers are currently trying to find a way to avoid facing a singularity in their equations. Such things are extremely unpleasant for a scientist to handle, because all the laws of physics and rules of nature break down. It is just not possible to tell what happens next. Does the matter all disappear? Does it re-form as something? Does it do something that we can't even imagine? There is no way to know, once this matter really goes to a singularity.

It may be amusing to learn that in the mid-1970s, when it appeared that collisions with singularities were inevitable in a black hole, physicists pointed out that physics, at least, was saved. The lost spaceship and its crew would have to face up to the collision, but external physicists would not, as there is no

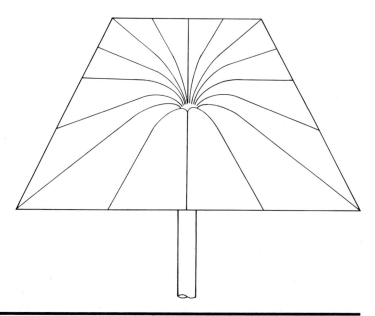

Fig. 17-13. A black hole can be pictured as a dent in a warped two-dimensional space that has a bottomless pit in it.

way that any information about the collision can escape, and therefore no need ever to worry about explaining it. The relativist Roger Penrose even wondered whether the Schwarzschild radius barrier was not nature's way of preventing us from seeing a "naked" singularity, and he therefore called it the "cosmic censor."

In the late 1970's however, progress is being made in treating the singularity problem in a new way, progress led largely by the young British physicist Stephen Hawking. He has tried a new approach that combines relativity with the techniques, called *quantum mechanics*, that are used to understand tiny particles and subatomic events. In 1974 Hawking showed that when quantum mechanical effects were added to the general relativistic equations describing black holes, a singularity might not be involved. Instead, the black holes might be able to create particles as they swallow them, and emit elementary particles such as electrons, neutrinos, and photons (light) in all directions. This loss of material would cause the mass of a black hole to diminish, thus increasing the rate of emission, until a runaway situation develops and the black hole explodes. Unfortunately for us as outside observers, "normal" black holes, say a few times or so as massive as the sun, will take a tremendously long time to explode in this way, far longer than the age of the universe, so we have no hope of detecting them now. For tiny black holes (see Sec. 17-6) the time scale is more favorable.

**17-4
OBSERVING
BLACK HOLES**

A neutron star's mass (Chap. 16) cannot be greater than approximately three solar masses. The fact that the estimated mass of HZ Herculis, which is based on the period of the binary system, is about 0.6 solar masses seems to conform to this calculated limit. If, on the other hand, a more massive object, say tens, hundreds, or thousands of times the sun's mass, should collapse, and if rotation is insufficient to halt the collapse, that massive object will find it impossible to stop its collapse. No amount of gas pressure, no amount of high-temperature nuclear reactions or any other agency is great enough to slow down the inward rushing of material toward the center of such an object. The result is that the collapse accelerates rapidly, and the surface of the collapsing star moves with greater and greater velocity, until finally it becomes a black hole, disappearing from our view.

It is of course impossible to see the black hole. Nevertheless, astronomers hope to detect them by one of two means. Because of their large mass, it may be possible to detect them by their gravitational influence on nearby objects. For example, it is proposed that if astronomers can determine the masses of all the stars in a star cluster accurately and then, after adding up these masses, find that the total mass of the star cluster (determined by the motions of the stars) is significantly greater than can be accounted for by the visible stars, then the invisible mass may be attributed to the presence of black holes in the cluster.

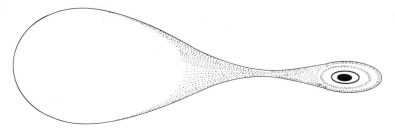

Fig. 17-14. Cygnus X-1 may be a double star, one component (right) of which is a black hole surrounded by a hot gaseous disc of material being pulled in, and the other component of which is an expanding star that is slowly giving up its material to its insatiable companion.

The centers of galaxies are considered possible residence places for black holes, because of the very high density of material there, and the likelihood that in the early history of a galaxy very massive objects may have existed and collapsed. If the centers of galaxies can be seen to contain fewer stars than expected on the basis of the measurement of the mass of material at the centers, then perhaps black holes are present. An example, announced in 1978, is the giant radio galaxy M 87 (Chap. 20) which apparently has more mass in its nucleus than can be accounted for by its starlight.

A second means of detecting a black hole is by discovering its effects on neighboring interstellar material. Because of its large mass, a black hole might be expected to pull any gas around it into its realm; calculations indicate that before this gas disappears into the black hole forever, it will temporarily rotate in a ring or shell around the black hole. Under such circumstances, the ring would be expected to have unusual optical properties that might be detected from the earth. Searches for such objects are going on, and one of them may have already been discovered. In the constellation Cygnus, there is a strong source of x-rays which astronomers find to be a double star, or at least a double something. A visible supergiant star is rotating around an invisible companion, which seems from its spectrum to have a rotating disc of hot gas around it. The x-rays come from a stream of gas that issues forth from the visible star, pulled rapidly to its invisible companion. By studying the orbit, astronomers have determined the period of revolution for the system and its semimajor axis, which allows them to use Newton's laws to gauge the mass of the system. The mass of the invisible object is apparently greater than 3 solar masses, so it cannot be a neutron star like HZ Herculis. Possibly it is a black hole, once a massive star but now visible only by its gravitational effect on its companion, and by the glow of gases that it pulls into its relentless trap.

Most stars rotate. The sun makes a complete roation in 29 days, and some stars rotate even faster. If a massive star should collapse to a black hole, the black hole will also rotate, as there will be little opportunity during collapse for the object to get rid of its angular momentum. If it rotates slowly as a star, this speeds up tremendously during collapse because of *conservation of angular*

**17-5
ROTATING
BLACK HOLES**

Fig. 17-15. A worm hole
is a hypothetical black
hole that connects to an-
other universe.

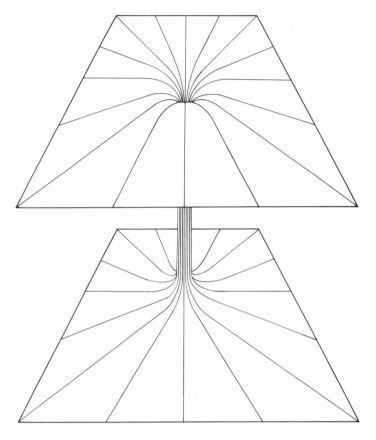

momentum, as discussed for the proto-sun in Chap. 2. Thus a black hole may be rotating with almost incredible speed by the time it disappears through its Schwarzschild radius. What are the consequences of this fact?

When physicists added roation to their equations, a surprising and exciting result emerged. Instead of proceeding to a singularity, the rotation seemed to allow the material to squirm out of such an impossible situation. Just before being squashed to infinite density, the material seemed to stop and begin to bounce back, opening out again, just reversing its collapse. People in a spaceship caught in the black hole would see their relativistic prison open up again slowly, and they would be thrown back out just as they came in. The heavens, which had closed up around them when they sped in through the Schwarzschild radius, would open up again and the sky would gradually unfold, revealing the stars.

But their friends and relatives outside would wait for them in vain. The equations seemed to say that it could not be our universe that they emerge into, but some other universe. The stars would be some other set of stars, if such a set exists; we could watch the black hole forever and it would never reappear to us.

The burst of light and matter that suddenly flashes into this other universe has been termed a "white hole" and the connection between the universes is called, somewhat facetiously, a "worm hole." Such holes, it was suggested after their discovery, could connect our universe to another, or perhaps they could connect two widely separated parts of our universe, if there is only one universe available. To see this boggling scenario in the mathematics took some imagination, of course, and it really lies as close to science fiction as it does to predictive science. There are often several solutions to a set of equations, and not all are necessarily real solutions that represent something in nature.

Science fiction, in fact, makes great use of "worm holes," especially when some crisis makes it necessary for the hero to get somewhere faster than the speed of light. He just steers his spaceship into a rotating black hole and emerges in a white hole somewhere else, hopefully where he wants to.

Unfortunately for those planning such a journey, it was shown in the mid-1970s that white holes may only occur if they had been planted in place in the beginning. A black hole would have to collapse at precisely the place where a singularity was established when the universe began. This kind of coincidence would be incredible, and so the worm hole idea may turn out to be unlikely, if not impossible.

Any star must be massive, at least three times more massive than the sun, if it is to collapse to a black hole; otherwise it will merely become a neutron star or white dwarf. But Hawking has pointed out that near the beginning of the expansion of the universe, just after the "big bang" (Chap. 24), the incredibly high pressures and the large density of matter was such that much smaller objects could collapse, forming tiny black holes. Typically, such a black hole might have a mass of 10^{15} grams (about 1 billion tons) and a radius of only 10^{-13} centimeters, about the size of a proton. Because of its small size, such a tiny black hole would not have much influence on its environment, and it would be difficult to detect it even a few hundred feet away. There could perhaps be some of these in the solar system, and our Galaxy might have as many as 10^{23} of them without our having discovered them.

If such tiny black holes were induced near the beginning of time, they may have swallowed up a large fraction of the mass of the universe. Perhaps the density of the universe is larger than we measure it to be because much of it is locked up invisibly in tiny black holes. Astronomers have often worried that what they can see of the universe may not be all there is; the age, shape and

**17-6
TINY
BLACK HOLES**

Fig. 17-16. Hawking's
tiny primordial black
holes may pervade
space. They are not very
massive and therefore
cause hardly any distor-
tion of space-time.

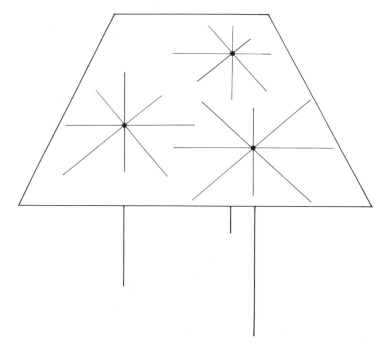

future of the universe can be determined by measuring its density (Chap. 24), but if we see only part of what is there, we will calculate the density of the universe wrong. Thus it is essential that we know whether some part of the universe is "missing." We do not know yet whether or not it is missing in the form of tiny black holes.

Some cosmologists believe that we may have a way of detecting Hawking's tiny black holes, if they exist. For a billion-ton object, the time until it explodes, if it does as Hawking predicts (Sec. 17-3), is only about 10 billion years, about the age of the universe. Thus if they were formed right after the birth of the universe, they should be popping off about now. Perhaps we will soon detect radiation from them, which might be most noticeable as gamma radiation detected from satellites. Hawking has made a prediction of how bright the gamma radiation will be, and our task now is to attempt to detect and measure the black holes. As for many of the theoretical results discussed in this chapter, many of which are freshly formulated and still very uncertain, the proof of the theory will come when we can go out there or look out there and check its predictions. Confirmation of Einstein's predictions about gravity and the theory of light led to acceptance of relativity in spite of its seeming at first to violate common sense. Perhaps the predictions about black holes and their connection with the universe will soon be confirmed, and these weird objects will turn out to be real and to have profound implications for the physical world.

1. What experimental fact led Einstein to develop the special theory of relativity?

2.* If astronauts are sent on a special mission to the Andromeda Galaxy (2 million light years away) in a vessel that can travel at almost the velocity of light, can they hope to reach their destination before they die of old age?

3.* For the astronauts in question 2, when (and if) they return, about how much time will have elapsed on the earth?

4. Why is a singularity an awkward entity to describe physically?

5. What are two possible methods of detecting black holes?

6. Why might a collapsing star that forms a black hole be expected to be rapidly rotating?

PART
THREE

CONCEPTS OF THE UNIVERSE

THE UNIVERSE OF GALAXIES

This chapter is intended to introduce you to the different types of normal galaxies that inhabit the universe. It points out why they all look so different, by explaining the kinds of stars and other material that each type of galaxy contains.

The universe is made up of large conglomerations of stars, gas, and dust called *galaxies.* Consisting of anywhere from several million to hundreds of billions of stars, galaxies contain most of the visible mass of the universe, and most of the mass of each galaxy is contained in its stars. The smallest galaxies, called *dwarf galaxies*, can be as small as 300 parsecs across, while the largest have diameters of 30,000 parsecs or more.

Fig. 18-1. An elliptical galaxy, NGC 3377. *(Lick Observatory.)*

Galaxies have a variety of forms; many, such as our local Galaxy, have a spiral shape, some others are irregular, but the most common have a smooth featureless structure. The astronomer Hubble was the first to organize galaxies into a scheme in which they were grouped according to appearance. Although not an evolutionary scheme, Hubble's classification of galaxies has helped immensely in our understanding of the evolution of galaxies. Hubble divided galaxies into three main types: the elliptical galaxies, the spiral galaxies, and the irregular galaxies.

18-2 ELLIPTICAL GALAXIES

Structure The elliptical galaxies are perfectly elliptical—their outlines are smooth and describe perfect geometrical ellipses (Fig. 18-2). A luminosity contour map of an elliptical galaxy shows a group of concentric ellipses, with the brightest contour in the center and with increasingly faint contours outward. All the ellipses have the same shape. The fact that there is no structure except for a smooth perfect form results from the fact that they are very old undisturbed galaxies—galaxies in which nothing has happened for a long time. Measures of ages of elliptical galaxies indicate that star formation in them ceased on the order of 10 billion years ago. If one measures how the light is distributed from the center outward, one finds that the intensity decreases according to a smooth curve, and that the curve describing this structure for all elliptical galaxies is the same, except for the scale and the degree of flatness. One formula can be written to describe all elliptical galaxies. They all have the same basic structure, except for two scale factors, one being related to how rich the galaxy is and the other telling how widely distributed the stars are. This fact is important because it indicates that it has been a long time since the system was

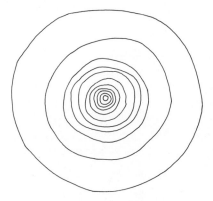

Fig. 18-2. A contour map of an elliptical galaxy, NGC 3379, showing lines of equal brightness.

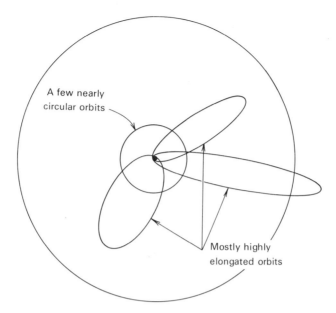

Fig. 18-3. Samples of orbits of stars in an elliptical galaxy.

A few nearly circular orbits

Mostly highly elongated orbits

formed, long enough so that there is no trace left of the initial conditions. The original shape of the galaxy has been completely erased during its intervening history by the encounters and the orbital smoothing of its stars.

Orbits of Stars

The elliptical galaxies normally are made up of billions of stars, all of which have orbits around the center of mass of the system. By comparing the structure of these galaxies with those of theoretical mathematically constructed models, we can determine the kinds of orbits that exist. Different galaxies have different arrangements of orbits, but most of the elliptical galaxies seem to have a large percentage of stars with extremely eccentric orbits. The earth's orbit around the sun is nearly circular, but in the case of most of the stars in most elliptical galaxies, the orbits are *not* circular but very elongated, though there are always a few stars that have nearly circular orbits.

Colors

Another important uniformity for elliptical galaxies is their color. All elliptical galaxies are red, and there is very little variation in color from one to another; there are small differences, but they are minor and not yet entirely understood. The measured colors of elliptical galaxies indicate that they consist primarily of old stars. As was shown in Chap. 11, young stars are blue, and of course a red galaxy cannot be made up of blue stars. The elliptical galaxies must therefore contain old stars.

Spectra The spectra of galaxies are rich sources of information. A galaxy spectrum is a superposition of the spectra of the stars in the system, giving a general idea of the different kinds of stars there. Galaxy spectra indicate that the stars in elliptical galaxies are old, as no young stars show up.

Insterstellar Content A further uniform property for elliptical galaxies is that they contain only stars and no detectable interstellar material, gas clouds, or dust (with very infrequent and minor exceptions). This implies that there cannot be any stars forming now in elliptical galaxies, because there is no gas and dust to provide the raw materials for star formation. The lack of such gas and dust means that star formation has ceased in elliptical galaxies, and that the stars are old and stable objects.

Subclasses In the Hubble classification, elliptical galaxies are divided up into eight groups, depending upon their appearance. This is not a physical distinction, but a geometrical one. Some elliptical galaxies appear nearly circular in outline, whereas others are highly elongated. The eccentricity of an ellipse is a measure of how stretched it is. It is defined in terms of the ratio of the minor to major axis of the ellipse that forms the outline of the system. Calling the major axis a and the minor axis b, the Hubble subclasses are defined according to the following scheme: an E, meaning elliptical, followed by a number equal to 10 $(1 - b/a)$. For a galaxy with a circular outline, where a and b are equal, $(1 - b/a)$ is 0, so it is called an E0. For a highly eccentric ellipse, for instance one with b that is $3/10$ of a, the classification would be E7. There are eight classes, E0 through E7.

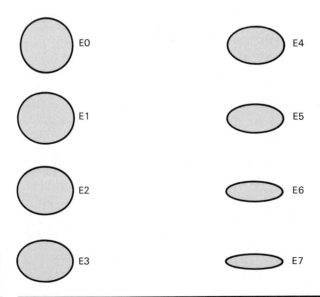

Fig. 18-4. Subclasses of elliptical galaxies depend upon their shapes.

Hubble found no galaxies which were flatter than E7 among the elliptical galaxies; apparently they do not exist.

Spiral galaxies, with their bright centers and star-filled spiral arms, are beautiful and spectacular objects. They are commonly large objects, with two or more arms. Less abundant than elliptical galaxies, there are only a dozen or so large

18-3
SPIRAL
GALAXIES

Fig. 18-5. A spiral galaxy with a peculiar companion, NGC 5194 and 5195. *(Kitt Peak National Observatory.)*

spiral galaxies within the nearest 2 million parsecs, but they are among the most interesting objects in our galactic neighborhood.

Shape and Rotation

They are quite flat, except for the central regions, a fact that can be deduced from the statistics of the orientations of spirals as seen in the sky. Enough are seen to be apparently quite thin for it to be inferred that most of them are intrinsically flat, with a maximum value of 0.8 for the quantity (1 - b/a). Their flatness is apparently the result of their fairly rapid rotation. Studies of the radial velocities of different parts of spiral galaxies show that they rotate with velocities up to 200 or 300 kilometers per second. The rotation is not like that of a solid body, but is most rapid near the outer parts of the system, slow near the center, and also slow at the greatest distances.

Because the stars and gas clouds are not held together rigidly, the outer parts revolve like the planets around the sun, the more distant ones moving the more slowly. In the inner regions, the stars revolve more nearly at the same angular speed (like a solid body) because of the fact that the gravitational pull that causes them to revolve is not just in the center of the galaxy, but is all around them.

Structure

Spiral galaxies can be thought of as having two components in their structure—an ellipsoidal component and a flat component. The central part consists mostly of the ellipsoidal component, with a structure like that of elliptical galaxies. The outer portions show mostly a flat distribution of stars with a very different structure. The ellipsoidal component has elliptical contours—old stars, no gas, and no dust—and its luminosity distribution follows the structural law of elliptical galaxies. The flat component contains the spiral arms. Most spiral galaxies contain two such arms, and sometimes they are multibranched. In some galaxies the spiral arms are open and spread out, and in others they are tightly wound. They contain large amounts of young stars, gas, and dust. The gas shows up in two forms, both as giant gas clouds (H II regions) that can be seen optically, and as invisible (except to radio telescopes) neutral hydrogen clouds that extend over the entire arm structure. Dust shows up conspicuously as dark blobs of matter that obscure the stars behind them. The young stars are conspicuous, at least in the nearer spiral galaxies, as bright blue stars.

From this evidence, it can be deduced that the arms are the locus of star formation in spiral galaxies. Stars exist there as young as a million years old or less, which is very young considering that these galaxies are 10 billion years old altogether. The raw materials—gas and dust in clouds—are also there, ready to make stars in the future.

The spiral arms may exist because of a somewhat remarkable physical effect described in C.C. Lin's *density wave theory*. This mathematician from M.I.T.

showed that a rotating disk of gas can generate a spiral-shaped wave of high density that will persist for a long time. Stars and interstellar material gather together in these arms for a certain time, moving slowly out again as new material is "captured" by the density wave. Formerly, it was thought that magnetic fields helped to form the arms, but it is now generally conceded that

Fig. 18-6. A photograph of a nearly edge-on spiral galaxy, NGC 4565, showing how flat it is. *(Kitt Peak National Observatory.)*

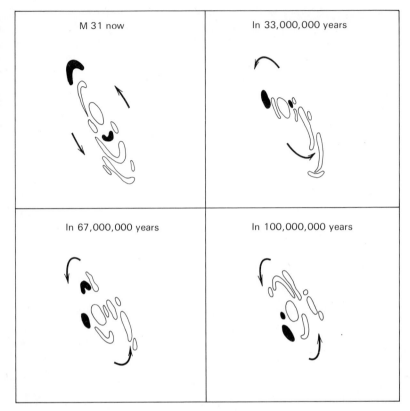

Fig. 18-7. A schematic illustration of the way in which the galaxy M 31 will rotate between now and 100 million years from now. Bright segments of spiral arms are delineated. Two are blacked in to make it easy to follow them around.

although magnetic fields do exist in spiral galaxies, they are much too weak to be effective in governing their structure. The magnetic field in our Galaxy, for instance, is found to be about 10^{-5} times the strength of earth's surface magnetic field.

Subclasses   Hubble divided the spiral galaxies into eight different subclasses. First are the S0 spirals, which have the characteristics listed above as far as structure is

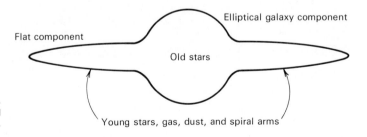

Fig. 18-8. The two components of a spiral galaxy.

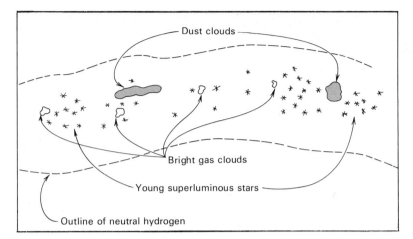

Fig. 18-9. Four components of spiral arms.

concerned. They have the flat component and the ellipsoidal component, and they often have dust, but they have no spiral arms. Some have a faint and indistinct outer structure that appears almost to be spiral, but not definitely so.

The next three subclasses are defined according to how tightly wound the arms are. An Sa spiral has quite tight arms, an Sb has less tight ones, while those

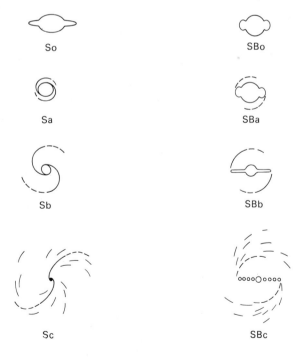

Fig. 18-10. Subclasses of spiral galaxies.

of an Sc are very loose and spread out. Sa galaxies usually have a large bright nuclear region (ellipsoidal component) and smooth bright arms. For Sb galaxies, the central ellipsoidal region is smaller and fainter and the arms more patchy in appearance, while Sc galaxies usually have a small nuclear area and very patchy arms.

Barred Spiral Galaxies

A separate and somewhat rarer type of spiral galaxy is characterized by a bar-shaped feature in its central regions. Instead of having a nuclear region that looks like an elliptical galaxy, barred spirals have a long, rather narrow structure that is bright and straight and centered on the nucleus. The arms usually emanate from the ends of the bar, rather than from the center of the galaxy. Hubble distinguished these objects by inserting a B between the S and the 0, a, b, or c. Otherwise, the subclasses are similarly defined. A barred spiral without arms is an SB0, one with tight arms is an SBa, and so on down the sequence. The "bar" of a barred spiral is still very much an enigma. It is not known how it can exist or how it forms, and a great deal of current research is going into solving this puzzle.

Fig. 18-12. An Sc galaxy, NGC 5457. *(Lick Observatory.)*

The third main type of galaxy is the *irregular* class. Generally, "Irr" galaxies have the same stellar makeup as spiral galaxies, but without well-defined spiral arms. They sometimes have bars, seldom have a clearly identifiable nucleus, and always contain large amounts of young stars, gas, and dust. Often, an irregular galaxy also contains many star clusters, many more than a spiral of the same total population would have. The outstanding character of a typical irregular galaxy is *youth*. Although all such galaxies appear to contain some old stars, most of the light from an irregular galaxy comes from its abundant and brilliant young stars and its luminous gas clouds. Irregular galaxies rotate like spiral galaxies, but for reasons that we do not yet know, spiral arms have not formed. It is true, however, that some irregular galaxies have what look like "primitive," distorted, or partly formed spiral arms, and there thus seems to be a close relationship between the Sc galaxies and the irregular ones.

Not all galaxies fit into Hubble's simple classification scheme, and most of the odd ones are very odd indeed. We now have reason to believe that certain

**18-4
IRREGULAR
GALAXIES**

Fig. 18-13. An SBb galaxy, NGC 4303. *(Lick Observatory.)*

Fig. 18-14. An irregular galaxy, the Large Magellanic Cloud. *(Harvard Observatory.)*

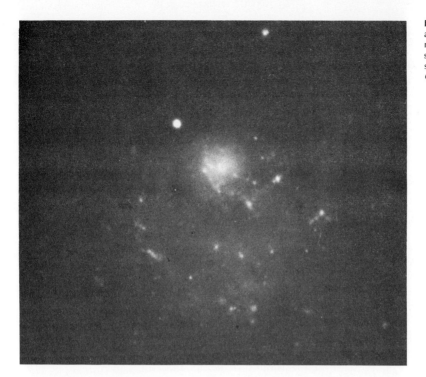

Fig. 18-15. A peculiar galaxy with an ellipticallike nucleus and segments of spiral arms on only one side, NGC 5474. *(Lick Observatory.)*

Fig. 18-16. A peculiar irregular galaxy, NGC 3077. *(Lick Observatory.)*

of these strange galaxies are experiencing huge, violent, disruptive, and explosive events, with the result being an unclassifiable jumble of stars, gas, and dust. Chapter 20 discusses these explosive galaxies in detail.

18-5
CLUSTERS OF
GALAXIES

The largest entities in the universe are the clusters of galaxies and the few identified groups of clusters. Ranging from small aggregates like the local group, with 20 to 30 members, to giant spherical systems of galaxies with tens of thousands of members, the clusters of galaxies are providing us with a great deal of information of cosmological importance. By studying the motions of the member galaxies we can deduce the mass of a cluster, applying the laws of gravity to the system. This tells us how much matter there is in the entire cluster. In many cases, we find that the mass of the cluster is greater than the masses of the galaxies that we see, and therefore we are forced to conclude that much of the matter in the cluster (and in the universe) is invisible to us, in the form of hot diffuse gas, small chunks of rock, black holes or some other dark form. The existence of intergalactic gas is proven by the detection of x-rays from clusters of galaxies, x-rays which seem to result from the high-speed travel of member galaxies through a gaseous medium. Further x-ray studies, using x-ray satellites designed with high-resolution detectors, may soon allow us to determine how much intergalactic gas there is in clusters and between clusters, and then we may be able to explain the "hidden matter" of the universe.

There is still some debate about whether clusters of clusters of galaxies exist. If they are real, and statistics suggest that they are, they are immense objects averaging about 60 million parsecs (200 million light years). We ourselves may be in a member group of a local cluster of clusters (or *supercluster*), which includes the local group, the giant Virgo cluster, and many other nearby groups.

QUESTIONS

1. What would be the Hubble class of a perfectly spherical elliptical galaxy? Of a very flat elliptical galaxy seen "face on"?
2.* Why do the inner stars, near the center of a galaxy, not revolve in orbits according to Kepler's laws?
3. What are the principal *observed* differences between elliptical galaxies and spiral galaxies? What basic differences can explain these observed effects? How?
4. What is the difference between Hubble's irregular galaxy class and peculiar galaxies in general?

OBSERVATIONS

• Galaxies, because of their immense distances, are normally disappointingly faint objects to observe visually. Nevertheless, many can be seen with a small telescope (three with just the naked eye) and it is often worth the effort to find and see them, if only to realize the fact that the light that you will be seeing was emitted by the galaxy millions of years ago. Listed below are some of the brighter galaxies, which can be seen with small telescopes. Try on dark moonless nights, away from city lights. The charts at the back of the book show where to look.

| Name | | | | | Distance |
M	NGC	Constellation	Hubble type	Magnitude	(million parsecs)
Fall objects:					
M31	224	Andromeda	Sb	4	0.6
M32	221	Andromeda	E	9	0.6
M33	598	Triangulum	Sc	6	0.7
Winter objects:					
M81	3031	Ursa Major	Sb	8	2
M82	3034	Ursa Major	radio	9	2
Spring objects:					
M49	4472	Virgo	E4	9	10
M104	4594	Virgo	Sb	9	10
M94	4736	Canes Venatici	Sb	9	4
M87	4486	Virgo	E (radio)	9	10
Summer objects:					
M63	5055	Canes Venatici	Sb	9	4
M51	5194	Canes Venatici	Sc	9	4
M101	5457	Ursa Major	Sc	8	7

19

OUR GALAXY

To appreciate the various properties of galaxies, in this chapter we look in detail at our own Galaxy, measure its size, determine how many stars it contains, look at its other contents, and figure out its structure.

Our local Galaxy, the Milky Way, is at the same time one of the best observed and one of the least understood galaxies in our local neighborhood of space. The fact that we are embedded in it makes it possible for us to make many measurements and to carry out surveys that would not be possible for a distant galaxy. On the other hand, this fact also means that we cannot get a good view of our Galaxy. We are entangled in the gas and dust of our system, which obscure much of it from view optically and prevent us even from properly mapping it from radio observations. It is difficult in our Galaxy to see the forest for the trees.

As was pointed out in Chap. 9, for many years astronomers had a completely wrong impression of the size of the stellar system to which the sun belongs. Until the interstellar dust was recognized and accounted for, it looked to astronomers as if our Galaxy were a rather small object compared to what we

Fig. 19-1. A section of the Milky Way that shows star clouds, clusters, and nebulae. *(A negative print, courtesy Lick Observatory.)*

know now about its size. The dimensions of the system were found by means of stellar statistics; stars were counted in various directions in the sky and their number was determined for various brightness limits. This gave a statistical, but not a precise, idea of the distribution of stars in space and showed that the number of stars appeared to decrease for faint luminosities, a fact which was interpreted as indicating that the census had reached the edge of the Galaxy. When interstellar dust was discovered, it was found instead that this apparent thinning out of stars was the result of the increased effects of absorption of light by the dust.

Shapley's Measurement

The true size of our Galaxy was first recognized by the astronomer Harlow Shapley in about 1917. It occurred to Shapley to look at the distribution in space of the globular star clusters, which he noted occurred all over the sky, not just in the galactic plane. He also noticed that their distribution in space was highly asymmetrical, with the majority of the globular clusters in the direction of the constellations Sagittarius and Scorpio. Shapley measured the individual distances for all the globular clusters for which he could do so, and he then examined their distribution in space, comparing the distribution with the model astronomers had developed for our Galaxy.

To get the distances to globular clusters, Shapley turned to the RR Lyrae variables (Chap. 12). Although no RR Lyrae variable is close enough for its parallax to be determined, there are enough of them close to the sun so that statistical parallaxes can be obtained for a sampling of the near ones. From this technique, it was known that RR Lyrae variables all have about the same absolute magnitude, about $M = 0$. Since RR Lyrae variables are fairly common in globular clusters, Shapley found that he could determine the distances to globular clusters from measurements of the apparent brightnesses of their RR Lyrae members. He also noted that the RR Lyrae variables are always about two magnitudes fainter than the brighter stars in the clusters, and that allowed him to measure the distances to globular clusters that did not contain RR Lyrae variables.

Fig. 19-2. Harlow Shapley, pioneer explorer of the Milky Way galaxy. (*Harvard Observatory.*)

Fig. 19-3. Globular cluster M 15. *(Manastash Observatory.)*

The result of Shapley's discovery was a revolutionary change in our understanding of our own Galaxy. The distribution of globular clusters in space was found to be far different from the distribution of stars that had previously been plotted. Instead of lying near the center of this distribution, the sun was found to be almost 10,000 parsecs away from the center, which Shapley found to lie in the direction of the corner shared by the constellations Sagittarius, Scorpio, and Ophiuchus. The total extent of our Galaxy was not 3,000 parsecs as had been previously estimated, but 30,000 parsecs from one side to the other.

Although more than 60 years old, Shapley's data has held up remarkably well. Modern measurements of the distances to the globular clusters confirm the picture that he produced of the size and shape of the cluster system. Since his pioneering work, it has been possible to understand the reason for the earlier mistakes about the size of the Galaxy in terms of the very large amounts of absorption caused by dust in the plane of the Galaxy. Because that dust is restricted to a thin layer along the Milky Way, Shapley's results were little affected by it because most of his globular clusters lay above or below that plane.

Attempts to pierce the plane by observing in what appeared to be holes in the

Modern Optical Determinations of the Size of the Galaxy

interstellar dust were carried out in subsequent years, particularly by the Mt. Wilson and Palomar Observatory astronomer Walter Baade. He found several holes that allowed examination of clusters of stars apparently very close to the center of our Galaxy near the plane. Baade's estimates and more recent determinations by other astronomers (for example, Sidney van den Bergh) all agree that the distance to the center of our Galaxy is approximately 10,000 parsecs, with an uncertainty of about 10 percent. Our sun is about two-thirds of the way toward the edge, so that the total diameter of our Galaxy is about 30,000 parsecs.

Fig. 19-5. Two globular clusters near the center of our Galaxy, NGC 6522 and NGC 6528. *(Kitt Peak National Observatory.)*

**19-2
CONTENT**

Stellar Content

The total number and kind of stars found in our Galaxy is well known only for that portion of our Galaxy that can be observed well optically. Astronomers have taken a census of stars that is reasonably complete down to very faint magnitudes only for the nearest 100 parsecs or so. For more luminous objects, for example the very bright stars called supergiants, the census extends to much greater distances, on the order of 1,000 to 2,000 parsecs. The result of such examinations show that in our Galaxy, the vast majority of stars are the stars of low total luminosity. Chapter 8 gave a result in terms of a stellar luminosity function that showed that the stars fainter than the sun far outnumber the stars brighter than the sun, even though the sun is about average in absolute brightness.

The total number of stars of all kinds in our Galaxy, although unknown, can be estimated very simply; it is possible to calculate the total volume of our Galaxy, as was done in Chap. 15. There we found the volume to be approximately 7×10^{11} cubic parsecs. The number of stars of all kinds within 20 parsecs of the sun is about 1,000. Since the volume of the sphere of space within 20 parsecs of the sun is about 3×10^4 cubic parsecs, the ratio of the volume of the space to the volume of our Galaxy is about 2×10^7 (for such rough calculations one can round off numbers to one significant figure). If we

Fig. 19-6. The density distribution in the Galaxy decreases away from the plane of the Milky Way.

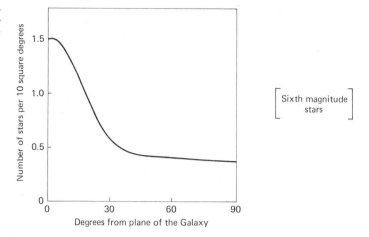

Fig. 19-6. The density distribution in the Galaxy decreases away from the plane of the Milky Way.

now assume that our sun is in a relatively typical area of our Galaxy as far as density is concerned (not a very good assumption, but very roughly true), it is then clear that the total number of stars in our Galaxy must be about 2×10^{10}.

A much better idea of the number of stars in our Galaxy can be obtained by a more realistic comparison of the mean density of stars near the sun to that for the whole Galaxy. As Fig. 19-7 shows, the sun is in a portion of our Galaxy that is distant and very sparsely populated by comparison with the inner, brighter part. It is therefore not in a good position to be considered in an average density

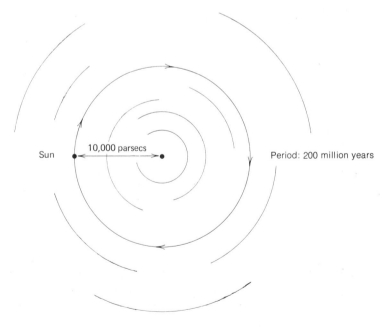

Fig. 19-7. The orbit of the sun in the Galaxy.

region. Using more accurate data, one calculates that the total number of stars in our Galaxy is about 2×10^{11}, ten times more than calculated above.

Another means of measuring the total amount of material in our Galaxy, somewhat more accurate than counting stars, is to determine the total mass of our Galaxy by means of observing the orbits of stars. In general principle, the method is similar to the way by which the masses of planets can be obtained from the orbits of their satellites (Chap. 2). In the case of the Galaxy, however, account must be taken of the fact that individual stars not only experience a gravitational pull toward the center of the Galaxy, but usually have considerable amounts of material outside their orbits that also influences their motions. Only in the far outer regions of our Galaxy do the orbits of the stars obey Kepler's laws. **Total Mass**

An approximate idea of the mass of our Galaxy can be obtained from the fact that the sun rotates in a nearly circular orbit around the center of our Galaxy with a known velocity. This velocity is obtained by measuring the apparent velocities of relatively fixed objects, such as nearby galaxies, in different parts of the sky, and by noting that in one direction in the sky they all appear to be moving away from the sun, while in the opposite direction, they move toward the sun. In this way, it is found that the sun's rotational velocity is about 250 kilometers per second. Its distance from the center of the Galaxy is known to be 10,000 parsecs, and it is possible to use these facts to determine the period of the sun around the center of the Galaxy. From the radius of the sun's orbit, it can be seen that the circumference must be about 60,000 parsecs. Converting its known velocity of 250 kilometers per second into parsecs per year, it is found that the sun moves about 1 parsec every 4,000 years. In order to cover an entire orbit of 60,000 parsecs, it takes the sun about 240 million years, which is the period of the sun in its orbit.

It is possible to calculate the approximate mass of our Galaxy by comparing the orbit of the sun around the galactic center with the orbit of the earth around the sun. In each case, the sum of the masses of the two bodies is given by the orbital radius (or semimajor axis) and the period; Newton's law of gravitation (Chap. 2) gave the following relationship:

$$m_1 + m_2 = \frac{r^3}{p^2}$$

where m_1 and m_2 are the masses in solar units of the two bodies (the earth and sun in one case, and the sun and the Galaxy in the other), r is the orbital radius in astronomical units, and p is the period in years. Thus, it is possible simply to calculate the mass of our Galaxy (plus the mass of the sun, which is negligibly small in comparison) in units of the mass of the sun. The radius of the sun's orbit is 10^4 parsecs, and a parsec is equal to approximately 2×10^5 astronomical units. Therefore, the radius of the sun's orbit in terms of the earth's orbit is 2×10^9 astronomical units. The period we have found to be approximately 2×10^8 years. Therefore, the mass for the Galaxy is equal to

$$M_G = \frac{r^3}{p^2} = \frac{(2 \times 10^9)^3}{(2 \times 10^8)^2} = 2 \times 10^{11} \text{ solar masses}$$

This is only an approximation of the mass, assuming that there is no effect on the sun's motion due to the matter outside its orbit. Since this assumption is not strictly correct, our calculated mass for the Galaxy is not precise. However, astronomers using complicated mass distribution models and solving precisely for the total mass for these models have reached the conclusion that the mass of our Galaxy is between 2 and 4×10^{11} solar masses, close to our simply calculated value.

The very large mass of the Galaxy indicates that in counting stars in the solar neighborhood and comparing volumes in the previous section, we did not adequately take into account the increased density of stars toward the central parts of our Galaxy. Nevertheless, both ways of looking at the total content of our Galaxy agree that it is a very massive object, containing at least 100 billion stars and having a mass of several hundred billion times the mass of the sun.

The Nonstellar Content

In addition to its stars, the Galaxy contains interstellar gas and dust. This interstellar material has been discussed in some detail in Chap. 9. Here we are most concerned about the total amount of such material in our Galaxy, and in the next section about its distribution. The only way to measure the total amount of neutral hydrogen gas in our Galaxy, which is the most massive nonstellar component according to present surveys, is to observe the radio radiation that it emits at 21-centimeter wavelength. Surveys of the entire Galaxy by means of radio telescopes have indicated that the total mass of neutral hydrogen is approximately 5×10^9 times the mass of the sun. This means that the neutral hydrogen gas comprises approximately 1 to 2 percent of the mass of our Galaxy.

The amount of dust in our Galaxy is less easily determined, because we can only measure it with great accuracy in the solar neighborhood. Neutral hydrogen radiation penetrates through the dust from all parts of our Galaxy, and is essentially undiminished by obscuration. The dust itself, however, prevents us from seeing and measuring the more distant dust clouds, and we have no information whatever on the dust content beyond 3,000 to 4,000 parsecs from the sun (in the plane of the Galaxy). Estimates from the local part of our Galaxy suggest that there is about 1 gram of dust for every 200 grams of hydrogen gas. On this basis, the total mass of dust in our Galaxy would be calculated to be the equivalent of about 25 million suns.

Fig. 19-8. Jan Oort, a pioneer in the understanding of our Galaxy. *(Leiden Observatory.)*

Other measurements of the dust and gas in our Galaxy come from studies of the clouds of material that can be distinguished as discrete objects, including the emission nebulae, the reflection nebulae, and the dark nebulae. The total masses of these objects, however, are small compared to the mass of the more evenly spread interstellar medium. A typical emission nebula has a mass of a few hundred or thousand times the mass of the sun, and a typical dust complex of clouds represents only a hundred or so times the mass of the sun. From emission nebulae, however, we do obtain information on some of the other chemical elements that exist in the interstellar material, and it is found that in general, the gas has about the same chemical abundances as the sun, with hydrogen the most abundant element, helium making up about 10 percent of the atoms, and the heavier elements all contributing a total of only 3 or 4 percent of the mass.

The Milky Way galaxy is found to have a structure typical of a spiral galaxy of Hubble class Sb or Sc. It shows a typical flat plane where most of the stars and all of the interstellar material are concentrated, a large central bulge (mostly made up of old stars), and a thin halo of old stars that includes the globular clusters. These same features are common for all spiral galaxies.

**19-3
THE STRUCTURE
OF OUR
GALAXY**

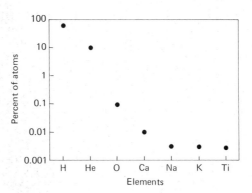

Fig. 19-9. Abundances of different atoms in the interstellar gas.

Fig. 19-10. Age versus position in our Galaxy.

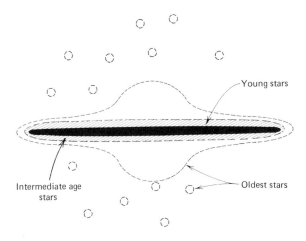

The Vertical Distribution of Stars in the Galaxy

Some idea of the nature of what our Galaxy looks like in profile, seen edge-on, can be obtained by stellar counts made in directions that are perpendicular to the plane of the Galaxy. We look out of the plane into the clear regions of intergalactic space and determine how fast the different kinds of stars thin out as our surveys move farther from the sun. In this direction, it is found that interstellar absorption is small and easily corrected. Therefore, we have a chance to carry out a fairly complete census without concern about missing any important elements.

Figure 19-10 shows the result of such a survey and illustrates the fact that different kinds of stars are distributed differently in the vertical direction. Stars of high temperature and large absolute luminosity are packed rather tightly to

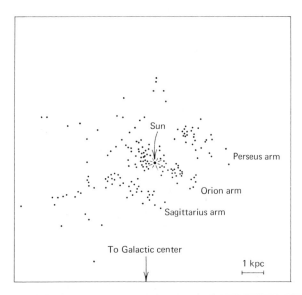

Fig. 19-11. Spiral arms from optical tracers near the sun. *(After B. J. Bok.)*

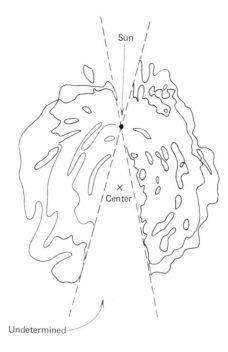

Fig. 19-12. The HI spiral structure from radio observations of our Galaxy.

the Galactic plane, as is the interstellar gas and dust. Fainter stars, which include the older stars, are much more loosely bound to the plane. This fact can be interpreted as resulting from two causes. First, a correlation of the distribution of objects above the plane with their age is probably an indication that in the earlier periods of our Galaxy's history, the system was less flat than it is now. Calculations of models of the evolution of our Galaxy confirm this conjecture and show that at a very early period of our Galaxy, during the first billion years or so, the Galaxy was collapsing from a more nearly spherical

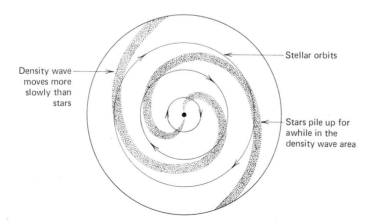

Fig. 19-13. The density wave model of spiral structure.

Fig. 19-14. The Milky Way approximately in the direction of the anticenter (180° away from the center). *(Lick Observatory.)*

object. Any stars, therefore, that formed during this early period were formed in a much more nearly spherical array. The collapse to the plane occurred fairly rapidly, however, and therefore most stars in our Galaxy have been formed in the thin plane of the system. The second effect that can occur has to do with the fact that even if a star is formed in the plane, it will not necessarily remain exactly there. If in the process of formation it is given a motion that is at even a very small angle to the plane, then this motion will carry it out of the plane for a certain distance. Its orbit will twist in and out of the plane in a snakelike manner, and therefore in general such stars will show a wider distribution above and below the plane than others. In particular, stars that have just been formed out of the interstellar material, those very young stars that have very high luminosities and temperatures, will not have had time for this effect to carry them out of the plane in their oscillatory path between the time of their formation and that of our observing them. Thus, these very young stars must be seen only in the flat plane, i.e., the part of our Galaxy where the interstellar gas and dust out of which they were formed exists.

The fact that our Galaxy is spiral in shape is established through a combination of neutral hydrogen measurements made of radio wavelengths and optical measurements of young objects. Examination of other galaxies shows that spiral arms are conspicuous because they contain a relatively large number of very luminous bright objects, including O and B type supergiant stars, large bright star clusters, stellar associations, and H II regions (bright gas clouds). Radio surveys of nearby galaxies also show that the spiral arms represent concentrations of neutral hydrogen. From these facts, astronomers have argued that the apparent concentrations of such objects in the immediate solar neighborhood show us the spiral structure of our Galaxy in our local area of space. Furthermore, radio observations covering our whole Galaxy suggest an overall spiral pattern that resembles that of other spiral galaxies.

The Spiral Structure of Our Galaxy

There are many hypotheses that have attempted to explain the spiral structure of our Galaxy and others. One of the most successful theories is one developed by C. C. Lin, who showed that a very large-scale disturbance in a galaxy could set up a *density wave,* somewhat like a wave on the ocean, which would then propagate through the galaxy, triggering star formation as it went (Chap. 18). The complicated mathematical details of the theory have led to predictions about the exact properties of spiral arms which seem, so far at least, to agree with measurements made at the telescope.

QUESTIONS

1. From your present knowledge, what methods in addition to Shapley's would you propose for measuring the size of our Galaxy?
2. Approximately what percent of the mass of our Galaxy is in the form of dust?
3. Approximately what percent of the *mass* of our Galaxy is helium?
4. What are the distinguishing characteristics of the material making up spiral arms?
5. From star maps at the back of this book, what are some of the constellations that lie in the plane of our Galaxy?

Fig. 19-15. An example of a Milky Way profile measured with a small telescope.

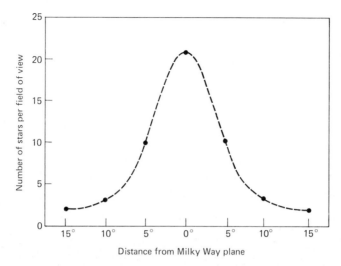

OBSERVATIONS

• The Milky Way is largely unresolved to the naked eye, and appears as a band of diffuse light. Binoculars will show many stars resolved in it, and telescopes even more. Sir William Herschel was one of the first to explore it telescopically. He used a simple method that you could try as an interesting experiment. Choose a section of the Milky Way and draw a line perpendicular to it on a star map. Using binoculars or a small telescope, count the number of stars that you can see in the field of view at various distances from the Milky Way plane. In this way, you can establish a star *profile* of the galaxy. Using both binoculars and a telescope, two profiles and two different average distances can be obtained, and these can be intercompared.

20

EXPLODING GALAXIES

20-0
GOALS

One of astronomy's grandest spectacles, exploding galaxies, is described and several of the mechanisms that astronomers have proposed to explain these gigantic events are introduced.

20-1
THE ENERGY
OF EXPLODING
GALAXIES

Among the more bewildering objects in the universe are the galaxies that seem to be in the process of explosion. Astronomers are not yet certain why these galaxies are experiencing violence, but scientists are continually gaining more evidence about the nature of the violent events. The most remarkable feature of these explosions is their size, as an explosion that can disrupt an entire galaxy (made up of a million, million stars) must be exceedingly energetic.

Explosions in galaxies were first detected by radio telescopes. Whereas ordinary galaxies normally emit most of their radiation in the form of normal light, it was discovered many years ago that some peculiar-looking galaxies also emit strong radio radiation. Over the last few years, detailed studies of these strange objects using *both* radio telescopes and large optical telescopes have shown that the most reasonable explanation of the source of radio emission is some type of explosive event.

In physics and astronomy, an energetic event like an explosion is usually measured in terms of the amount of energy released. The unit used in most cases is called the *erg*, a very small unit that measures the capacity to do work. One ton of TNT, when exploded, releases 4×10^{16} ergs of energy, while a 20-kiloton atomic bomb gives off 10^{21} ergs. A major volcanic eruption can give off energy equivalent to many atomic bombs.

Fig. 20-1. An erupting volcano, Mt. Mayon, in the Philippines in 1968. *(U.S. Air Force.)*

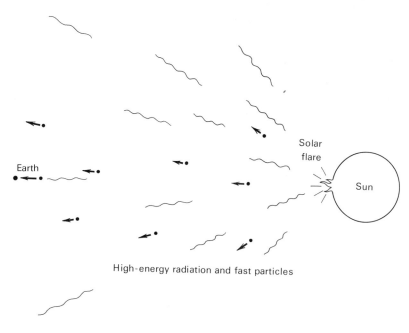

Fig. 20-2. A solar flare can influence the earth's outer atmosphere.

Earth

Solar flare

Sun

High-energy radiation and fast particles

We can measure, for example for an erupting volcano, both the rate of release of energy (in ergs per second) and the total energy given off over the entire time of the eruption. The rate of energy released from the eruption of Surtsey, the volcano that recently formed a new island off the coast of Iceland, averaged about 10^{18} ergs per second. The eruption lasted for months, so that the total energy given off was about 10^{25} ergs.

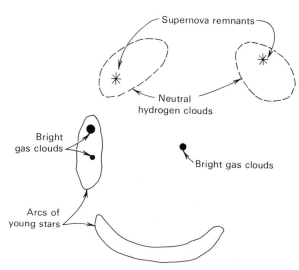

Supernova remnants

Neutral hydrogen clouds

Bright gas clouds

Bright gas clouds

Arcs of young stars

Fig. 20-3. The remnants of a super-supernova tentatively identified in a nearby galaxy (the LMC). The explosion pushed the gas in the galaxy back into arcs to make the observed areas of young stars and gas, which contain supernova remnants and bright gas clouds. *(After Westerlund and Mathewson.)*

Even more energy is released in some of the types of explosive events among the stars. The least energetic of these is immensely more powerful than the biggest volcanic eruption witnessed on the earth. These are the solar flares, which eject from the sun vast numbers of nuclear particles and energetic light and radio waves. The average high-intensity solar flare explodes with a total energy that is so large that it would take the equivalent of over a million volcanoes to equal it, averaging some 10^{31} ergs total.

A much more energetic explosion of stellar material is that of the nova, which while ejecting part of its outer layers explosively becomes thousands of times brighter than normal. An ordinary nova explosion is about 10^{14} (100 million, million) times as powerful as a solar flare. This means a total energy of 10^{45} ergs. Supernovae, which involve the entire mass of a star, are even more energetic. On the average, the amount of energy produced in a supernova explosion is about 10,000 times that of a nova explosion, some 10^{49} ergs.

Even more energetic than the supernova is a very recently discovered phenomenon called the *super-supernova*. We only know of a few such explosions, all of which occurred perhaps millions of years ago and for which we only see the circumstantial evidence made up of remnants. Nevertheless, from the evidence that remains, it is possible to learn that super-supernova explosions, whatever their cause, are so energetic that they throw out vast amounts of material, the equivalent of about a million suns. They can completely disrupt portions of galaxies. The estimate of the energy in a super-supernova explosion is approximately 100,000 times that of an ordinary supernova, some 10^{54} ergs. The range of the energies of the explosions described here is immense, but this range is extended considerably by the total energies involved in the explosive galaxies. Typically, the total amount of energy involved in the most energetic radio galaxies is about 10 million times that of a super-supernova (about 10^{61} ergs). Explosive galaxies *average* approximately 10^{58} ergs, the equivalent of a billion supernovae.

In general, the explosive objects that we can identify as extragalactic (beyond the edges of our galaxy, the Milky Way) can be divided into two classes: the quasistellar objects, which include the quasars and the quasistellar blue objects; and the radio galaxies. The quasars are remarkable and as yet unexplained objects that emit very intense radio emission and that look optically very much like stars. Since their discovery in 1961, a considerable controversy has surrounded them.

20-2
TYPES OF
EXPLOSIVE
GALAXIES

The quasistellar objects have a very small apparent size, both in optical photographs and according to radio telescopes' measurements. They appear very much smaller than known galaxies, and most would not be distinguishable from stars by ordinary techniques. They have very broad emission lines in their spectra, which indicates that they are made up of hot, thin gas clouds with large turbulent velocities. Their radial velocities, as measured from the Doppler

20-3
QUASARS

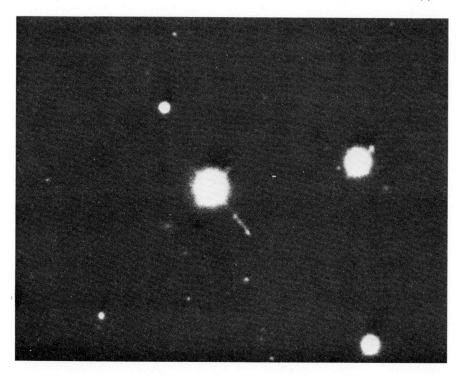

Fig. 20-5. The brightest quasar, 3C 273 *(Kitt Peak National Observatory.)*

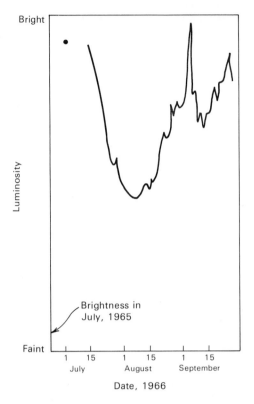

Fig. 20-6. Light curve of a quasar (3C 446) as measured in the summer of 1966. *(After T. D. Kinman.)*

Bright

Luminosity

Brightness in
July, 1965

Faint

1   15   1   15   1   15
July   August   September

Date, 1966

shifts in their emission lines, are very large, ranging from several thousand miles per second up to velocities of about 90 percent of the velocity of light. They sometimes show rapid and usually irregular fluctuations in their intensities over a few days or weeks, which indicates that the objects are intrinsically very small. If they were many light years across, the whole object could not vary in luminosity over a short period, because of the light travel time from one side to the other. If, for instance, an object 1,000 light years across is varying in brightness like a quasar, and if it suddenly changes in brightness (all over) in some synchronous way, then we would see this occur first due to the light from the nearest part of the galaxy and then a thousand years later we would see the variation occur at the farthest part. The entire variation would be spread over a period of a thousand years as seen by us. Therefore, it is argued that we cannot in the case of quasars be looking at objects that are more than a few light days across. And yet the quasars are so bright and so energetic at both visual and radio wavelengths, that it is very difficult to understand how so much energy can be emitted by such a small object. Some astronomers have argued that the quasars are not as distant as is indicated by the use of normal distance indicators, such as red shift (Chap. 22), and this may someday be proven to be the case but at the time of the writing of this chapter we are left with conflicting

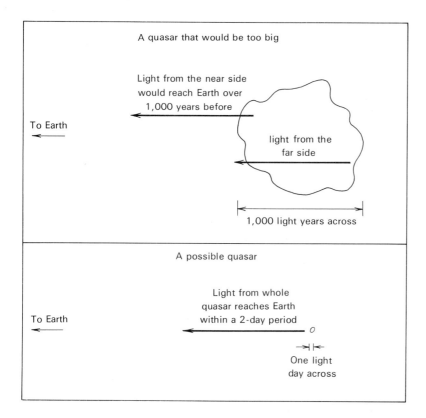

A quasar that would be too big

Light from the near side would reach Earth over 1,000 years before

To Earth

light from the far side

1,000 light years across

A possible quasar

Light from whole quasar reaches Earth within a 2-day period

To Earth

One light day across

Fig. 20-7. Why a quasar must be small if we see it change in brightness within a day or two.

evidence. They are either much too close for their red shifts, or they are much too energetic for their size.

20-4 RADIO GALAXIES

The other type of extragalactic explosive object is the radio galaxy. These galaxies show very strong emission at radio wavelengths, but otherwise look something like ordinary galaxies with certain optically evident peculiarities. In general, most of the radio galaxies that have been identified with optical objects appear to be somewhat disrupted, show bright emission of light from the nucleus, and often show vast quantities of dust and gas strewn about in a chaotic pattern. In some cases, the image appears to be that of an ordinary elliptical galaxy, but with a chaotic or unusual arrangement of dust and gas superimposed on it.

20-5 CAUSES OF THE EXPLOSIONS

The total amount of energy involved in the type of explosion that causes the radio galaxies and the quasars is so big, that in order to produce this energy by ordinary nuclear reactions it would be necessary to completely annihilate

Fig. 20-8. M 82, a galaxy that may have exploded only a few million years ago. *(Hale Observatories.)*

Fig. 20-9. The galaxy NGC 6946 is an "active" galaxy that has had several supernovae recorded in the twentieth century. *(Kitt Peak National Observatory.)*

Fig. 20-10. A quasar spectrum. *(Courtesy of Roger Lynds, Kitt Peak National Observatory.)*

billions of suns. There have been many suggestions of explanations for how this might occur, and in the next paragraphs some of these theories are reviewed.

When radio galaxies were first discovered, it was believed that the cause of the disruption was the collision of two galaxies. Since galaxies probably collide occasionally, it was hypothesized that when such collisions occur, a very disruptive and energetic event results. However, the stars in galaxies are spread out so thinly, and the sizes of the stars are so small compared to their distances apart, that two galaxies made up only of stars would pass through each other

Collisions of Galaxies

Fig. 20-11. The giant radio galaxy NGC 5128. *(Cerro Tololo Interamerican Observatory.)*

cleanly without any violent collision occurring. The stars would appear for a while to be twice too numerous to an observer in such a colliding galaxy, and then they would thin out again to their normal density as the two galaxies passed through each other. The probability that even one star of one galaxy would collide with one star of another during such a collision of galaxies is small. However, if such colliding galaxies contain gas and dust clouds that are very large compared to the sizes of stars, then a violent collision would occur. The amount of energy produced, however, is not large enough to explain the most energetic radio sources and, furthermore, very few of the nearby radio galaxies have the appearance of galaxies in collision.

Antimatter   Another early theory suggested to explain the high-energy output from radio sources is the idea that they represent an encounter of matter with antimatter. We know from physics that there is a fairly strict symmetry in nature between two different forms of material, one of which we call ordinary matter—that is, the kind of material that we, our earth, the sun, and everything that we come in contact with is known to be made of—and the other is called *antimatter*, which looks the same, except that it is exactly opposite in charge and in other fundamental properties, so that it has the property of annihilation when it comes into contact with matter. For instance, the proton is an ordinary

Galaxies approaching

Dust and gas clouds

Galaxies colliding

Galaxies after the collision

Fig. 20-13 Two galaxies that collide leave behind their clouds of gas and dust, but their stars do not collide.

elementary particle of matter. Several years ago, it was found that there is another particle with the same mass as the proton, but with negative instead of positive charge, and this was called the *antiproton*. Experimentally, it was found that a proton and an antiproton, when brought together, annihilate each other and produce energy in their place. Although there is no evidence that antimatter exists in nature (we have always had to produce it in the laboratory), there is still the idea that perhaps elsewhere in the universe there are galaxies made up of stars of antimatter. If one of those galaxies should collide with one of the galaxies of matter, it has been suggested that perhaps a quasar or extremely strong radio galaxy might result. At the present, this theory is not held by very many people because of the difficulties involved in understanding why

Fig. 20-14. The collision of a galaxy of antimatter with a galaxy of matter.

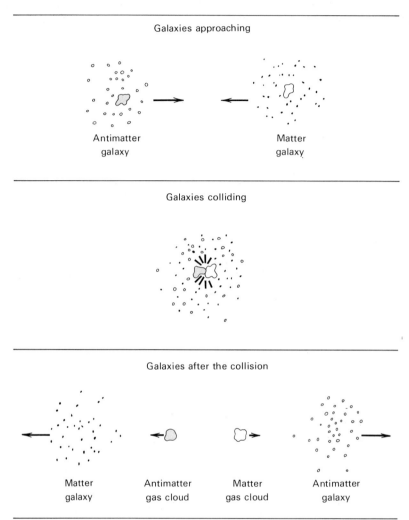

Galaxies approaching

Antimatter galaxy   Matter galaxy

Galaxies colliding

Galaxies after the collision

Matter galaxy   Antimatter gas cloud   Matter gas cloud   Antimatter galaxy

such a collision would proceed for very long. As soon as two clouds of gas, one made up of matter and the other made up of antimatter, should come close to each other, the first few atoms that would collide and annihilate each other would cause a heating up of that portion of the collision edge of the two clouds, to the point where the two clouds might then separate and no large collision would result.

Accretion   Another idea that has been put forward to explain radio sources is the idea of accretion of matter. It was suggested that large amounts of matter between the galaxies might fall into a galaxy under the galaxy's gravitational pull, and that the nucleus of the galaxy would eventually experience a tremendous overpopulation of matter, which would cause an explosion.

A variation on this theme was put forward in 1977 by Philip Morrison and his collaborators at M.I.T., who took a second hard look at the galaxy M 82 (Fig. 20-8). Rather than an explosion, they found that the same distorted and distended appearance could come about if M 82 were a normal spiral galaxy which happened, some millions of years ago, to drift into a huge, thin cloud of gas and dust inhabiting the intergalactic regions of a loose group of galaxies around M 81 (Fig. 11-4). It could have triggered massive star formation in the central areas, giving rise to high temperatures and a bright, spotty appearance there. The emersion in the gas and dust could also have swept the outer parts of M 82 clean of its own gas, so that the spiral arms would have had no materials left to maintain themselves with new star formation. Also in the group and not far from M 82 is the peculiar galaxy NGC 3077 (Fig. 18-16), which may owe its dusty and lumpy appearance to a similar effect.

This idea, still only a tentative explanation of M 82, does not explain most other radio galaxies. M 82 is not an especially intense radio source, hardly very different from the center of our own Galaxy. For the giant radio galaxies and the quasars, some explosive mechanism seems necessary to explain the incredible amount of energy being released.

Black Holes

Probably the most promising hypothesis for explaining quasars and radio galaxies is that of gravitational collapse to a massive black hole (Chap. 17). Many astronomers believe that the only possible way to generate so much energy is for an object millions or perhaps even billions of times the sun's mass to collapse at the center of a galaxy. Then matter exterior to the black hole will continually fall into it, and will emit brilliant light just before disappearing. This light could violently push away the material just beyond the critical radius of the black hole, making a galaxy appear to have an exploding nucleus.

Quasars, some astronomers argue, may be a phase of this collapse seen just prior to the disappearance of the massive object into its black-hole cocoon. It would be like a giant neutron star (Chap. 16), probably spinning and emitting radio waves, x-rays, and flashing light. However, because it is so much larger than a pulsar, the flashes occur in weeks or months instead of in small fractions of a second. It is calculated that a body with mass of about a hundred million solar masses would (temporarily) form a neutron star about a light-month across. Interferometer measures by intercontinental radio-telescope arrays (Appendix 5) show that quasars are about this small, and of course their rapid light fluctuations also argue this (Sec. 20-3).

It was discovered in 1978 that the center of the bright radio galaxy M 87 contains a very massive object that is not made up of normal stars. Measure-

ments indicate that its mass may be as much as 5 billion suns. Such an object may well be a black hole, in which case the radio emission from this galaxy is undoubtedly due to the energy released by this collapsed object at its center.

As the above theories have shown, astronomers and physicists have given a great deal of time and thought in attempting to explain the explosive galaxies, but it cannot be said at this time that they have succeeded in completely explaining these strange objects, as none of the theories is yet fully satisfactory. More needs to be known about the characteristics of these explosive galaxies, their distances, and the mechanisms whereby the energy is released. Hopefully, when such knowledge is attained, we will be better able to identify the immense forces that are at work when an entire galaxy is disrupted or blown apart.

QUESTIONS

1. What fraction of the energy of a typical radio galaxy outburst do you expend in climbing a staircase? (Confer with Fig. 8-2.)
2.* How much energy would be emitted if one billion (10^9) suns were converted entirely into energy?
3. Our galaxy has an ''active nucleus,'' that is, it has a small area at its center

that emits intense radio waves. Why can we not see this active region in order to study it optically?

Only two radio galaxies are visible with small telescopes, and neither is bright enough for its optical peculiarity to be visible. These are M 82 and M 87, both tabulated in the *Observations* section following Chap. 18.

OBSERVATIONS

21

WHAT IS THE UNIVERSE

This chapter forms a base for exploring the universe at large. It tries to come to terms with a definition of the word "universe," and it deals with the possible curvature of space in our universe.

The term *universe* is used to designate the entirety of space and all of its contents. *Cosmology* is the term for the scientific study of the universe as a whole. The universe is made up of galaxies, light, and intergalactic matter. The efforts of cosmologists have resulted in some understanding of the nature and history of these things. This chapter deals with two large-scale aspects of the universe that are often confusing and difficult to visualize. They are, however, important for an understanding of the material in the chapters that follow.

One of the key questions in cosmology is the question of whether or not the universe is bounded, that is whether it has an edge or border where it stops.

**The Observable
Boundary**

The observable universe is bounded by the effectiveness of our equipment, of course, but this observable boundary is not a permanent one, because every year astronomers work out new methods of improving telescopes and other astronomical instruments. The largest optical telescope in America, the 200-inch telescope on Palomar Mountain, is an instrument that was designed especially for the purpose of exploring the most distant realms of the universe. Also, the limits of observation are being extended by other kinds of techniques, particularly radio telescopes, and every time more knowledge is added to what we know about the most distant objects we can see, more distant ones are found. The universe tends to grow, as far as its *observable* boundary is concerned, and recently it has doubled in size every five or so years.

**The Geometrical
Boundary**

There is a very different kind of boundary that is not a human or physical limitation, but a geometrical one. It is possible that the universe is bounded by

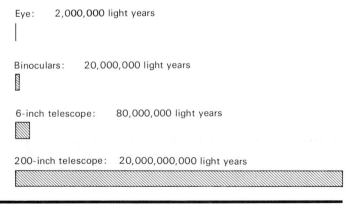

Eye: 2,000,000 light years

Binoculars: 20,000,000 light years

6-inch telescope: 80,000,000 light years

200-inch telescope: 20,000,000,000 light years

Fig. 21-1. The size of the observable universe as seen by the eye and by telescopes of various sizes.

Fig. 21-2. Dome of the 100-inch telescope at Mt. Wilson, where the depths of the universe were first probed.

Fig. 21-3. NGC 4594, an Sb galaxy with large amounts of obscuration by dust. *(Kitt Peak National Observatory.)*

geometry alone. The universe can be considered to be infinite and without boundary, as visualized by Newton, for example. There are difficulties with this view, however, as there are observations that do not seem to agree with the predictions that can be made from the newtonian hypothesis of an infinite universe. For instance, recent measurements indicate that the most distant galaxies have velocities and luminosities that are not quite right for a newtonian type of infinite universe.

One can avoid this difficulty by hypothesizing that the universe is not shaped the way our intuition tells us, a suggestion that comes from some work of Lobachevsky, a Russian mathematician. If one rejects Euclid's axiom about parallel lines (that two straight lines that are parallel to each other will extend to infinity and never intersect, always remaining at the same distance), a much looser axiom about parallel lines is possible, i.e., one that allows the lines to diverge or converge. This gives a whole new system of geometry that is perfectly self-consistent and logical, as is Euclid's.

Einstein found that by using such a noneuclidean geometry, he could derive a cosmology of the universe that is finite in volume. The reason for a finite volume is that a straight line extended indefinitely in what is called spherical geometry will eventually return to the place from which it started. That is, a person projected in a straight line by a space vehicle in some direction will eventually return to his or her starting place from the opposite direction. A

Fig. 21-4. The flat earth and the round earth, examples of how two-dimensional space curvature acts.

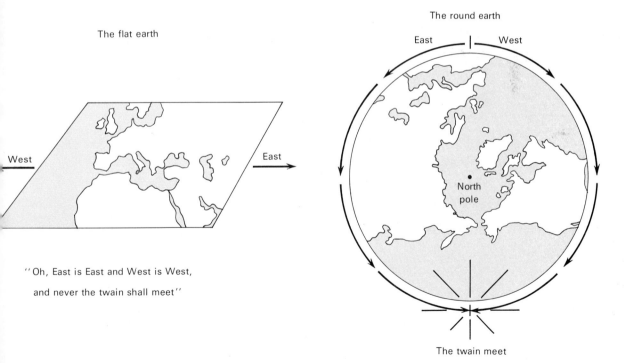

The flat earth

West — East

"Oh, East is East and West is West,

and never the twain shall meet"

The round earth

East — West

North pole

The twain meet

Fig. 21-5. The sums of the angles in a triangle on a flat surface and on a sphere.

$$90° + 90° + 90° \neq 180°$$

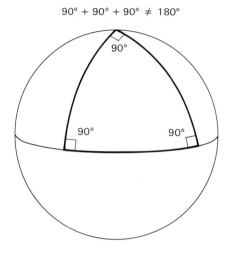

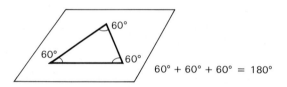

$$60° + 60° + 60° = 180°$$

1,000,000,000 years

100,000,000 years

10,000,000 years

1,000,000 years

Earth

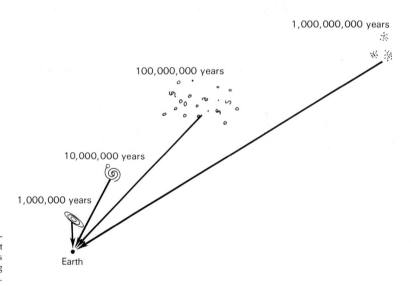

Fig. 21-6. The time problem: the most distant clusters are seen by us as they looked long, long ago (not to scale).

Fig. 21-7. The Hercules cluster of galaxies is a relatively open cluster. *(Kitt Peak National Observatory.)*

straight line "bends" with a certain radius, the radius being a description of the size of the universe. In this kind of universe, there would be no edges, there would be no empty space, and the whole universe could be filled uniformly with galaxies and stars. At the same time, it would have a finite number of galaxies and stars and a finite volume. Noneuclidean geometry, therefore,

Fig. 21-8. The clustering hierarchy.

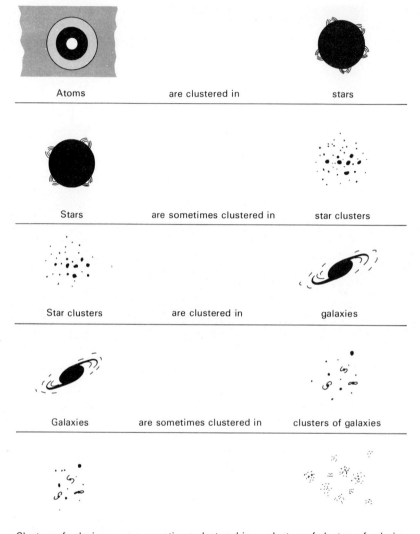

Atoms	are clustered in	stars
Stars	are sometimes clustered in	star clusters
Star clusters	are clustered in	galaxies
Galaxies	are sometimes clustered in	clusters of galaxies
Clusters of galaxies	are sometimes clustered in	clusters of clusters of galaxies

provides a possible way of explaining the known features of what on some grounds seems to be a finite universe.

These special types of geometry are virtually impossible to visualize, because we are so completely used to thinking in terms of euclidean space. But consider an historical analogy. Not many years ago, there were still some people who thought that the earth was flat. When a scientist would say "the earth is round, so if you go west far enough, you will end up in the east," these people just could not understand. They pointed out that the earth "looks" flat and that it was unnatural to think otherwise, just as one might now argue that the world

Fig. 21-9. The Hydra cluster of galaxies, a rich distant cluster. *(Kitt Peak National Observatory.)*

looks euclidean and that it is unnatural to imagine the alternatives. Just as we now know that the earth is round, that its surface has no edges, and that its surface is finite in area, so did Einstein argue that the universe is spherical, that it has no edges, and that its volume is finite. The main difference is that for the earth, we are talking about two-dimensional curvature, while Einstein talked about curvature of three-dimensional space. Another difference is that for the earth, we have demonstrated the truth of this idea, while no one has yet proven beyond doubt the existence of curvature for the universe.

The Time Quandary

When we observe the universe, we are observing light that is emitted by various parts of the universe at various distances, but since light doesn't travel instantaneously, when we look at very distant objects we aren't seeing what they look like now, but rather how they appeared many years ago, according to their distance. The most distant galaxies that we can examine are seen as they looked billions of years ago. We have a picture of the universe that isn't really a snapshot; it is not a picture of what it looks like ''now,'' but is a historical parade of events and conditions, from the beginning (if there was one) to the present.

Corrections have to be made for the time problem. For instance, when measuring a galaxy at a very large distance, it is necessary to calculate what it

Fig. 21-10. The giant galaxy NGC 7331. *(Kitt Peak National Observatory.)*

looks like now, knowing something about how galaxies evolve. A galaxy, for instance, that sent out its light 4 or 5 billion years ago, is now possibly rather different from what it was when it sent out the light that we see. Cosmological calculations must include corrections for this evolutionary change.

Another aspect of the time problem has led to a famous paradox. We have found above that we can't know what is happening way out there "now" because the information can never reach us sooner than the light-travel time. But relativity opens up an opportunity for trying to get around this problem (see Chap. 17). Suppose some day we have spaceships that are capable of traveling *almost* at the speed of light. We could send such a ship, with appropriate astronomers on board, to a distant galaxy (say 2 million light years away) at so high a velocity that time on board would be greatly slowed down. It would seem to the travelers to only take, say, a few weeks to get there, even though a light beamed by their colleagues to the galaxy would take 2 million years to make the trip. Thus the astronomers on the spaceship could find out the current status of the galaxy (Is it exploding? Does it have new stars? etc.) long before the folks at home know. If they wish to transmit their finding to earth, however, they have a problem. If they radio it back, the signal will take 2 million years to reach earth. So they might as well just bring it with them. Using the same high velocity and slowing of time, they could return to earth again in just a few weeks, anxious to tell their friends the news. Unfortunately, however, they would discover that time here on earth did not slow down while they were gone, and 4 million earth years would have passed during their absence. If there were any earthlings still left on earth, they would not be too surprised by the news brought back about the galaxy, as the light from the galaxy sent out while the spaceship was there would have arrived at earth a few weeks earlier. This problem is often referred to as the "twin paradox," but it is not a true paradox, because although it would seem to violate the usual experience of twins (one of whom travels and one of whom stays home), it does not violate the known laws of nature, represented by the theory of relativity.

QUESTIONS

1. What shape of surface can you think of that would have the property that the sum of the angles of a triangle drawn on it would be less than 180°?
2. From your knowledge of stellar evolution, how do you think a distant elliptical galaxy might be different now from the view that we have of it as a result of the time of light travel?

22

THE SIZE OF THE UNIVERSE

To establish the size of the universe, it is necessary to measure distances to galaxies in a huge realm of space. This chapter is intended to make clear to you the step-by-step process that is employed.

The extragalactic scale of distances is established by a stepwise process. Because of the great difficulty in measuring the distances to distant galaxies, it is first necessary to establish methods of measuring nearby galaxies' distances, and then of calibrating new methods for another step in distance, and so on.

In practice, the distances to the nearest galaxies are established by the

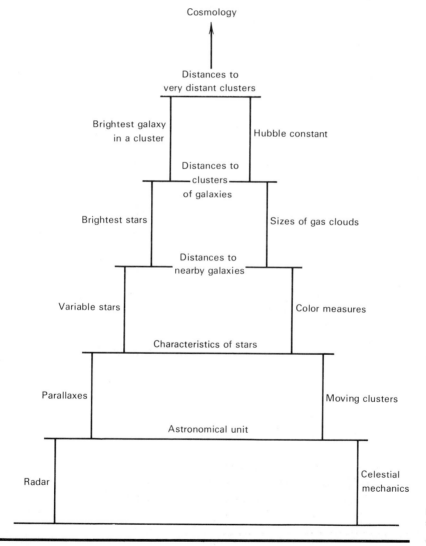

Fig. 22-1. The distance-scale pyramid.

period-luminosity relationship for the Cepheid variable stars (Chap. 12). These nearest galaxies, mostly in the local group, are then used to calibrate the brightness of the brightest stars in a galaxy so that they can be used as a criterion in more distant objects. Similarly, the nearby galaxies can be used to calibrate the sizes of the largest H II regions (Chap. 9) so as to allow them also to be used in the next step outwards. Novae (Chap. 14) provide a further method of calibrating the first two steps in the distance scale, and there are other secondary methods that will be mentioned later in this chapter. Beyond distances of about 10 million parsecs, individual objects in galaxies can no longer be distinguished, and so it is necessary to establish distances by cruder means. From the Hubble relationship between apparent magnitude and radial velocity (Chap. 23), radial velocity is directly proportional to distance, and distances can simply be calculated from the measured radial velocity. For some purposes, however, it is necessary to obtain a distance measure independent of the red shift, and so such methods as the determination of luminosities of individual galaxies and the calibration of the luminosities of the brightest galaxies in clusters are used.

22-1 CEPHEID VARIABLE STARS

The Period-Luminosity Relation

In the early part of this century, Henrietta Leavitt at Harvard discovered a curious relationship between the periods and luminosities of variable stars in the Magellanic Clouds (Chap. 12). It was some years, however, before it was realized that this apparent curiosity in the Magellanic Clouds could become a fundamentally important means of measuring distances to galaxies. In fact, it was not until E. Hertzsprung of Denmark pointed out the probable identity of these objects as Cepheid variables similar to the high-luminosity objects in our Galaxy that it was even realized that Leavitt's observation implied a considerable distance for the Magellanic Clouds.

Extensive use of this relationship between period and luminosity for Cepheid variables was first made by Harlow Shapley, who attempted to calibrate the period-luminosity relationship in absolute units by comparison with variables of similar range of period in globular star clusters and in the field near the sun. The calibration in modern times uses Cepheid variables that are members of galactic star clusters, the distances to which are established by precise photometry of the main-sequence stars.

Is It Universal?

There is no reason to suspect that the Cepheid period-luminosity relationship should not be identical from one galaxy to another, and tests within the local group of galaxies indicate that any differences must be very small and probably

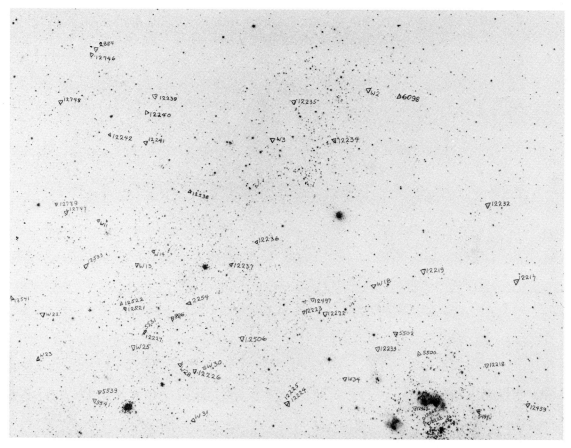

are not significant for the distance problem. At first, there were many reports of important differences between Cepheids in the Magellanic Clouds and in our Galaxy, but most of these differences have since been found to result from selection effects and photometric difficulties. There is an apparent difference in period distribution between galaxies, but this is believed most likely to result from differences in star formation rates. Therefore, it is believed that the basic similarity of Cepheids in the galaxies is fairly safely established. Differences resulting from small differences in chemical composition can be accounted for theoretically if these differences in composition are known, but presently any such differences are not established well enough to be used in any systematic way. They seem to be smaller than the differences found within an individual galaxy.

Fig. 22-2 Cepheid variables identified in a portion of a nearby galaxy, the Large Magellanic Cloud.

Figures 12-3 and 22-3 show light curves for several Cepheid variables of the Large Magellanic Cloud. The period-luminosity relationship for these variables

Distances from Cepheids

Fig. 22-3. The light curve of a Cepheid variable in the Large Magellanic Cloud. Its period is a little over 4 days.

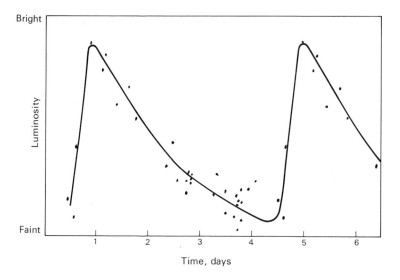

Bright

Luminosity

Faint

1 2 3 4 5 6

Time, days

Fig. 22-3. The light curve of a Cepheid variable in the Large Magellanic Cloud. Its period is a little over 4 days.

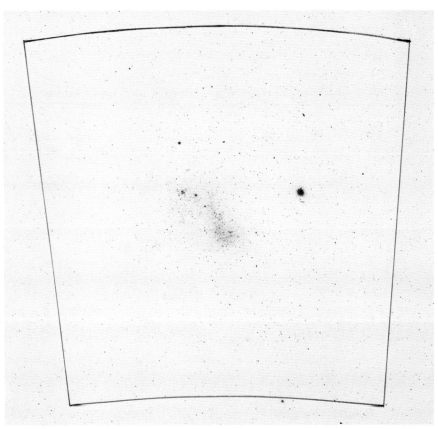

Fig. 22-4. The Small Magellanic Cloud. *(Mt. Stromlo Observatory.)*

in apparent magnitudes can be compared with the period-luminosity relationship in terms of absolute magnitudes established by A. R. Sandage of the United States and Gustaf Tammann of Switzerland (Fig. 12-12). A comparison of the observations with the Sandage-Tammann curve allows an exact determination of the distance to the Large Magellanic Cloud, when reddening and absorption in the galaxy and in our stellar system has been subtracted from the data.

The distance to a galaxy can be expressed in two ways, either as a linear distance given in units such as kiloparsecs or magaparsecs, or in terms of magnitudes in the form of a difference called the *distance modulus*, which is defined as the difference between the apparent magnitude of an object in the galaxy and its absolute magnitude. The comparison discussed above for the Large Magellanic Cloud leads immediately to a distance modulus for this galaxy, which, corrected for dust absorption, is 18.90. From the definition of magnitudes, it can be calculated that this distance modulus means a linear distance of 54,000 parsecs.

The Andromeda Galaxy, M 31, is the most completely studied spiral galaxy and, being the nearest large spiral to us, the best one in which to observe Cepheid variables. Walter Baade and Henrietta Swope of the Palomar Observatory measured the period-luminosity relationship for 20 Cepheids in a distant arm of M 31. Their observations (see Fig. 22-5) give a relationship with a slope very similar to that found for the Magellanic Clouds, and a comparison with the calibrated standard period-luminosity relationship leads to an apparent distance modulus of $m - M = 24.65$. By comparing their measured colors with colors of Cepheid variables in our Galaxy, Baade and Swope determined that in the mean their Cepheids are 0.16 magnitudes too red in B-V, due to intervening dust. This means that there is a total absorption in V of approximately 0.48 magnitudes, so that the true modulus of M 31 is $m - M = 24.17$ (a distance of 670,000 parsecs).

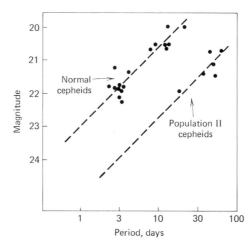

Fig. 22-5. Period-luminosity diagram for variable stars in M 31, the Andromeda Galaxy. *(After Baade and Swope.)*

In 1925, Edwin Hubble published his first full study of an extragalactic object at Mt. Wilson Observatory—the small irregular galaxy NGC 6822 (Fig. 22-6). Here he found Cepheid variables similar to those in our Galaxy and in the Magellanic Clouds, and he was able to obtain a period-luminosity relationship for them. More recently, Susan Kayser has published a more complete study of the period-luminosity relationship for this galaxy (see Fig. 22-7). Of the 32 variables studied, she found 13 Cepheids, which gave a clear period-luminosity relationship and a distance of 550,000 parsecs.

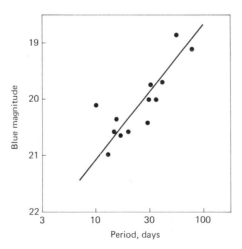

Fig. 22-7. The period-luminosity relation for NGC 6822. *(After Kayser.)*

Fig. 22-8. Draco, a near-by dwarf elliptical galaxy, the faint members of which are scattered over this negative photograph. *(Lick Observatory.)*

22-2
RR LYRAE
VARIABLES

Because they are relatively less intrinsically bright than the Cepheids, the RR Lyrae variables (Chap. 12) are less frequently used for distance measurements of galaxies. Nevertheless, there are some galaxies that contain no normal Cepheids and that are near enough so that we can measure the RR Lyrae members. Galaxies of this sort are the Draco system, (Fig. 22-8), the Leo I system (Fig. 22-9), the Ursa Minor system, and the Sculptor system, all of which are local-group, dwarf elliptical galaxies. Figure 22-10 shows light curves of RR Lyrae variables in the Draco system.

The only other galaxies in which RR Lyrae variables have been studied are the Magellanic Clouds. Generally, although precise photometry is difficult for these objects, which are at apparent magnitude between 19.0 and 20.0, the results are in close agreement with distances measured by means of Cepheid variables.

22-3
NOVAE

Novae (Chap. 14) similar in properties to those in our local Galaxy have been observed in the Magellanic Clouds and in M 31, as well as in a few more distant galaxies. It was found in studying the novae in M 31 that there is a relationship between the maximum brightness of the nova and its rate of decline after maximum, and a similar relationship exists for novae in our own Galaxy (Fig. 14-10). Therefore, it is possible by observing any given nova, both at maximum and at intervals following maximum, to establish its distance by comparison with standard novae of known distance in our Galaxy. In the case of M 31, hundreds of novae have been observed. For the Magellanic Clouds,

only about a dozen have been observed, and only a few of these have been studied reasonably accurately.

22-4 BRIGHTEST STARS

Beyond the local group, it is possible to measure the luminosities of the brightest stars in galaxies out to about 10 million parsecs. However, difficulty is encountered because of the fact that the brightness of the brightest stars not

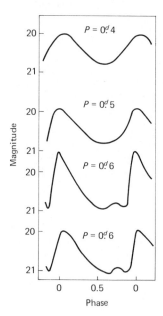

Fig. 22-10. Light curves of RR Lyrae variables in Draco. *(After Baade and Swope.)*

Fig. 22-11. A nearby galaxy, M 33, the brightest stars of which can be measured and used to calibrate this distance criterion. *(Kitt Peak National Observatory.)*

only depends upon the galaxy type and total mass, but is also extremely difficult to measure precisely because the brightest stars are usually found in stellar associations and clusters. Allan Sandage has found that the brightest stars in galaxies have an absolute magnitude of approximately -9.3, but that the range for calibration galaxies is between -7 and -10.

**22-5
H II REGION
DIAMETERS**

The Argentine astronomer Jose Sersic pioneered in utilizing the apparent diameters of the largest H II regions in galaxies as distance indicators. If the assumption is made that physical conditions in galaxies are sufficiently similar so that the largest H II regions always have the same intrinsic diameter, then a simple measurement of their angular diameters gives an immediate value for the distance to the galaxy. Allan Sandage has also utilized this method widely and Robert Kennicutt has recently refined it.

Fig. 22-12. Large gas clouds in a distant galaxy, NGC 2403. *(Hale Observatories.)*

A recently employed method of obtaining distances to objects in the range of the distance of the Virgo Cloud (about 10^7 parsecs) has been the measurement of the integrated luminosity of the globular star clusters. For example, the Canadian astronomer Rene Racine has measured luminosities for hundreds of the globular star clusters in the giant Virgo Cloud elliptical galaxy M 87. Using these data, Sandage obtained a distance for M 87 by comparing globular star cluster luminosities with those of local group galaxies, including our own Galaxy and M 31. The difficulty with using this method of distance determination lies in the assumption, unconfirmed by recent evidence, that the distribu-

22-6 GLOBULAR STAR CLUSTER LUMINOSITIES

Fig. 22-13. A globular cluster, NGC 1978, in the Large Magellanic Cloud. *(Harvard Observatory.)*

Fig. 22-14. The giant elliptical galaxy M 87 in the Virgo cluster of galaxies. Some of its thousands of globular clusters are visible. *(Kitt Peak National Observatory.)*

tion of luminosities of globular star clusters is everywhere the same, and that the brightest globular star clusters in galaxies of different total mass or size are the same. It is more reasonable to expect that very massive galaxies will have very much larger numbers of globular star clusters than less massive galaxies. M 87 itself has several thousand recognized globular star clusters, and this does not even count the faint clusters that are not registered on 200-inch telescope plates. By comparison, our local Galaxy and M 31 have estimated total populations of 200 and 400 globular star clusters, respectively.

22-7
THE
LUMINOSITY
CLASSIFICATION
OF GALAXIES

Sidney van den Bergh has devised a luminosity classification of galaxies, which if used with sufficient care, makes it possible to estimate distances by the galaxy class alone. The classification is based on the appearance of certain features in the galaxy, including the development of the arms and the surface brightness. For example, M 31 is classified by van den Bergh as luminosity class I-II. Comparison with the calibration given by him for the luminosity classification system shows that this agrees well with the measured total luminosity for M 31 (absolute magnitude is $M = -20.05$). Used for more distant galaxies, this method gives moderately reliable estimates for individual galaxies, though not as surely as by the above methods.

Fig. 22-15. M 31, the Andromeda galaxy. *(Hale Observatories.)*

The most luminous stellar objects in galaxies are the infrequently occurring supernovae (Chap. 16). They, of course, occur only once every century or so in any one galaxy. Supernovae are of two types (see samples in Fig. 16-3), one of which reaches much larger total luminosity than the other. If the supernova in a given galaxy can be identified as to type, then using calibrations based on supernovae that have been observed in nearby galaxies, including our Galaxy and M 31 (for example, the Crab Nebula, which was observed in 1054 in our Galaxy, and S Andromedae, which was observed in 1894 in M 31), it is possible to calibrate the distance to the galaxy. This is a method that has not been used very frequently, primarily because of the large apparent spread in

22-8
SUPERNOVAE

supernova absolute luminosities and because of the difficulty of catching supernovae exactly at maximum light.

1. How would the presence of dust between galaxies affect the use of the period-luminosity relationship for determining distances to nearby galaxies? How would this dust be detected?
2. What is the distance modulus of NGC 6822?
3. Why can RR Lyrae variables be used to determine distances only for the nearest few galaxies?
4. Why can the classical Cepheid period-luminosity relation not be used to measure the distance to the Draco dwarf galaxy?
5. What methods would you use to determine the distance to a spiral galaxy if it were 50,000 parsecs distant? 500,000 parsecs? 5,000,000 parsecs?

23

THE EXPANSION OF THE UNIVERSE

23-0
GOALS

The purpose of this chapter is to make clear what evidence there is for saying that the universe is expanding, and to explain how the properties of the expansion tell us about the beginning and the age of the universe.

23-1
DOPPLER
SHIFTS OF
GALAXIES

For many years, it was thought that the universe was static. Aside from the motions of the planets and stars in their orbits, no motions on a large scale were expected, certainly not for the universe as a whole. Yet in about 1920, when the first velocities of galaxies were measured, it was found that the observable universe has motion in the form of rapid outward expansion.

From the shifts in their spectra (see Appendix D), it is now known that galaxies are moving with respect to one another with very high velocities. Even in the local neighborhood of our Galaxy, there are large motions, on the order of 300 to 400 kilometers per second (about 1,000,000 kilometers per hour).

All that can be measured with the Doppler effect is that component of the motion that is in the "line of sight," that is, in the direction that we are looking. In the cases of stars, we can do more, especially for stars that are near the sun. In addition to the Doppler shift, we can measure the velocity across the sky (perpendicular to the line of sight) by taking a photograph now and another one several years later and measuring the displacement. However, galaxies are so much farther away that to measure any such displacement for them would take millions of years. Therefore, all we know for galaxies is the velocity in the line of sight direction.

Looking farther and farther away at Doppler shifts of more and more distant galaxies, the spectrum lines move systematically farther and farther to the red. This effect is called the *red shift* of the galaxies. As red is a long wavelength, this trend means that the distant galaxies are moving away from us. At distances beyond about 10 million parsecs, all galaxies are moving away from us, and beyond this the velocities get greater with greater distances. At smaller distances, the individual ("peculiar") motions of the galaxies are larger than the systematic outward velocities, so that the red shift is difficult to detect.

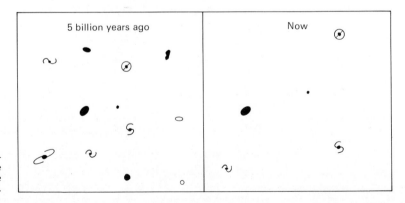

Fig. 23-1. In an expanding universe, only the distance between the galaxies grows.

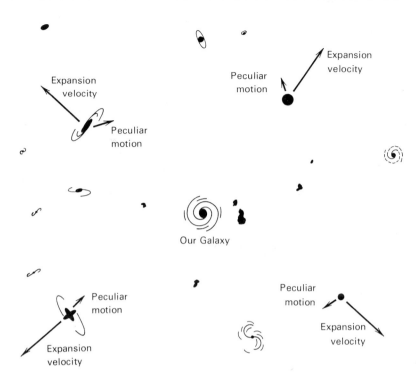

Fig. 23-2. Peculiar motions of galaxies are small and random in orientation.

Although accurate measurement of distances to galaxies is difficult beyond about 10 million parsecs, we can use some of the less accurate distance indicators to continue out farther. We find that the relationship is extremely good, and what exists is a law of nature—a close correlation between velocity and distance for galaxies.

For nearby objects, where astronomers can use the brightest stars and the variable stars for distances, they find that individual galaxies have velocities that are only a few hundred kilometers per second. For the distant clusters, where the brightest galaxy in a cluster is used as a distance criterion, very large velocities are observed, up to 100,000 kilometers per second or so. At the limit of present measurement capabilities the velocity becomes immense, almost as large as the velocity of light. The velocity-distance relationship, called *Hub-*

23-2
THE HUBBLE
LAW

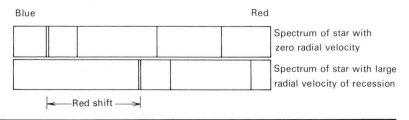

Fig. 23-3. How the radial velocity of a star is measured.

Fig. 23-4. The line-of-sight (radial) velocity is not necessarily the true velocity.

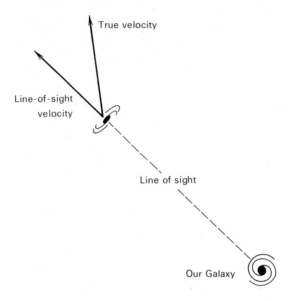

True velocity

Line-of-sight velocity

Line of sight

Our Galaxy

ble's law, is described by a very simple formula: The velocity is equal to the distance multiplied by a constant (abbreviated by *H*). This constant, called the *Hubble constant*, was derived for the first time by Edwin Hubble, astronomer at Mt. Wilson Observatory in California.

Calibration of the Hubble Constant

As an example, consider the large cluster of galaxies in the constellation of Virgo. We measure its distance by the methods discussed in Chap. 22 and find that it has a distance of 20 million parsecs, a value based on the luminosity of the brightest stars and the sizes of the H II regions in the individual galaxies in the cluster. The spectra of the Virgo cluster galaxies show a red shift from which we calculate a velocity away from us of 1,100 kilometers per second, on the average. Now, using the Virgo cluster, we can calibrate Hubble's law by determining what the value of *H* is, namely, about 55 kilometers per second per million parsecs.

Now we can apply Hubble's law to any other galaxy. Take, for instance, a cluster of galaxies that is so far away that even the 200-inch Palomar telescope, the largest in the world, has difficulty in obtaining spectra of the members. If we have no other measure of its distance, we can use Hubble's law to get one by measuring its Doppler shift. The lines on a spectrum of such a galaxy will be shifted some 1,000 Angstroms to the red, indicating a velocity of about 60,000 kilometers per second. The distance to this cluster therefore is one billion (10^9) parsecs.

Because the Doppler formula given above is really only the approximation of

Fig. 23-5. Two spiral galaxies in the Virgo cluster are among the many that can be used to find the cluster's distance. *(Lick Observatory.)*

a more complicated formula, if one observes much faster velocities than this, this simple version can no longer be used. Also, the geometry of the universe enters into the calculations, and the formulas become very much more complicated, and even to a certain extent unknown.

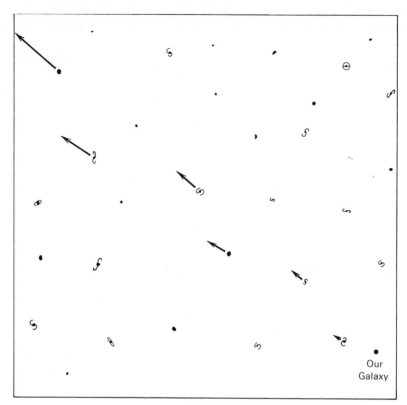

Fig. 23-6. The more distant a galaxy, the faster its recession.

Our
Galaxy

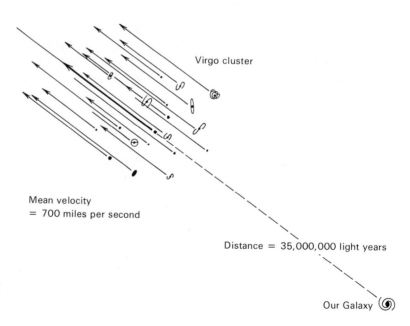

Virgo cluster

Mean velocity
= 700 miles per second

Distance = 35,000,000 light years

Fig. 23-7. The mean velocity of the galaxies in the Virgo cluster is a calibration of the Hubble constant.

Our Galaxy

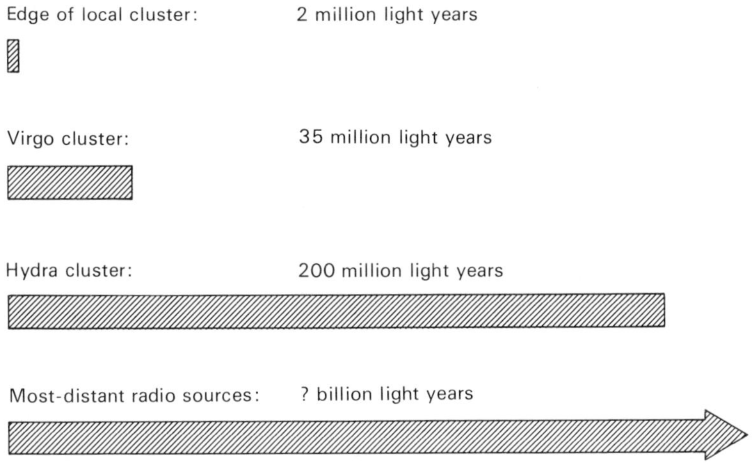

Edge of local cluster: 2 million light years

Virgo cluster: 35 million light years

Hydra cluster: 200 million light years

Most-distant radio sources: ? billion light years

Fig. 23-8 Distances of clusters of galaxies.

There are important observational limits to how far away astronomers can measure velocities to explore the general expansion of the universe. Among the farthest galaxies for which red shifts can be measured is the Hydra cluster, at a distance of about 1 billion parsecs. This is about the most distant cluster for which it has been possible to obtain spectra of *normal* galaxies. The reason that spectra of more distant ones cannot easily be obtained is that the night sky, even at the best observatory sites, has a certain luminosity of its own. It looks dark, but it is not completely black, and when an astronomer uses a very powerful telescope with a powerful spectrograph, and tries to photograph something extremely faint, the night sky appears brighter than this faint object, giving a spectrum of itself as well, and the spectrum of the galaxy is lost. For the largest telescopes, the night sky fogs the plate after about 5 to 10 hours, before it has reached much beyond about 1 billion parsecs.

An obvious solution is to get above the night sky, that is, above the upper atmosphere where the night sky light originates. The way to do this, of course, is to put a telescope out in space. For this work, a very large telescope on the order of 100 inches in diameter is needed. Smaller telescopes have been orbited, and plans to orbit a telescope of this size indicate that it may be done by about 1983.

The Most Distant Galaxies

A small percentage of galaxies is experiencing very unusual but violent events (Chap. 20). These explosions result in the presence of extremely hot gas clouds in the galaxies, which cause very bright emission lines in their spectra. Emission lines are easy to detect and measure, much more easy than absorption lines, so by looking at these peculiar emission-line galaxies, red shifts can be detected out to a distance at least twice the distance otherwise possible. The

Red Shifts of Radio Galaxies and Quasars

Fig. 23-9. The Pegasus cluster of galaxies. *(Hale Observatories.)*

most distant emission-line galaxy measured has a velocity of 120,000 kilometers per second, thus placing it at a distance of about 2 billion parsecs.

Though a tremendous distance, this is still not far enough to tell unambiguously about the geometry of the universe. It is only about half the velocity of

Fig. 23-10. The faint smudges at the center of this photograph are the brightest members of a very distant cluster of galaxies *(Lick Observatory.)*

Fig. 23-11. This famous group of galaxies pictures in the current redshift controversy. The spiral galaxy in the lower left has a far smaller redshift than the other four, leading to a conventional interpretation that puts it much closer, just a chance superposition. H. C. Arp and others, however, find evidence that suggests that all these galaxies are interacting, and that therefore the redshift is mysteriously *not* a true gauge of distance. *(Kitt Peak National Observatory.)*

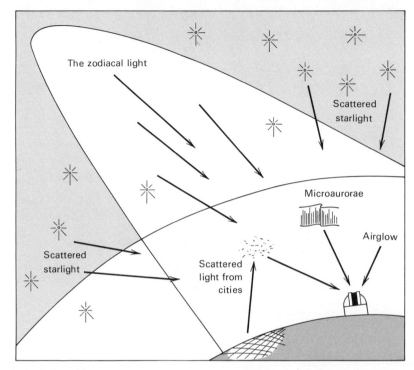

The zodiacal light

Scattered starlight

Microaurorae

Airglow

Scattered starlight

Scattered light from cities

light, and one would like to see eight-tenths or nine-tenths the velocity of light in order to solve the problem of cosmology. For these velocities, we must turn to the quasars.

Although most astronomers believe that quasars obey the Hubble relationship between velocity and distance, it is also possible that they are instead much closer objects that have high red shifts because of some heretofore unexplained phenomenon. At present the evidence, though still confusing and contradictory, favors the idea that the quasars are very distant objects, some of which are receding with enormous velocities. Therefore, quasars may be able

Fig. 23-13. The Coma cluster of galaxies is the nearest giant cluster. Almost all the images on this print are elliptical and SO galaxies in the cluster. *(Hale Observatories.)*

to extend knowledge of the universal expansion to vastly greater distances. To date, the largest velocity measured anywhere is for one of the quasars, something between 0.9 and 0.95 times the velocity of light. For this system, the red shift in the spectrum is over three times as big as the wavelength itself. Such objects are rather difficult to work on for the reason that we are looking at a part of the spectrum that we have never seen before, because it is normally in the invisible ultraviolet.

These quasistellar objects are possibly the most distant things we can see, and we still find that among the faintest ones are those that are moving away the fastest. This is an important fact because it indicates that the entire universe that can be measured and seen is in a state of expansion.

Questions that come to mind in thinking about an expanding universe include: What is it expanding from? Where is the center? And what caused the expansion? These are questions that cannot yet be answered. We do not know where the center is, and there may not be one. Even if a center exists, we may not be able to find it, because when there is a one-to-one correlation between

**23-3
THE NATURE
OF THE
EXPANSION**

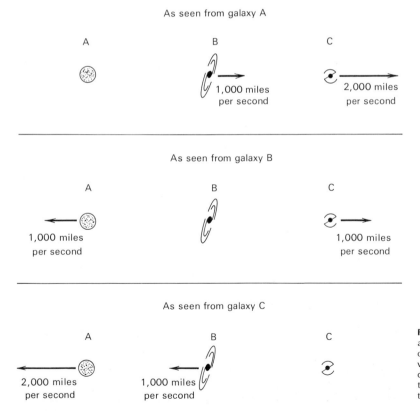

Fig. 23-14. When galaxies are receding from each other with an apparent velocity proportional to distance, it is impossible to determine which is at the center.

velocity and distance, one cannot detect a center. For example, consider three galaxies: ours; one "north" of us at, say, 1 million parsecs distance; and one "south" of us at 1 million parsecs. If the others were together at one time near our galaxy and are now moving away from us as the center point, we would see what we should do; both would have the same distance and the same velocity. But we would also see this if either the north one or the south one had been at the center, because then our galaxy would be moving away with the appropriate velocity, and the other noncentral one with twice that velocity.

The Center of the Universe

Astronomers are not even sure there is or was a center. If the universe is closed, if it exists in curved space with a positive curvature so that the universe is finite in volume, then there is no center. Perhaps it is clearer to say that the term "center" does not have a definition. The *surface* of the earth doesn't have a center, and it is a two-dimensional analog of three-dimensional curved space.

The Cause of the Expansion

What caused the expansion is also something we cannot answer now. In fact, we may never know what went on at the time of the beginning of the expansion, because all the clues may have been erased by the violence of the event itself. What happened before that may also be entirely unknowable. Perhaps nothing existed, not even space and time. In some cosmological theories, on the other hand, the universe is cyclic, varying from periods of high density followed by expansion, to periods of low density followed by contraction.

23-4 AGE OF THE UNIVERSE

It is possible to calculate the age of the universe from the expansion. From the expansion rate, one can extrapolate back to the time when all the galaxies were together at one point, calling that time the beginning of the universe. The

Measured ages

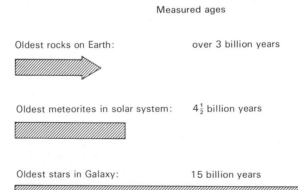

Oldest rocks on Earth: over 3 billion years

Oldest meteorites in solar system: $4\frac{1}{2}$ billion years

Oldest stars in Galaxy: 15 billion years

Fig. 23-15. A comparison of the ages of the earth, the solar system, and our Galaxy.

Billions of years ago

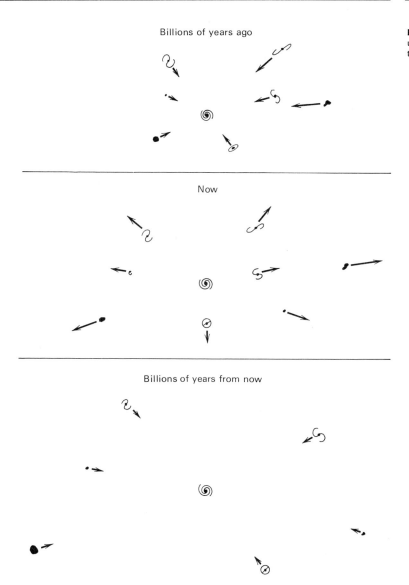

Now

Billions of years from now

Fig. 23-16. An oscillating universe seen at three times in its history.

time interval from that date to the present is simply the age of the universe. However, one first must know the geometry of the universe, and then must use the appropriate equations for the relationships between the cosmology and the properties of the universe. Because each cosmological model gives its own answer, we will not know the age of the universe until we discover which cosmological theory is correct.

The simplest model of all is the euclidean model of the universe, with flat space. For such a model, the age is just 2 divided by three times the Hubble constant, which gives about 11 billion years. This value is larger than the age of the earth (4 to 5 billion years), and age measurements of the meteorites show that the solar system itself has an age of about 4.6 billion years. It is thus not unreasonable to think that the universe as a whole should be somewhat older than these values.

One possible difficulty arises, though, if an age of only 11 billion years is accepted for the universe. When the ages of the oldest stars in our Galaxy are determined (Chap. 13), values of between 12 and 15 billion years are obtained, but this is more than the age of the universe. Apparently, then, some error or misinterpretation is present. Several possibilities exist. One way of avoiding this discrepancy might be to say that the universe is not euclidean. If we choose some other geometry and recalculate the age, it is possible to obtain values as large as 15 billion years for the age. Another possibility is that there is some error in the calculations of the ages of the oldest stars, an error yielding values that are somewhat too large. A third alternative is that the universe did not originate from a point, but is, in fact, oscillating. We just happen to live at a time when it is expanding, but it also contracts. A model that contracts to a volume that is not greatly smaller than its present size, when it reverses and expands again, can have stars that are much older than the measured contraction or expansion age. A fourth method of getting around the age difficulty is provided by the assumption that the universe is infinite in its history, and that galaxies are continuously formed to fill up the space left by those that are moving away. This is the so-called *steady state* universe, and it is one of the various cosmological models that will be described in the next chapter.

QUESTIONS

1. Why is the Doppler effect for galaxies usually referred to as the *red shift?* Why not the *blue shift?*
2. What is the distance to a cluster of galaxies that has a recessional velocity of 10,000 kilometers per second?
3. Why do astronomers find radio galaxies and quasars to have larger measurable red shifts than normal galaxies?
4. Explain why we do not seem to be able to locate the center from which the universe is expanding.

MODELS OF THE UNIVERSE

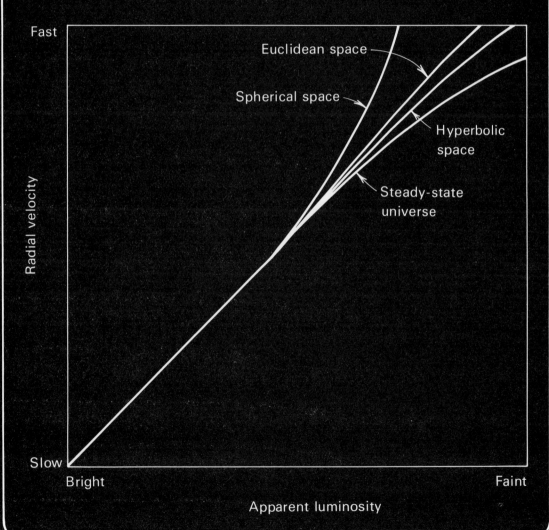

24-0
GOALS

This chapter explains some of the various models that cosmologists have worked out for the universe. Its aim is to make clear the ways in which what we see and measure allows us to zero in on the cosmological problem, the grandest unanswered question in all astronomy.

24-1
BASIC COSMOLOGICAL ASSUMPTIONS

Cosmology is the study of the nature of the universe as a whole. It involves the construction of hypothetical models of the universe and the testing of these models by comparison with observations. Because of the tremendously complicated nature of the universe, mathematics is the only language capable of describing it adequately. For that reason, most cosmologists are first and foremost mathematicians or mathematical physicists, and the approach that they take, of course, is a highly theoretical one. Furthermore, instead of trying to take into account all the known facts and every detail, they simplify the properties of the universe to the point where the universe can be handled reasonably well mathematically. Among these simplifications are a number of standard assumptions.

Homogeneity

First, there is the assumption that the universe is *homogeneous*, which means that the matter in the universe is spread out uniformly and that there are no large-scale clumps or empty spaces. In detail, of course, this is not strictly true. There are large-scale clumps such as the galaxies and clusters of galaxies. But the cosmologists usually take the point of view that they need not worry about these. If they consider a big enough volume, then it is very nearly homogeneous.

Isotropy

The second assumption is that the universe is *isotropic*. That means that the galaxies are distributed and are moving, whatever their velocities, in a smooth and uniform manner. All the velocities, for instance, are velocities of expansion. The universe near us sees the same pattern of velocities as at other places. So far measurements support this assumption, except for a relatively small effect caused by our Galaxy's peculiar motion among galaxies (see Sec. 24-3).

Incoherence

The third assumption is that the universe is *incoherent*, meaning that the different parts of the universe are independent of events in other parts, except for light and gravitation. It is not bound together by pressure like a solid or a gas, and the effect of any local change or perturbation is only felt locally. This idea is well supported by our knowledge of the very low matter densities prevailing in space.

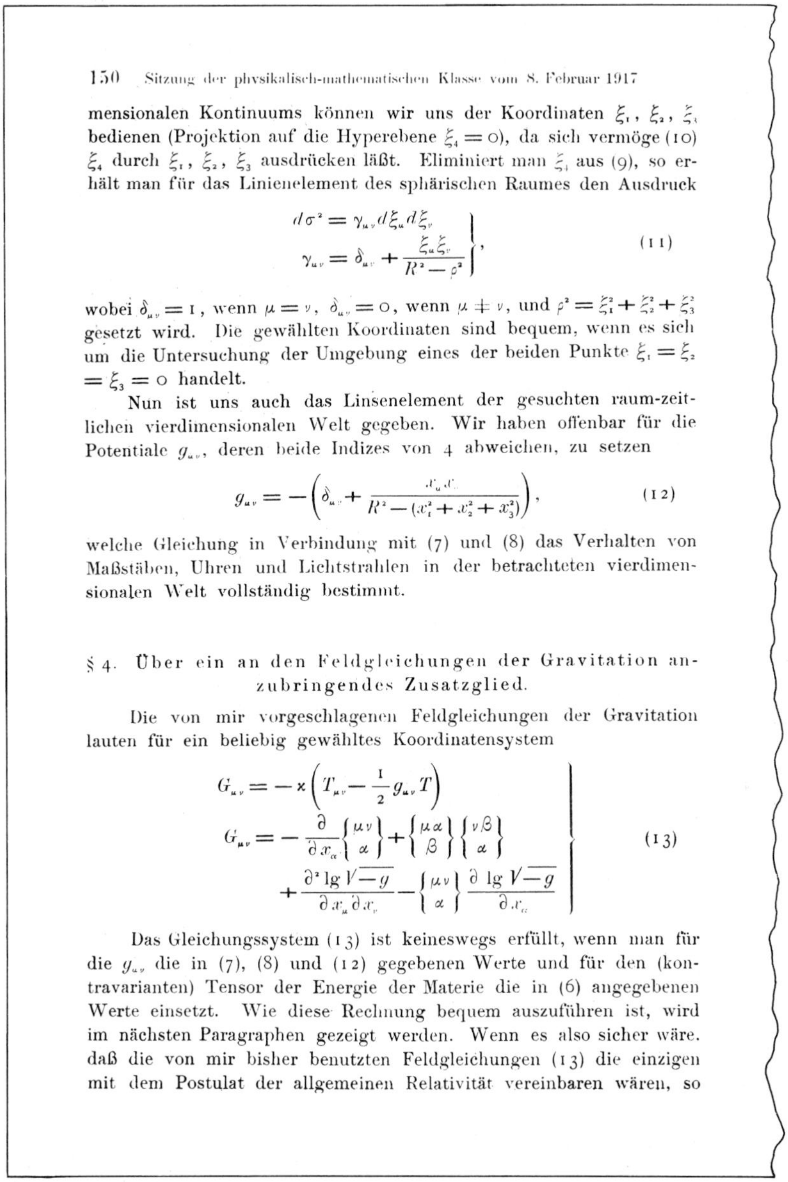

150 Sitzung der physikalisch-mathematischen Klasse vom 8. Februar 1917

mensionalen Kontinuums können wir uns der Koordinaten ξ_1, ξ_2, ξ_3 bedienen (Projektion auf die Hyperebene $\xi_4 = 0$), da sich vermöge (10) ξ_4 durch ξ_1, ξ_2, ξ_3 ausdrücken läßt. Eliminiert man ξ_4 aus (9), so erhält man für das Linienelement des sphärischen Raumes den Ausdruck

$$d\sigma^2 = \gamma_{\mu\nu} d\xi_\mu d\xi_\nu \left.\right\} $$
$$\gamma_{\mu\nu} = \delta_{\mu\nu} + \frac{\xi_\mu \xi_\nu}{R^2 - \rho^2} \left.\right\}, \qquad (11)$$

wobei $\delta_{\mu\nu} = 1$, wenn $\mu = \nu$, $\delta_{\mu\nu} = 0$, wenn $\mu \neq \nu$, und $\rho^2 = \xi_1^2 + \xi_2^2 + \xi_3^2$ gesetzt wird. Die gewählten Koordinaten sind bequem, wenn es sich um die Untersuchung der Umgebung eines der beiden Punkte $\xi_1 = \xi_2 = \xi_3 = 0$ handelt.

Nun ist uns auch das Linsenelement der gesuchten raum-zeitlichen vierdimensionalen Welt gegeben. Wir haben offenbar für die Potentiale $g_{\mu\nu}$, deren beide Indizes von 4 abweichen, zu setzen

$$g_{\mu\nu} = -\left(\delta_{\mu\nu} + \frac{x_\mu x_\nu}{R^2 - (x_1^2 + x_2^2 + x_3^2)} \right), \qquad (12)$$

welche Gleichung in Verbindung mit (7) und (8) das Verhalten von Maßstäben, Uhren und Lichtstrahlen in der betrachteten vierdimensionalen Welt vollständig bestimmt.

§ 4. Über ein an den Feldgleichungen der Gravitation anzubringendes Zusatzglied.

Die von mir vorgeschlagenen Feldgleichungen der Gravitation lauten für ein beliebig gewähltes Koordinatensystem

$$G_{\mu\nu} = -\varkappa \left(T_{\mu\nu} - \frac{1}{2} g_{\mu\nu} T \right)$$
$$G_{\mu\nu} = -\frac{\partial}{\partial x_\alpha} \begin{Bmatrix} \mu\nu \\ \alpha \end{Bmatrix} + \begin{Bmatrix} \mu\alpha \\ \beta \end{Bmatrix} \begin{Bmatrix} \nu\beta \\ \alpha \end{Bmatrix} \qquad (13)$$
$$+ \frac{\partial^2 \lg \sqrt{-g}}{\partial x_\mu \partial x_\nu} - \begin{Bmatrix} \mu\nu \\ \alpha \end{Bmatrix} \frac{\partial \lg \sqrt{-g}}{\partial x_\alpha}$$

Das Gleichungssystem (13) ist keineswegs erfüllt, wenn man für die $g_{\mu\nu}$ die in (7), (8) und (12) gegebenen Werte und für den (kontravarianten) Tensor der Energie der Materie die in (6) angegebenen Werte einsetzt. Wie diese Rechnung bequem auszuführen ist, wird im nächsten Paragraphen gezeigt werden. Wenn es also sicher wäre, daß die von mir bisher benutzten Feldgleichungen (13) die einzigen mit dem Postulat der allgemeinen Relativität vereinbaren wären, so

Fig. 24-1. A page from Einstein's paper (1917) describing his general relativistic cosmological model.

The fourth assumption is that of *uniformity*. It is assumed that the universe is uniform in its properties at extremely large distances; the galaxies are different in no fundamental way from those that we are measuring in our neighborhood. Only predictable differences such as the evolution of galaxies (which we can take into account) exist. This assumption is supported by the uniform physical and chemical properties found for galaxies wherever they are observed.

Uniformity

Fig. 24-2. Homogeneity.

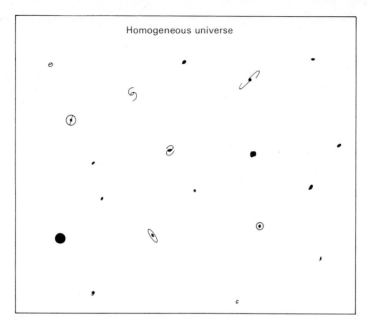

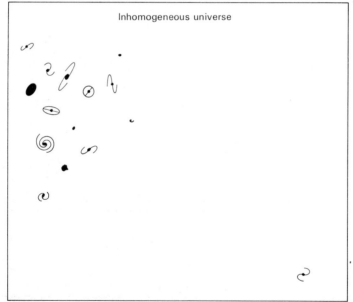

Universality   The fifth assumption is *universality*, which means that the laws of physics discovered here on the earth apply throughout the universe. Some cosmologists also assume that the laws of physics also apply to the universe as a unit, an even bigger assumption.

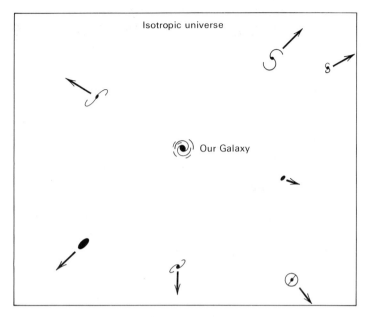

Fig. 24-3. Velocity isotropy.

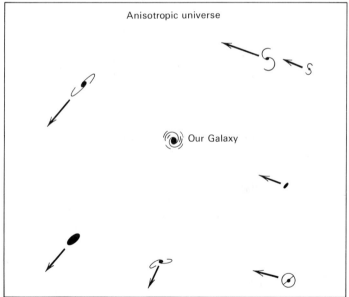

There have been many different cosmological models put forward in recent years by mathematically oriented astronomers and physicists, and they all involve something about the geometry. It was shown in Chap. 21 that many kinds of geometry are possible, and that the particular geometry that our universe possesses is not necessarily the plane geometry of Euclid. Different

**24-2
COSMOLOGICAL
MODELS**

Fig. 24-4. Coherence.

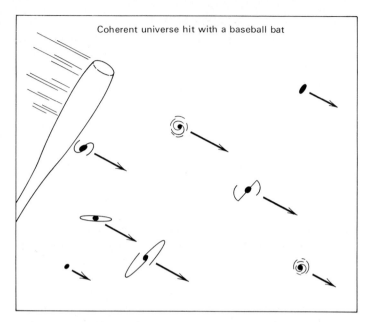

Coherent universe hit with a baseball bat

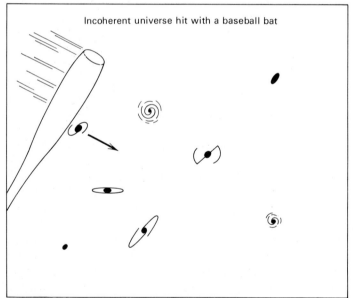

Incoherent universe hit with a baseball bat

cosmological models treat the problem of geometry differently. Some leave the question open, and some assume a certain geometry, requiring it to be correct if the cosmological model is correct.

In the following paragraphs, several modern models of the universe are described. Some of these are based on Newton's laws of physics, some are based on Einstein's general relativity, and others are based on still different laws of nature. At the present time, it is not yet possible to choose between the various models. However, it should be emphasized that each of these cosmological models has been built on a very firm and usually very complicated foundation of mathematical physics, and that the brief qualitative descriptions given here do not give a true idea of the nature of the theories. It is not enough in science simply to think about the universe and write down a description of some model of it. No model of the universe can be taken seriously by scientists until it has been completely worked out in mathematical and physical detail.

Newton was one of the first scientists to take what can be considered a scientific approach to the cosmological question. He attempted to apply his theory of gravitation to the entire universe, which he assumed to be euclidean, and found certain difficulties in achieving this. The worst problem was that the universe defined by the space containing matter must be infinite in extent, because if it is finite, then Newton's law of gravity would predict that it would immediately contract to a central point. It would not be a stable universe if finite in size.

Newtonian Cosmology

More recently, other difficulties of Newton's cosmological model have been pointed out. One is called *Olbers' paradox*, because the astronomer Olbers showed that if the universe is infinite in extent, the sky in all directions would not be dark, but would be extremely bright—either infinitely bright or at least as bright as the average surface brightness of a star. This would mean that in an infinite universe, all of the sky would have about the brightness of the disc of the sun. In such a universe, we would hardly even detect the sun, and there would be no night or day. It would be too bright and too hot on the earth for life as we know it. As this is obviously not the case, we might conclude that the universe is either not infinite or else is not homogeneous. Perhaps, one might suggest, the stars become scarcer or fainter at very great distances from the sun, or perhaps large amounts of dust absorb distant starlight. Neither of these suggestions is attractive to the cosmologist, because they necessitate assuming that the sun is in a very special or preferred location in the universe. Many lessons throughout history have taught us that making the assumption that we are at the center of the universe is dangerous.

Olbers' Paradox

Another point concerning Newton's cosmology is that it and subsequent newtonian models dealt only with static (nonmoving) universes. They did not and could not take into account the fact that the universe is expanding. The expansion of the universe was not known, of course, in Newton's time. But we now know that if the universe is not static, and is expanding as we know it is, then Olbers' paradox can be avoided. At very great distances, the galaxies will be moving at such high velocities that most of their radiation will not reach us as visible light, and therefore we will not expect a bright sky to result from them. However, even when Newton's cosmologies are revised to take into account the expanding nature of the universe, the results are not entirely satisfactory because they cannot account for the long-established effects of relativity.

Einstein's Model   In 1917 Einstein attempted to work out his first cosmological model based on general relativity. The red shifts of the galaxies were still not known, so he assumed a static universe without expansion. He found that the universe could not be infinite in extent, as Newton found, nor could it be finite in extent and surrounded by an empty infinite universe. Therefore, he suggested that space is not euclidean in nature, but is finite and positively curved. When he worked out the details of the mathematics, he found that if his ideas were correct, then the radius of the curved finite space can immediately be found from the density of the universe. Astronomers have attempted to measure the density of the universe by measuring masses of galaxies and then by counting them and measuring their separations. From the best estimates presently available of the density of the universe, the radius calculated using Einstein's equations would turn out to be about 20 billion (2×10^{10}) parsecs.

Relativistic Expanding Models   After the discovery of the expansion of the universe, Einstein and many others subsequently attempted to develop a relativistic cosmological model with expansion. The most general type of expanding relativistic models of the universe differ primarily in their curvature, so that if the assumptions underlying them are correct, the main thing that observations need to determine is the geometry of the universe. Generally speaking, most cosmologists today believe that one of these expanding, general relativistic models is probably correct, and that the curvature of the universe is the chief unknown quantity.

However, there are some doubts about the initial assumptions outlined in the beginning of this chapter, and several recent models have been proposed that reject the assumption, for instance, of isotropy. In these models, for example, the universe might be rotating and expanding. Complications like this are very difficult to treat exactly, and there is no complete agreement among cosmologists about the implications of properties such as the possible rotation of the universe, although it can be shown from the measured anisotropy of the background radiation (Sec. 24-3) that any rotation must be very slow.

Steady-State Models   A very different idea was proposed about 20 years ago by cosmologists Bondi and Gold, who rejected the entire approach taken by their predecessors in

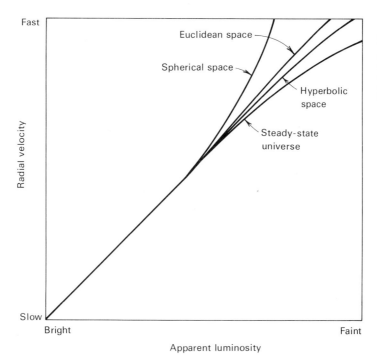

Fast

Euclidean space

Spherical space

Hyperbolic space

Steady-state universe

Radial velocity

Slow

Bright

Faint

Apparent luminosity

Fig. 24-5. When the radial velocities and the distances of very distant galaxies can be determined independently of each other, we can find out what kind of space we live in.

cosmology. They hypothesized that an overriding principle should take precedence over all other principles of physics, and they called this the *perfect cosmological principle*. It leads to the conclusion that the universe and its large-scale nature are exactly the same everywhere *and at all times.* The implications of the perfect cosmological principle lead to the model of the universe called the *steady-state cosmology,* in which the universe is infinite in extent and age. As we know from measurements, the universe expands, so that in order for its density not to change as it expands, Bondi and Gold, and later Hoyle, proposed that matter is continuously created in the universe to fill up the space left by the expansion. This matter is created out of nothing, and although it is created on an extremely slow scale, much too sparsely for us ever to detect, many scientists object to the idea of creation out of nothing because of the way it violates physical principles involving conservation.

Other Models

In addition to the steady-state model, there have been many other exotic ideas that have led to models of the universe. For instance, some scientists have suggested that the red shifts of galaxies are not Doppler shifts and that the universe is not truly expanding. These scientists have had to hypothesize a completely new and unspecified effect which mimics the Doppler shift, and the lack of any confirmation of such an effect makes it difficult to accept this type of cosmology without considerable reservation. Other scientists have suggested

Fig. 24-6. The steady-state universe, though expanding, always has the same density of galaxies because new galaxies are formed out of nothing.

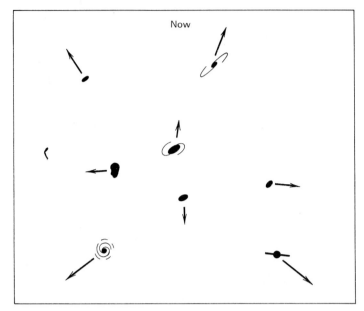

Now

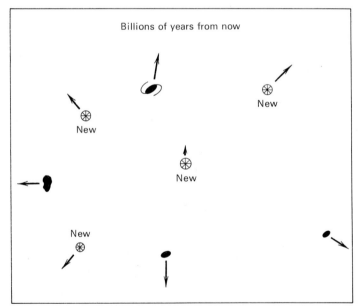

Billions of years from now

New

New

New

New

that the universe might be decipherable by looking more closely at the various constants of physics and by exploring the nature of the nucleus and the atom. It is found, for instance, that certain fundamental properties of elementary particles, like the electron and the proton, can be related in a way that suggests a close numerical relationship with the properties of the universe as a whole.

Although such juggling with physical parameters is very suggestive, its significance is not generally understood or agreed upon.

Perhaps the most important discovery in cosmology in recent years is that of the so-called *background radiation* at radio wavelengths. This has been shown to very probably be radiation that is "left over" from a time when the universe was very small and very hot, billions of years ago. The background radiation can best be explained as greatly red-shifted light rays that were produced shortly after the universe began (the *big bang*) and that are now so far away, so diffuse, and so rapidly receding that we can detect them only at radio and infra-red wavelengths. If this interpretation is correct, then the steady-state theory is impossible.

**24-3
THE COSMIC
BACKGROUND
RADIATION**

Discovered in 1965, the cosmic background radiation has now been studied so thoroughly that there is little doubt of its identity. The blackbody spectrum has an equivalent temperature of 2.76K, which is in excellent agreement with big-bang cosmological models. In 1977 further experiments showed that the radiation is *anisotropic*, that is, is slightly different in one direction from the other, which amounts to a temperature difference of about 3×10^{-3} K. This anisotropy can be explained if the solar system is moving slowly (cosmically speaking) with respect to the rest of the universe. The earth's velocity is found to be about 400 kilometers per second in the direction of the constellation Leo. A further fact that can be gleaned from these measurements is that the universe cannot be rotating very rapidly; any rotation must be less than one-billionth of a second of arc per century.

In spite of a great deal of effort on the part of astronomers to check all of the many different cosmological models, it is still true that we do not know which is correct, if any. It is obvious that the problem of the nature of the universe is one of the most difficult in all science, and that it will be many years before it is solved.

1. What are the differences among homogeneity, isotropy, uniformity, and universality?
2. Can you think of possible assumptions about the universe that are different from those usually made by cosmologists?
3. Why did Einstein's 1917 cosmological model not include the expansion of the universe?
4. To what extent is Einstein's method of measuring the radius of the universe dependent upon the value of the Hubble constant?
5. For the surface of a sphere, a two-dimension analog to curved space, what corresponds to the Einstein radius?
6. Why does the cosmic background radiation (if we are interpreting it correctly) rule out the steady-state theory of the universe?

QUESTIONS

Appendix A

GLOSSARY

ablation. Loss of material from a meteorite as it passes through the atmosphere. Includes such processes as melting and fragmentation.

absolute luminosity. A measure of the true brightness of an object. Determined as the amount of energy emitted per unit time.

absolute magnitude. The real brightness of a star, measured in units of magnitudes and defined as the apparent magnitude a star would have if its distance were 10 parsecs.

absolute zero. The lowest possible temperature that anything can have, when almost all molecular motion has stopped. It is -273°C or 0 K.

absorption spectrum. A spectrum of a bright object in front of which is a cooler gas, which produces dark lines across the continuous spectrum of the object.

acceleration. A change in velocity, such as that produced by gravitation or by another force acting on a body.

achromatic. Without color, such as a lens for which chromatic aberration has been corrected.

albedo. The amount of light reflected by an object. Usually defined as the ratio of incident to reflected light.

almanac. A collection of tables of astronomical data, especially about day-to-day changes in the positions of heavenly bodies.

altitude. The angular distance of an object measured perpendicularly above the horizon.

amplitude. The amount of change, for example in velocity for a binary star, or in light from a pulsating star.

angstrom (Å). A unit of length used in spectroscopy. One angstrom is 10^{-8} centimeters.

angular size. The extent of an object. Measured by the apparent angle that it subtends as seen by an observer. Its true size can be derived from its angular size if its distance is known.

angular momentum. The amount of mass and motion involved in a body's rotation or revolution. For a body in a circular orbit, it is the product of its mass, its velocity, and the radius of its orbit.

annular eclipse. An eclipse of one object by another in which the eclipsing body does not quite cover the other body entirely. Usually refers to eclipses of the sun.

antimatter. Material made up of particles that have the exact opposite properties of ordinary particles, e.g. antiprotons, antineutrons, and positrons.

aphelion. The point in an object's orbit about the sun where its distance from the sun is at its maximum value.

apogee. The point in a satellite's orbit about the earth where its distance from the earth is at its maximum value.

apparent magnitude. The brightness of a star or other object as seen from the earth and measured in units of magnitudes.

association. An assemblage of stars of common age, usually very young and not gravitationally bound as in a star cluster.

asteroid. One of thousands of small, solid bodies with planetlike orbits in the solar system. Most lie between Mars and Jupiter.

astrology. A nonscientific method of divining the past and future, which uses mystical interpretations of the apparent motions of the planets, moon, and stars.

astrometric binary. A double star in which only one star is observed; its companion is detected by the periodic motion it causes the brighter star to have as they orbit each other.

astrometry. The study of the positions and motions of the stars and planets.

astronomical unit (AU). The mean distance of the earth from the sun (1.495985×10^8 kilometers). Used as a unit of distance in the solar system.

astronomy. The study of the physical universe, including the galaxies, stars, planets, and other cosmic bodies, their nature, their origins, and their evolution.

astrophysics. That part of astronomy that deals with the application of physics to astronomical problems. Almost all current astronomy could be so termed.

atom. The part of an element, consisting of a nucleus surrounded by electrons, that is the smallest division possible without the element's losing its identity as such.

atomic number. The number of protons in the atom of a particular element.

atomic transition. The process of an atom's gaining or losing energy by emission or absorption of light or by collision with another particle.

atomic weight. The combined number of protons and neutrons in an atom of a particular element.

aurora. The "Northern (or Southern) Lights" caused by collision of particles from the sun with the earth's upper atmosphere.

autumnal equinox. The place where the ecliptic intersects the celestial equator, with the sun moving from north to south. Happens on or about September 21.

azimuth. The angle, measured eastward along the horizon from due north, which together with altitude defines an object's position in the sky.

Balmer lines. The lines of atomic hydrogen that appear in the visible part of the spectrum. They arise from atomic transitions from or to the second energy level of the electron.

bands. Spectroscopically, closely spaced series of lines in the spectrum, usually of molecules or radicals.

barred spiral. A galaxy with its central structure dominated by a bar-shaped linear feature.

big bang. The beginning of the expansion of the universe. Thought to have been a giant explosion involving all cosmic matter.

binary. A double star, with the two components revolving around each other.

blackbody. An object, most nearly approximated in real life by a dull black closed box, that absorbs then reemits all radiation that falls on it.

black hole. An object that has collapsed to such a high-density state that no radiation or anything else can escape from it.

Bohr atom. An early model of the atom that pictures the electrons as having circular orbits (of different energy) around the atomic nucleus.

bolide. An extremely bright meteor. Usually accompanied by a loud report of sound waves.

bolometric magnitude. The magnitude of an object that takes into account its brightness at all wavelengths.

brightness temperature. The temperature of a body calculated by comparing its brightness to that of a theoretical blackbody.

carbon cycle. A series of nuclear reactions that converts hydrogen to helium, by using carbon along the way to help.

cassegrain. Referring to a telescope that uses a second small and convex mirror to focus the light near (usually behind) the primary mirror.

Cassini's division. A conspicuous gap in Saturn's rings.

cD galaxy. A large, diffuse galaxy, often the largest in a cluster and a radio source.

celestial equator. The projection onto the sky of the earth's equator.

celestial mechanics. Application of the law of gravity to the motions of the planets and stars.

celestial navigation. Use of the positions of the stars, sun, and moon, along with an almanac, to find your position on earth or in space.

celestial sphere. The apparent sphere of the heavens. Considered at infinite distance and centered on the eye of the beholder.

center of gravity. Same as the center of mass.

center of mass. The point between two objects (or the midpoint of more objects), around which they move.

centrifugal force. Not really a force, but a concept introduced to describe the fact that a body moving along a curved path appears to experience a force outward from the center of curvature.

centripetal force. That force that causes a body to move on a curved path instead of in its natural straight-line direction.

Cepheid variable. A pulsating star with a period of variability of between about 1 and 100 days.

chromatic aberration. The separation of colors of light that passes through a glass lens, causing the image to be focused at different positions for different colors.

chromosphere. The atmosphere of the sun just above the visible solar surface.

chronometer. A very accurate clock.

circumpolar. That part of the heavens near the celestial poles that never sets for a given observer on the earth.

cluster. A group, in astronomy usually referring to a gravitationally stable group of stars or galaxies.

coelostat. A system of moving mirrors that allows a telescope to remain stationary as the sky turns.

collimator. A lens in an optical system that changes the light beam to make the light rays parallel.

color excess. The amount of reddening of light that is caused by dust between object and observer.

color index. A measure of color. Defined as the difference between the magnitudes of an object measured at different wavelengths.

color-magnitude (CM) diagram. The collection of magnitudes and colors of a group of stars (a cluster, association, galaxy, etc.)

color temperature. The measure of the temperature of an object taken as the temperature of a blackbody of the same color.

coma. Distortion of the image shape in a reflecting telescope, away from the center of the observed field. Caused by the parabolic shape of the mirror. Also used to refer to the bright envelope of gas around the nucleus of a comet.

comet. A solar-system object, made up primarily of icy substances and considerably smaller than a planet and larger than a meteoroid.

commensurability. Relating to a simple mathematical relation between periods or other characteristics of two or more orbits.

compound. A material that contains two or more elements chemically bound together.

conjunction. The situation in which two celestial bodies are lined up (with the same celestial longitude) as seen from the earth.

constellation. An arrangement of stars that appear near each other in the sky. Usually given a name by the ancients.

continental drift. The gradual motion of the earth's continents over its surface.

continuous spectrum. The light emitted by an object, including all wavelengths without gaps.

convection. The means of transferring energy from a hot place to a cold place by movement of heated material in a current to the cold place, where the energy is deposited, and then back to the hot place.

coriolis effect. The apparent deflection of an object, moving about the surface of the earth, caused by the earth's rotational motion under it.

corona. The extremely hot outer atmosphere of the sun.

coronagraph. A specially designed telescope for photographing the corona.

cosmic rays. High-energy particles from the sun and beyond. Mostly protons accelerated to very high speeds.

cosmogony. The study of the origin of the solar system, or sometimes of the universe.

coudé. A focus position of a telescope that involves several mirrors to render the image stationary at some point in the telescope building.

crater. Any depression of circular outline. Caused by volcanism, impact, or some other process.

crescent. The phase of the moon occurring when the moon is closer to the sun than is the earth.

dark nebula. A dust cloud between the stars that obscures more distant stars.

deceleration. Change in velocity from higher to lower.

declination. The coordinate on the celestial sphere that measures the distance of an object, north or south from the celestial equator.

degenerate gas. Gas under such high pressure that all the energy states of the atoms are filled by electrons.

density. The concentration of matter in an object. Measured by the ratio of its mass to its volume.

density-wave theory. A possible mechanism for explaining the spiral arms of a galaxy.

diffraction. The wavelike behavior of light as it passes the edge of an obscuring body, wherein the light is spread out.

diffuse nebula. A cloud of gas or dust that is illuminated by starlight.

dispersion. In spectroscopy, the degree of spreading out of the different wavelengths.

distance modulus. A measure of the distance to an object, given as the difference between its apparent and its absolute magnitude.

diurnal. Relating to one day or the daily rotation of the earth.

Doppler shift. The change in apparent wavelength or frequency of a source moving with respect to the observer.

dwarf. In astronomy, a star on the main sequence.

dynamical parallax. A measure of the distance of a binary star based on the mass-luminosity relation and a determination of its orbit.

eccentricity. A measure of how flat an ellipse is. Defined as the ratio of the distance between the foci to the length of the major axis.

eclipse. Any obscuration of a body that is caused by another body's moving into place between the first body and the observer.

ecliptic. The path of the sun in the sky; the ecliptic plane is the plane of the earth's orbit.

effective temperature. The temperature of a body as determined by the color of its peak radiation.

electromagnetic radiation. All lightlike radiation, ranging in wavelength from gamma rays to radio waves.

electron. A subatomic particle that inhabits the outer parts of an atom and carries a negative charge.

element. The simplest substance that has unique chemical properties; it has a single atomic nucleus and is surrounded by a cloud of electrons.

ellipse. A curve formed by cutting completely through a circular cone at some angle to the cone's axis.

elliptical galaxy. A galaxy with regular elliptical contours and smooth outward-decreasing luminosity gradient.

elongation. The angular separation between a planet and the sun in the sky. Measured along the ecliptic.

emission line. A bright line of a particular color in the spectrum of an incandescent gas.

emission nebula. A gas cloud with its light being given off as emission lines.

ephemeris. A collection of tables of astronomical data giving positions of various bodies at various times.

epoch. A date chosen for reference in organizing observations, as for variable stars.

equation of state. The relationship between the temperature, density, and pressure for a substance.

equation of time. The difference (in time) between the apparent position of the sun and the position it would have were the earth's orbit a perfect circle.

equator. The great circle that is 90° from the poles of a rotating object.

equatorial mounting. A telescope mounting for which one axis is parallel to the earth's axis.

equinox. Either of the intersections between the ecliptic and the celestial equator.

equivalent width. A measure of the strength of a spectral line. Given in terms of the width of a hypothetical, equally strong absorption line that has zero intensity over its whole width.

erg. A small metric unit of energy that is equal to the work done by a force required to accelerate one gram by an amount of 1 centimeter per second per second through a distance of 1 centimeter.

ether. The hypothetical all-pervading medium of space through which light was once thought to be transported as a wave.

euclidean. With regard to space, indicating a geometry that conforms to Euclid's axioms; this is so-called flat space.

event horizon. A surface surrounding a collapsed black hole, and marking the position where its velocity of escape equalled the velocity of light.

excitation. The acquiring of energy by atoms so that their electrons move to higher energy levels than the least-energy state.

extinction. The reduction of intensity of light from a source that results from intervening gas or dust.

extragalactic. Beyond the edges of the Milky Way galaxy.

f ratio. The ratio of a telescope's (or camera's) focal length to its aperture, giving its speed.

faculae. Brighter than average regions seen near the edge of the sun.

fireball. An unusually luminous meteor. In cosmology, the explosion of light associated with the big bang.

flare. A brief explosive event on the sun.

flare star. A kind of star whose brightness suddenly increases briefly and unpredictably.

flocculi. Bright regions seen on the sun in monochromatic light; more commonly called *plages.*

fluorescence. The process involving absorption of light of one wavelength and then reemission at a different wavelength.

flux. The amount of energy through a surface of unit area per unit time.

flux unit. A measure of the flux of radiation, equal to 10^{-26} watts per square meter per Hertz.

focal length. The distance between a lens or mirror and the point where all the light rays come together.

focus. The point where the light rays, which are bent by a lens or mirror, come together.

forbidden lines. Spectrum lines that result from electron transitions that are very improbable.

force. Agent responsible for the change in momentum of a body.

Foucault pendulum. A large pendulum that is capable of demonstrating the turning of the earth.

Fraunhofer lines. The more conspicuous lines of the solar spectrum.

frequency. The number of vibrations per unit time.

fusion. The building of heavier atoms by joining lighter atoms together.

galactic cluster. An open cluster in our Galaxy.

galactic coordinates. A grid of positions with its equator along the Milky Way.

galaxy. A giant collection of stars that is gravitationally independent of other objects and usually separated from its neighbors by millions of light years.

Galaxy. When capitalized, this word refers to our own galaxy, the Milky Way.

gamma rays. Very short wavelength electromagnetic radiation.

gauss. A unit of strength of a magnetic field.

gegenschein. A faint patch of light in the sky 180° from the sun's position and resulting from the reflection of sunlight from dust particles.

geodesic. A path through space-time.

giant star. Any luminous star, brighter than a main-sequence star of the same temperature.

gibbous phase. The phase of the moon exhibited when the moon lies farther from the sun than does the earth.

globular cluster. A large assemblage of about a million stars in a compact cluster. The stars are uniformly very old.

globule. A small very dark nebula.

granulation. The grainy structure of the sun's visible surface.

gravitation. The force that causes matter to attract matter.

gravitational redshift. The shift in wavelength of light that is caused by a strong gravitational field.

great circle. The path on a sphere that is traced by its intersection with a plane which has passed through its center.

greatest elongation. The largest apparent distance in the sky along the ecliptic between the sun and an inferior planet.

Greenwich. The observatory site in England through which the 0° longitude line on the earth passes.

Gregorian calendar. The calendar now in use and introduced in 1582 by Pope Gregory XIII.

H I. Neutral hydrogen.

H II. Ionized hydrogen.

H II region. An interstellar cloud that contains ionized and excited hydrogen.

hadron. A class of subnuclear particles with masses approximately equal to that of the proton.

half-life. The time required for half of the radioactive atoms of a material to disintegrate.

halo. For a galaxy, the outer spherical component of stars or gas.

Hayashi line. The path on the temperature-luminosity diagram of a completely convective star.

heliocentric. Centered on the sun.

helium flash. The sudden onset of helium burning in the core of a red-giant star.

Hertz. A unit of frequency, equal to one cycle per second.

Hertzsprung gap. The area above the upper main sequence where stars are not found.

Hertzsprung-Russell diagram. A plot of absolute magnitude versus spectral type for stars.

high-velocity star. A star whose velocity with respect to the sun is high, and which therefore has a different kind of motion around the Galactic nucleus.

horizontal branch. A part of the CM diagram of globular clusters lying at absolute magnitude about 0.7.

hour angle. A great circle passing through an object and the poles of the equatorial coordinate system.

Hubble constant. The multiplying constant that relates the distance of a galaxy to its recessional velocity; it is abbreviated as H and is about 55 km per second per megaparsec.

Hubble law. The proportionality between galaxies' distances and their velocities of recession.

hydrostatic equilibrium. The balance between the internal pressures in a substance and the weight of its various layers.

hypothesis. A tentative idea that explains certain observations and that is subject to further verification by more experimental comparison.

IC. The Index Catalog, a supplement to the NGC.

image. The optical formation of a representation of an object by a lens or mirror.

image tube. An electronic device for forming images greatly amplifed in brightness.

inclination. The angle between the plane of an orbit and some reference frame, such as the earth's orbit.

inferior conjunction. A conjunction with an inferior planet in which the planet is on our side of the sun.

inferior planet. A planet whose orbit lies inside that of the earth.

infrared. Light of wavelength just longer than the longest visible radiation.

interferometer. An optical or radio instrument that uses wave interference to increase the quality of information gained from the source.

interplanetary medium. The gas and dust between the planets.

interstellar medium. The gas and dust between the stars.

interstellar lines. Spectral lines in a star's spectrum that are produced by the interstellar medium between us and the star.

ion. An atom that has lost or gained one or more electrons and is electrically charged.

ionization. The process of forming an ion from an atom.

ionosphere. The upper atmosphere of the earth or other planet, where many of the atoms are ionized.

isotope. Any of the forms of an element that differ only in the number of neutrons in the atom's nucleus.

isotropic. Having properties that do not depend on direction.

Jovian. Having to do with Jupiter or the other outer planets.

Julian Day. The number of days elapsed since January 1, 4713 B.C.

Kepler's laws. Three laws of planetary motion derived by Kepler from extensive observations that were made largely by Tycho Brahe.

kiloparsec. Abbreviated kpc, 1,000 parsecs.

kinetic energy. That part of the energy of a body that is associated with its motion.

kinetic temperature. A measure of the temperature determined from the velocity distribution (for a gas).

Kirkwood's gaps. Gaps in the arrangement of bodies in the asteroid belt. Caused by commensurabilities with Jupiter's orbit.

Lagrangian points. Five places in a two-body system where the forces on a tiny test mass are zero.

laser. An instrument that amplifies a light signal of monochromatic radiation and gives a coherent beam of light of great intensity.

latitude. The angular distance from the equator to the poles. Measured perpendicular to the equator.

law. A simple statement of the behavior of some aspect of the physical world.

lepton. A class of subatomic particles of small mass, including the electron.

light curve. A graphic display of the variation in time of the brightness of a star or other object.

light year. The distance that light travels in one year (about 10^{13} kilometer).

limb. The edge of the moon, planet or sun as seen from earth.

limb darkening. The darkening of an object at the limb due to absorption of light in its atmosphere.

line broadening. The effect on spectrum lines that is caused by various physical processes, which broaden the lines.

line profile. The shape of the intensity curve of a line. Plotted against wavelength.

local group. The family of about twenty-five nearby galaxies, including our own.

long-period variable. Cool giant stars that show large brightness variations with periods longer than seventy five days.

luminosity class. Classification of the intrinsic brightness of a star, based on its spectrum.

luminosity function. A curve or table that tells how many stars of each absolute luminosity there are.

Lyman lines. A series of spectrum lines of hydrogen, like the Balmer series, except at far ultraviolet wavelengths.

Magellanic Clouds. Two irregular galaxies near the Milky Way Galaxy.

magnetosphere. The envelope around a planet or other body that encloses its detectable magnetic field.

magnification. The power of a telescope to magnify an apparent image, as compared to the unaided eye's view.

magnitude. A measure of the brightness of an object that uses a logarithmic scale.

main sequence. The sequence of stars in the CM diagram that contains most stars and that runs from hot bright stars to cool faint ones.

major axis. The length of the maximum diameter of an elliptical orbit.

major planet. One of the larger planets, which include Jupiter, Saturn, Uranus, and Neptune.

mare. Dark, flat lunar basins and similar features on Mars and Mercury.

mascons. Mass concentrations under the surface of certain mare on the moon.

maser. An instrument that amplifies microwave radio radiation of a given frequency into a coherent beam.

mass. The measure of the amount of matter in an object as determined by its gravitational field.

mass-luminosity relation. The relationship between the mass and the luminosity of stars, which holds for most stars.

mean solar time. The local hour angle of the mean sun.

mean sun. A hypothetical sun that moves uniformly with respect to the celestial equator during the year.

megaparsec (Mpc). One million parsecs.

meridian. In the celestial sphere, a great circle that passes through the celestial poles and the observer's zenith; on the earth, a great circle that passes through the poles and the observer.

Messier catalog. A list of 107 nebulae and clusters. Made in the eighteenth century by Charles Messier.

meteor. The streak of light observed when a meteoroid passes into the earth's atmosphere.

meteor shower. A large number of meteors made by meteoroids with similar orbits, all striking the earth at about the same time.

meteorite. The part, if any, of a meteoroid that survives to land on the ground.

meteoroid. A small body in space.

micrometeorite. A small meteorite that has been able, because of its small size, to fall to the ground unablated.

microwave. Short-wavelength radio waves, with wavelengths from about 1 millimeter to 30 centimeter.

Milky Way. The bright band of stars that marks the plane of the Galaxy.

minor axis. The length of the minimum diameter of an elliptical orbit.

minor planet. An asteroid.

Mira variable. A long-period variable.

molecule. Two or more atoms bound together to form a chemically distinct substance.

momentum. The state of motion of a body, calculated as the product of its mass and its velocity.

monochromatic. Pertaining to one color or one wavelength.

N galaxy. A galaxy with an unusually prominent nucleus.

nadir. The point directly opposite from the zenith in the sky.

nebula. A diffuse cloud of gas or dust.

neutrino. A very small subatomic particle with no charge and no rest mass.

neutron. A subatomic particle with no charge and with its mass nearly that of the proton.

neutron star. A collapsed star so dense that all its matter is squeezed into the form of neutrons.

new moon. The moon's phase when it is between the earth and the sun.

Newtonian focus. In a reflecting telescope, the arrangement in which the image is deflected to the side of the telescope by a flat mirror.

NGC. The New General Catalog, a listing of over 7,000 star clusters, nebulae, and galaxies.

node. The intersection of the plane of an object's orbit with a reference plane.

north celestial pole. The extension onto the sky of the earth's north polar axis.

nova. A star that suddenly brightens by several magnitudes and then slowly declines.

nucleus. Referring to the central part of an atom, a comet, or a galaxy.

nutation. The movement in direction of the earth's axis due to tidal action by the moon on its equatorial bulge.

O-association. A stellar association with stars of type O.

objective. The main lens or mirror of a telescope.

objective prism. A large prism that fits over the objective of a telescope to give spectral images for each star in the field.

oblateness. The degree of flattening of a nearly spherical body.

obliquity of the ecliptic. The angle between the poles of rotation of the earth and the poles of its orbit.

occular. The eyepiece of a telescope.

occultation. The eclipse of a smaller body by a larger one.

opacity. The degree of opaqueness to light; the ability of a material to impede the passage of light.

open cluster. A relatively small and loosely structured cluster of stars, found in the Milky Way plane or similarly placed in another galaxy.

opposition. The position of a planet that is opposite the sun as seen from the earth at a particular time.

optical depth. The degree to which light has been absorbed in a given amount of material.

optical double. A double star that is only apparently double; the two stars are seen in front of one another, but actually are far apart.

orbit. The path of one object around another, or around a point, due to gravitational force.

parabola. A conic section formed by cutting a circular cone with a plane that is parallel to the side of the cone.

parallax. The apparent back-and-forth motion of a distant object resulting from the real motion of the observer, as in the orbital motion of the earth.

parsec. A unit of distance that is defined as the distance to a star that has a parallax of 1 second of arc.

peculiar velocity. The velocity of an object that is different from the mean velocity of its neighbors.

penumbra. That part of a body's shadow in which only part of the sun's light is occulted; the outer part of a sun spot.

perigee. The point in a satellite's orbit that is nearest the earth.

perihelion. The point in a body's orbit around the sun that is nearest the sun.

period-luminosity relation. The relation between the period of light variation and the absolute luminosity of a variable star.

perturbation. The gravitational effect of a body on another that distorts its orbit, as around the sun.

phases. The changes in the amount of illumination for the visible side of a body such as the moon or an inferior planet.

photometry. Measurements of light intensity and/or color.

photon. A tiny discrete unit of radiation.

photosphere. The visible surface of the sun.

plage. A brighter-than-average region on the sun seen in monochromatic light.

Planck's constant. A number that relates the energy of a photon to its frequency.

planetary nebula. An ejected shell of gas surrounding a highly evolved star.

plasma. A gas whose particles are ionized.

polar axis. The line around which a body rotates; in a telescope, that axis that is parallel to the earth's polar axis.

polarization. The separation of the various alignments of the planes of vibration of light waves.

Population I and II. The two broad divisions of types of stars in our Galaxy.

position angle The orientation of a line in the sky, such as that connecting the components of a binary. Measured as the angle from the north, taken starting first toward the east.

positron. An antielectron, like an electron except that it is positively charged.

potential energy. The energy inherent in a configuration. Can be released in another form of energy.

Poynting-Robertson effect. The process whereby dust particles spiral in toward the sun under the influence of solar radiation.

precession. The slow turning of the earth's polar axis, caused principally by tidal action on its equatorial bulge.

prime focus. The focal point of a reflecting telescope's primary mirror.

prime meridian. The meridian of the earth that passes through a marker in Greenwich, England, and that defines 0° longitude.

prominence. A loop or streamer of hot gas seen just beyond the limb of the sun.

proper motion. The annual angular motion of a star in the sky as measured with reference to background stars or galaxies.

proton. The positively charged subatomic particle that exists in atomic nuclei.

proton-proton reaction. A process for the fusion of hydrogen nuclei to helium nuclei.

proto-. A prefix meaning the material as it existed before it formed into a particular object, such as a star or a planet.

pulsar. A neutron star that emits radio and light pulses in rapid succession.

quadrature. The position of a planet with respect to the earth, such that its elongation is 90°.

quasar. An extremely bright, small, and distant extragalactic object, often a radio source.

R Cor Bor stars. Eruptive variable stars that have abrupt irregular decreases in luminosity. Named after the bright example, R Coronae Borealis.

RR Lyrae variable. A short-period (less than 1 day) pulsating star of Population II.

radial velocity. The component of an object's velocity that is directly toward or away from the observer.

radiant. For a meteor shower, that point in the sky from which all the meteors appear to come.

radiation pressure. The transfer of momentum from photons to objects with which they collide.

radioactivity. The decay of atomic nuclei from an unstable form to a stable form. Particles and/or gamma rays are emitted as the nuclei decay.

ray. For the moon, streaks of lighter-colored material that radiate from certain craters.

red giant. A large, cool, evolved star.

reddening. The selective absorption of bluer radiation by the interstellar medium. The absorption causes an object to appear redder than it really is.

redshift. The Doppler shift for receding objects.

relativity. The formulations of natural law, due primarily to Einstein, that incorporate the constancy of the velocity of light and the relation between gravity and the curvature of space-time.

resolution. For a telescope, the degree to which fine details can be perceived.

resolving power. The closest resolution of a given instrument.

retrograde. Motion in the solar system that is opposite to the common direction of motion.

revolution. The motion of a body in its orbit.

right ascension. The coordinate on the celestial sphere that measures the east-west position of an object, as does longitude in the earth's coordinate system.

rille. A valley or long depression on the moon's surface.

Roche limit. The distance from a massive object at which tidal forces on a smaller object begin to pull it apart.

rotation. Movement of an object around its axis.

saros. The 18-year interval between eclipses that have similar properties.

Schwarzschild radius. The distance from the center of a black hole at which the escape velocity equals the velocity of light.

secular. Nonperiodic.

seeing. The result of the turbulent atmosphere above an observatory, which spreads out and moves the image of a star or planet.

semimajor axis. Half of the major axis.

semiregular variable. Star that varies in brightness, with a characteristic cycle but no regular period.

Seyfert galaxy. One of a class of galaxies that have bright, hot, gaseous nuclei.

shell star. A star that is surrounded by a thin shell of gas.

sidereal. Measured or determined with respect to the stars.

singularity. A point at which events or material have infinitesimal or infinite dimensions.

solar apex. The direction toward which the sun's peculiar motion is aimed.

solar constant. The average amount of energy received from the sun at the earth's distance from the sun, per unit area and per unit time.

solar motion. The velocity of the sun measured with respect to a given sample of stars or other objects.

solar wind. The particles, which make up an expanding plasma, expelled outward from the sun through the solar system.

solstice. On the ecliptic, the points farthest from the celestial equator.

spectral type. A classification of the spectra of stars according to the prominence of the lines of various elements.

spectrogram. A photograph of a spectrum.

spectrograph. A device for photographing a spectrum.

spectrophotometry. Measurement of the intensities in a spectrum of an object by means of careful, precise techniques.

spectroscope. A device that allows a spectrum to be viewed directly.

spectroscopic binary. An unresolved double star whose duplicity is discovered because the combined spectra show orbital motion due to the Doppler shifts in the spectra.

spectroscopic parallax. The distance to a star. Gauged by determining its probable intrinsic luminosity from its spectrum.

spectroscopy. The study of astronomical objects by means of their spectra.

spectrum. The light from a source, spread out into its different colors or wavelengths.

spherical aberration. The imperfection in an image formed by a lens as a result of the failure of light from different parts of the objective to come to a common focus.

spicule. Narrow needlelike rays of material in the sun's chromosphere.

sporadic meteor. A meteor that is not part of a shower.

star. A sphere of gas that is luminous because of its own energy.

stellar model. A mathematical and tabular description of the interior conditions in a star. Based on application of the relevant physical laws.

stimulated emission. Radiation of light from atoms as a result of incident light of the same wavelength.

Stromgren sphere. An H II region, ionized gas around a hot star.

subdwarf. A star that lies below the main sequence.

subgiant. A star that lies just above the main sequence.

sun spot. A temporary, small, cool region on the sun's surface.

sun spot cycle. The period of about 11 years of variation in the number of sun spots on the sun.

supergiant. A star that lies far above the main sequence.

superior conjuction. A planetary configuration in which the other planet lies opposite the sun as seen from the earth.

superior planet. A planet that is more distant from the sun than the earth.

supernova. An exceedingly bright, sudden increase in a star's luminosity, due to a near-complete explosive destruction of the star.

surface gravity. The strength of the force of gravity at the surface of a body measured in terms of the weight of the unit of mass there.

synchrotron radiation. Light emitted by extremely high-velocity particles moving in a magnetic field.

synodic. Referring to the interval between successive similar configurations with respect to the sun, as seen from the earth.

syzygy. The configuration of the earth-sun-moon system when the moon is full or new.

T Tauri. An irregular, very young star, prototype to a class of erratic variables located in gas and dust clouds.

tektite. A glassy object apparently formed in conjunction with one of a number of giant meteorite falls on the earth.

telescope. Any optical instrument used to increase perceptibility of distant objects.

telluric. Derived from the earth.

temperature. A measure of the internal energy in a body or other system of particles.

terminator. The lines of sunset and sunrise on the moon or a planet.

thermonuclear energy. Energy derived from the nucleus of the atom.

tide. Any distortion of a body by means of the differential gravitational force exerted on it by another body.

transit. The passage of a small body across the face of a larger body. Also, an astronomical instrument that times the passage of stars across the meridian.

triangulation. The use of trigonometry for determining the distance to an object from measures of two angles and one side of a triangle.

Trojan asteroid. One of two families of asteroids that move in Jupiter's orbit.

tropical year. The length of the earth's year measured with respect to the vernal equinox.

umbra. The portion of a body's shadow for which the light source is completely obscured; or the dark center of a sun spot.

universal time. The local mean time at Greenwich.

van Allen belt. The complex region above the earth's atmosphere where charged particles are trapped by the magnetic field.

velocity of escape. The velocity that a body must have in order to escape completely from the gravitational pull of a larger body such as a planet.

vernal equinox. The point on the celestial equator that is occupied by the sun as it moves from south to north on the ecliptic.

visual binary. A binary, both components of which can be detected through the telescope.

W Virginis variable. A pulsating star like a Cepheid, but of Population II.

white dwarf. A star that has used up its nuclear fuel and has collapsed to a small, hot sphere of degenerate gas.

white hole. A hypothetical (and probably impossible) opposite of a black hole in which material rushes back from a singularity.

Wolf-Rayet star. A class of extremely hot, massive stars that are ejecting high-velocity gas shells.

worm hole. A hypothetical bridge between two universes or two parts of one universe, by using a black hole and a white hole.

x-rays. Electromagnetic radiation with wavelengths shorter than visual and ultraviolet light, but longer than gamma rays.

Zeeman effect. Splitting of spectral lines into two or more close components, because of a magnetic field in the source.

zenith. The point directly above an observer.

zodiac. The band of constellations through which the ecliptic passes.

zodiacal light. A faint glow along the ecliptic. A result of sunlight reflected from interplanetary dust particles.

Appendix B

SOME MATHEMATICS

In astronomy it is especially important to be able to use powers-of-ten notation, since astronomical numbers are often cumbersome because they are so large. Any number can be written as some simple number between 1 and 10, multiplied by 10 raised to an appropriate power (a number raised to a power is just that number multiplied by itself as many times as the power). Thus, the number 100 is 1×10^2, or 1×10 raised to the power 2, which is $10 \times 10 = 100$. The number 200 is 2×10^2, and so forth. The luminosity of the sun in ordinary notion is 3,900,000,000,000,000,000,000,000,000,000,000 ergs per second. This is much more easily written 3.90×10^{33} ergs per second. The distance to the Andromeda galaxy is 2,200,000 light years, or 2.2×10^6 light years.

Negative powers of ten can be used to express small numbers. The number 10^{+1} is 10, 10^0 is 1, and 10^{-1} is $1/10$ or 0.1. This sequence continues: 10^{-2} is 0.01, 10^{-3} is 0.001, etc. One Angstrom is 0.00000001 cm or 10^{-8} cm. The mass of the electron is 9.1096×10^{-28} gm.

It is useful to keep in mind some of the properties of simple geometrical figures.
1. Triangles. The sum of the interior angles of a triangle in flat, euclidean space is 180° (but see Chap. 21 for other spaces). The area of a triangle is $1/2$ hb, where b is the length of the base of the triangle and h is a perpendicular line to b drawn from the opposite corner.
2. Rectangles. The four corners of a rectangle are 90° angles. The perimeter of a rectangle is $2l + 2h$, where l and h are the length and height of the rectangle. The area of a rectangle is lh. For a square of side s, the area is s^2.
3. Circles. The circumference of a circle is just $2\pi r$, where r is the radius and π is 3.1416. The area is πr^2.
4. Spheres. The area of the surface of a sphere is $4\pi r^2$. The volume of a sphere is $4/3 \pi r^3$.

This book does not use trigonometry, but it is important to have a clear idea of the properties of angles and their measurement. Most angles in astronomy use measurements made in units of arc seconds, minutes, and degrees. A circle is divided into 360 degrees, each degree into 60 arc minutes, and each minute into 60 arc seconds. Sometimes scientists use an alternative system called *radian* measure, in which the circle is divided into 2π radians, and smaller divisions are just fractions of radians. Astronomers also occasionally measure angular distances in the sky in terms of the time it takes to cross the meridian.

For instance, a 15° arc along the celestial equator is equivalent to 1 hour of time. A 1-arc minute separation along the equator is 4 seconds of time.

With triangulation to find the distance to a distant object, simple geometry can be used. One merely makes a scaled-down triangle in which the angles are kept the same as in the real triangle, and then the scaled-down distance to the object is measured off. It is much easier, however, to use simple trigonometry, which is the branch of mathematics that is based on the properties of triangles. For example, the following fact will enable you to compute the distance to any star whose parallax is measured. Consider a right triangle made by the earth-sun-star, with the 90° angle at the sun. Then the distance to the star is related to the angle, s, so that the star appears to move as a result of the earth's motion from the above position to a position in line with the sun. The base of the triangle is 1 AU and the star's distance is computed to be $\frac{1 \text{ AU}}{\text{tangent of } s}$ where the tangent of any angle is a trigonometric function, and can be looked up in any reference book of mathematical tables (or can be found by using many pocket calculators). The formula in general is

$$\text{distance} = \frac{\text{baseline}}{\text{tangent of angle opposite baseline}}$$

The angle used here is just ½ the parallax angle (that subtended by the earth's motion all the way from one side of its orbit to the other).

Appendix C

UNITS AND CONSTANTS

Metric and astronomical units

Length:
  meter (m)
  centimeter (cm) = 10^{-2}m
  kilometer (km) = 10^{3}m
  micron (μ) = 10^{-6}m
  Angstrom (Å) = 10^{-10}m
  Astronomical Unit (AU) = $1.49597870 \times 10^{11}$m
  parsec (pc) = 206264.806 AU = 3.085678×10^{18}cm = 3.26 ly
  light year (ly) = 9.460530×10^{17}cm

Mass:
  gram (gm)
  kilogram (kg) = 10^{3}gm
  ton (metric) = 10^{6}gm
  solar mass ($M_\odot$) = 1.989×10^{33}gm

Time:
  second (sec)
  day = 8.64400×10^{4}sec
  sidereal year = 3.155815×10^{7}sec

English unit conversions

1 in. = 2.54 cm
1 ft. = 0.305 m
1 yd. = 0.915 m
1 mi. = 1.609 km
1 lb. = 0.454 kg

Constants

Physical constants:

Speed of light	c	= 299,792,458 m/sec
Constant of gravitation	G	= 6.672×10^{-11}m^3/kg·sec^2
Planck's constant	h	= 6.626×10^{-27}erg·sec
Boltzmann's constant	k	= 1.3806×10^{-16}erg/kelvin
Stefan-Boltzmann constant	σ	= 5.66956×10^{-5}erg/cm^2·deg^4·sec
Wien displacement constant	$\lambda_{max}T$	= 0.289789 cm·K
Mass of hydrogen atom	m_H	= 1.6735×10^{-24}gm
Mass of neutron	m_n	= 1.6749×10^{-24}gm
Mass of proton	m_p	= 1.6726×10^{-24}gm
Mass of electron	m_e	= 9.1096×10^{-28}gm
Rydberg's constant	R	= 1.09677×10^{5}/cm

Astronomical constants:

Mass of sun	$M_\odot$ = 1.9891 × 10³³gm
Radius of sun	$R_\odot$ = 696,000 km
Luminosity of sun	$L_\odot$ = 3.827 × 10³³erg/sec
Solar constant	S = 135.3 mW/cm²
Mass of earth	$M_\oplus$ = 5.9742 × 10²⁷gm
Equatorial radius of earth	$R_\oplus$ = 6378.140 km
Mean distance center of earth to center of moon	D_M = 384,403 km
Radius of moon	R_M = 1738 km
Mass of moon	M_M = 7.35 × 10²⁵gm

Appendix D

LIGHT AND OPTICS

Most of our information about the universe comes to us in the form of radiation. Much of this radiation is the kind most familiar to us, light visible to the human eye. However, light is only one small part of an entire spectrum of related radiation. Because of its importance to astronomy, it is important to have some understanding of the nature of this radiation, including light.

The first physicist to give serious consideration to the nature of light in modern times was Sir Isaac Newton, who demonstrated that sunlight was made up of a mixture of light of all different colors, from red to blue. He evolved the idea that light consisted of particles with certain characteristics, which the human eye perceived as color.

In subsequent years and centuries, it was shown that light has properties very similar to those exhibited by a phenomenon called waves. Waves became recognized as a general way that nature has for transferring energy or matter from one portion of a medium to another. Ocean waves, sound waves, waves through the earth caused by earthquakes, and light waves all seem to share many of the same properties of behavior. These waves appear similar in that they can be characterized by a wavelength (the distance between crests of ocean waves, for example) and a velocity. Other similar characteristics include the way by which waves reflect from flat surfaces, the way in which they diffract around small objects placed in their path, the way in which their paths can be changed in direction by passing through media of different density, and the way in which they can experience interference, one by the other.

All the different kinds of waves found in nature also share one other feature, namely, that they are waves of something. Scientists came to realize that there had to be a *medium*, a substance that is vibrating or oscillating. In the case of water waves, there is the water itself vibrating up and down as the wave moves across the surface. In the case of sound waves, it is the air that compresses and expands as the wave of higher density moves swiftly through it. In the case of light, however, experiments showed that a medium was undetectable.

If a container is evacuated in the laboratory so that virtually no air remains in it, then sound waves cannot exist in it. A common experiment is to place an electrically operated bell in a glass jar and to listen to it as the vacuum pump proceeds to evacuate the jar. The bell is heard to ring loudly at first, and then gradually the ringing becomes fainter and fainter, and finally disappears—even though the bell is continuing to operate. On the other hand, if light is allowed to pass through the glass jar, it continues to pass through undiminished, even when the jar is completely evacuated. Physicists of the past explained this apparent contradiction by hypothesizing that there was a medium even less dense and elusive than any gas known to science; the medium existed in the jar even after all the air had been pumped out. This medium, which they called the *ether*, was hypothesized to exist throughout space, filling the universe in order that we might see light propagated even from the very most distant galaxies.

In 1889, however, physicists at the Case Institute helped to show that the ether does not exist. The Michelson-Morley experiment was set up to measure the velocity of light both along the direction of the earth's motion through the ether and perpendicular to it. There should have been a difference in the measured velocity because of the relative motion of the earth through the medium, but Michelson and Morley found no difference and were forced to the conclusion that the concept of a universal, weightless, ether was not a useful concept and should be dropped.

In the succeeding twenty years, physicists found other ways in which light waves did not agree with the other types of waves mentioned above. The most important way was discussed by Einstein, who found that light could behave under certain circumstances very much as if it were made up of individual particles of energy, somewhat similar to what Newton first suggested. The *photoelectric* effect explored by Einstein demonstrated that at very low light levels, one could detect individual "particles" of light by their action on certain materials, which were found to give up electrons when a particle of light hit them. Only light of certain wavelengths (certain energies) was found to be able to affect the electrons, and no matter how much light of less energy (longer wavelengths) was allowed to fall on the surface, electrons could not be ejected. This led to the realization that light under some circumstances acts like particles, which are called *photons*, and under other circumstances it acts like waves. The modern understanding is that the wave phenomenon of light is a manifestation of the laws that the photons must follow because of their exceedingly small "size", which is in the realm of nature where everything behaves according to certain laws of probability, rather than mechanistically.

Modern physicists understand the nature of light in this way. But it is impossible for them to describe in familiar terms what a photon "looks like," because light and other objects in this realm of microphysics behave in ways that are highly unfamiliar to us in the macroworld. Modern physicists find that light can be understood as consisting of small bundles of energy (*quanta*) which are discrete objects. These behave in a way described by probability rules, called *quantum mechanics*. It is these probability rules that make light appear under certain circumstances to act like waves. These rules also govern the behavior of the electron, proton, and other small particles that make up matter.

Along with this new understanding of light came the realization that light was a much more general phenomenon than mere visible radiation. We now realize that visible light is an exceedingly small portion of the wide spectrum of radiation, which extends all the way from gamma rays (with very small wavelengths) to radio waves (which have very large wavelengths). Figure A4-1 identifies the different names given to the different regions of wavelengths.

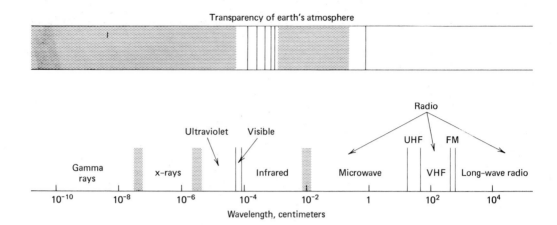

Fig. A 4-1. Electromagnetic radiation spans a huge range in wavelength, from gamma rays to radio waves. As indicated schematically at the top of this figure, the earth's atmosphere blocks much of this radiation, keeping it from reaching the ground.

Almost all these different types of radiation are useful in astronomy, because astronomical objects of various types radiate them. For example, satellites have detected a large number of sources of x-rays in the sky, and large radio antennae have for years supplied astronomers with information on cosmic radio-wave emitters.

Light at all wavelengths has the same velocity in a vacuum, and this velocity has considerable fundamental importance as one of the constants of nature. The theory of relativity is based partly on the discovery of the fact that the velocity of light is the same regardless of any motion of the emitter or receiver, and that nothing can exceed the speed of light in our universe. But for the purposes of this discussion, it is important to realize that the velocity of light can be slowed when it passes through something other than a vacuum. When light enters the earth's atmosphere the atmospheric medium bends the light slightly, because the velocity is slowed down slightly. A flashlight beam shone obliquely into a fish tank is another example of the way in which the velocity of light is altered when it passes from a medium of low density to one of a much higher density. The flashlight beam is bent conspicuously on entering the surface of the water. A similar effect is experienced by a swimmer who has difficulty judging distances at first in water because of the bending of light rays in that medium.

Early scientists who discovered this effect were very quick to realize that this bending of light could be put to their advantage. Lenses were developed in the sixteenth century and consisted of curved discs of glass, some thicker in the center than in the outer portions, and some in the reverse design. It was found that for the first case, the lenses could be used as magnifiers. But when a lens was thinner in the center than in the outer parts, it made things appear smaller than natural. The magnification of a lens results from the bending of the light rays passing through it, so that the human eye sees objects being examined as either occupying a larger area or a smaller area than in actual fact.

Appendix E

ASTRONOMICAL TELESCOPES

Although other scientists invented the telescope, Galileo was the first scientist to realize that an instrument built by combining two lenses of different curvature together could be used in astronomy to make distant objects in the sky appear closer. Galileo's first telescope was made by using a large positive lens (thicker in the center) at one end of the tube and a much smaller and more highly curved negative lens (thinner in the center) at the other end. By looking through the smaller of these lenses he could focus on distant objects and achieve a magnification of several times. Because the bending of light by a material is called *refraction*, Galileo's telescope was termed a *refracting telescope*.

Mirrors also can be used to alter the direction of light. This is certainly familiar to anyone in the case of flat mirrors, which merely reflect an exact image of the object in front of it. It is perhaps less obvious to most people that curved mirrors that have similar properties to lenses can be built. Shaving mirrors usually have a curvature that allows a man to examine an image of his chin enlarged over its appearance in a flat mirror. Similarly, curved mirrors at a fair can be used to cause great distortion of one's appearance. It was found by early astronomers, especially Newton, that a curved mirror thinner in the center than in the outer part could be used to create a telescope very much like Galileo's. Because reflection rather than refraction is the main operation, such telescopes are called *reflecting telescopes*.

Refracting Telescopes There are two basic ways to design a refracting telescope. Both designs have a large positive lens as the main light-gathering element, called the *objective*. The difference in the two designs is in the nature of the lens at the other end of the tube, called the *eyepiece*. It is the purpose of the eyepiece to take the light rays that have been gathered and bent together by the objective and to straighten them out again so that the human eye sees them as apparently coming parallel again, as before they entered the telescope. The eyepiece can be a negative lens, as used by Galileo, or it can be a positive lens. In the former case, the lens is placed at the proper distance from the objective so that it straightens out the light rays before they come together to a point (called the

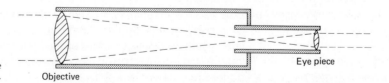

Fig. A 5-1. Parts of the telescope.

Objective

Eye piece

focus or *focal point* of the objective). When the eyepiece is a positive lens, it is placed beyond the focal point of the objective, after the light rays have come together to a point and begun to spread out again. The positive lens refracts the light rays so that they come out of it traveling parallel.

Each of these basic designs that use simple lenses has a drawback. For the galilean-type telescope, the drawback is that the field of view (the area of the image that can be seen) is limited to a rather small size. When such a telescope is made very large, and it is pointed toward the moon for example, only a very small portion of the moon can be looked at at a time. This is an especially important problem for large versions of this kind of telescope, and for that reason very few such telescopes have ever been built. The design is now almost exclusively used for very low-power telescopes and opera glasses.

The disadvantage of the refracting telescope that uses two positive lenses is an entirely different matter. The field of view for such a telescope can be quite large, even for very large telescopes and high magnifications. However, these telescopes suffer from a defect that is avoided in Galileo's design. Light waves are bent when entering a lens by an amount that depends somewhat upon the wavelengths of the light. For that reason, the objective lens of a telescope tends to bend the light of different colors into different angles; therefore, the image formed by a telescope that is made up only of two positive lenses is focused at slightly different positions for the different colors. The red portion of the image of a distant star will be out of focus if the eye focuses on the blue portion of that light. The galilean telescope avoids the problem because the negative lens on the eyepiece almost cancels the effect; but for a telescope made of positive lenses only, the effect remains. This is called *chromatic aberration* and is a serious design limitation for telescopes of the refracting type.

It was discovered in the 1800s that chromatic aberration could be largely avoided by employing double lenses made of two different kinds of glass with different refracting properties. The objective of such a telescope consists of a positive lens of one kind of glass cemented together with a slightly negative lens of another kind of glass. The negative lens is chosen to have a smaller refractive effect (*refractive index*) than the positive lens of the pair. The effect of the negative lens on bending the light is small, yet its thickness is large enough to correct the chromatic aberration. A telescope made with lenses of this type is called *achromatic*, and virtually all astronomical telescopes of the refracting type made and used today are achromatic refractors.

The largest refractors in the world are made up of objectives 3 feet or more in diameter, with the eyepiece located some 40 or 50 feet away at the bottom end of a very long tube. The Yerkes Observatory of the University of Chicago, located at Williams Bay, Wisconsin, has the world's largest objective—40 inches in diameter. It is achromatic, bringing the light of different colors almost exactly to the same focus. Large refractors of this type have not been built for many years, because the size of the Yerkes objective has been found to be just about the largest practical size for a telescope of this sort. With lenses very much larger than this, the weight of the glass tends to cause the lens to sag under its own weight, distorting the lens and the image. Also, reflecting telescopes of similar and larger sizes are less expensive and less difficult to construct.

Reflecting Telescopes

Reflecting telescopes are basically similar to refractors. The different types of reflectors differ in the way in which the focus is brought out of the beam of the telescope. The newtonian design uses a small flat mirror tilted at a 45° angle at the upper end of the telescope tube. This brings the rays of light from the objective or primary mirror out of the tube to one side of it, where it is possible to place an eyepiece or other astronomical receiving instrument. This design is the simplest and most inexpensive for small reflecting telescopes, and is the one used most often by amateur astronomers, who normally make their own. A great many thousands of such newtonian reflecting telescopes exist around the world, mostly in the hands of men and women who use them to explore the stars and planets as a hobby.

The largest newtonian telescope in the world is the 100-inch reflector at Mt. Wilson Observatory in California. It was installed in 1918. Its mirror is more than 8 feet in diameter and $1\frac{1}{2}$ feet thick. The telescope weighs so much that it was built to float in pools of mercury in order to smooth its movement. The entire weight of the moving parts of the Mt. Wilson 100-inch reflector is many tons. The newtonian focus is at the top end of the tube, which is normally three of four stories above the floor of the dome. An astronomer reaches it on a moving platform attached to the ceiling of the dome by cables.

The most common design for more modern large telescopes is somewhat different from the newtonian design. Instead of a flat mirror, a circular curved mirror is used to reflect the image back toward the primary mirror, which in this case has a hole cut in the center to allow the focus to lie out of the tube at the

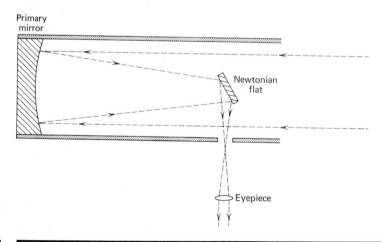

Fig. A 5-2. The newtonian telescope design, using a lens and a flat secondary mirror.

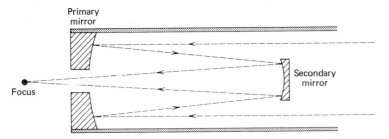

Fig. A 5-3. The cassegrain telescope design, using two curved mirrors, the largest of which has a hole cut to accommodate the light path.

bottom end, where an eyepiece or other instrument can be placed. The hole in the primary mirror does not affect the image, except inasmuch as it decreases the total area of the mirror. The final image is therefore somewhat fainter than would otherwise be the case. Similarly, the secondary mirror, which is usually convex (in which case it is called a *cassegrain* design, after a French telescope maker of the seventeenth century), also blocks a small portion of the light. The cassegrain design has an advantage over the newtonian design in that the image is brought to a position at the bottom of the telescope where it is more accessible to the astronomer, and where it is less difficult to attach large auxiliary instrumentation.

A third design for reflecting telescopes avoids the problem of any secondary mirror by placing the astronomer at the *prime focus*, the main focus of the primary mirror at the top end of the telescope. This design is only effective for exceedingly large telescopes, because otherwise the astronomer obscures too large a percentage of the light. The largest telescopes in the world—the 200-inch Mount Palomar telescope in California, the 234-inch telescope in the Soviet Union, and the two 158-inch telescopes located in Arizona and Chile—all have provisions for a prime focus observer's "cage" located at the

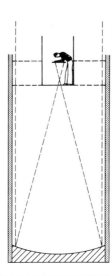

Fig. A 5-4. The prime focus, used only for very large telescopes.

top of the tube. To reach it, the astronomer must climb a ladder or take an elevator to the very top of the tube, climb into the cage, and then remain in this small enclosure as the telescope swings around to point to the objects which he or she wishes to measure or photograph. In order to obscure as little light as possible, the prime focus cage of such a telescope is made as small as possible. Astronomers who have used such instruments find that the highly cramped quarters and isolation from the rest of the world lend something of the quality of working in a space capsule.

A fourth position for the focus of a reflecting telescope is a very special one, requiring many reflections and extra mirrors. Called the *coudé focus*, after the French word for elbow, this focus has the special advantage of being perfectly stationary. A series of mirrors in the telescope and in the telescope mounting brings the focus of the telescope down a tube, through the mounting, and into a room below the domed observing room. Very large and heavy instruments can thus be set up to analyze the image, without having to be attached to the telescope tube. Ordinarily, large and cumbersome spectrographs, which are too great in size and too delicate in design to be able to swing around at the cassegrain or newtonian focus, are placed at the coudé focus in an isolated temperature-controlled room below the telescope dome. Only very large telescopes have been built using the coudé focus; the smallest is a NASA telescope in California with a 24-inch primary mirror, designed expressly for spectroscopic studies of bright planets. Most very large reflectors in the world

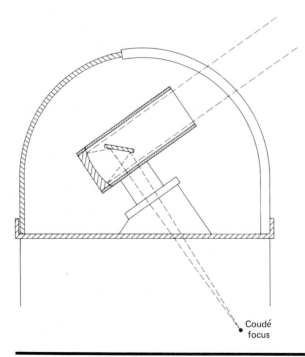

Fig. A 5-5. An example of a simple version of the coudé design. Others, involving several more mirrors, are more complicated.

Coudé
focus

have been designed with a coudé focus that can be activated by changing mirrors from the otherwise-employed cassegrain or prime focus setup.

Reflecting telescopes are completely achromatic; all colors of light are reflected from the surface of the mirrors at the same angles. Therefore, these telescopes do not suffer from any chromatic aberration such as those that plague refracting telescopes. Nevertheless, there are limitations of field size that are difficult to overcome for large reflectors. Image distortions occur at the edges of the field observed, and these distortions result from what is called *coma*, caused by the fact that the edges of the field are not brought together at a single focus. Correcting lenses, which are achromatic as for a refractor, can be used to correct for the coma effectively.

One extreme example of this technique is that developed by Bernard Schmidt in 1930. He placed a very thin correcting lens, curved exactly so as to correct the spherical aberration at the top end of a reflecting telescope. The Schmidt design is primarily used for very large photographic telescopes, usually called *Schmidt cameras*, which have exceedingly fast speeds and large fields of view. The largest is the 52-inch Schmidt in Germany.

The three features of a telescope that make it useful in astronomy are its ability to enlarge the apparent size of a distant object (*magnifying power*), its ability to make distant objects appear brighter (*light-gathering power*), and its ability to discern very small separations or small-scale features in a distant object (*resolving power* or *resolution*). **Magnifying Power**

The magnification of a telescope depends upon the strength of the lens. How much a lens bends the light can be measured either by describing the amount of curvature of the lens, or more conveniently, by measuring the distance from the lens to the point in space where it forms an image (called the *focal length*). If the curvature is very small, this distance is large. For example, if we use a lens that has no curvature at all, but is merely a piece of flat glass, no image is formed. The focal length is then infinitely large, and the strength of the lens is zero. On the other hand, if the lens is very highly curved, it brings the light to a focus very close to it, and the resulting focal length is short. In order to achieve large magnification, the telescope must have the ability to make the eye seem to see an object subtend a very much larger angle than it does without the instrument. This can be achieved by using a strong (short focal length) eyepiece. This image will be magnified, however, only if the focal length of the objective is comparatively long. The eyepiece thus receives an image that subtends only a very small angle, as seen by the objective. If both eyepiece and objective were to have the same focal length, then the objective would bend the light a certain amount, and the eyepiece would unbend it by the same amount. The result would be no magnification at all. To achieve maximum magnification for a telescope, the objective must have as long a focal length as possible, and the eyepiece as short a focal length as possible.

A student looking at the moon through a college's small telescope, which might have a focal length of 50 inches for the objective and 1 inch for the eyepiece, would see the image of the moon magnified by a factor of 50. On the other hand, if an astronomer working at the prime focus of the 200-inch telescope on Palomar Mountain (which has a mirror with a focal length of 700 inches) were to look at a distant galaxy by using an eyepiece of 1-inch focal length, he or she would see the image magnified 700 times. (Normally, he or she would only look through the telescope briefly to set it before making measurements, as very little visual observing is ever done with large telescopes.) To calculate the magnification of any telescope, it is merely necessary to divide the focal length of the objective by the focal length of the eyepiece.

Light-Gathering Power

Magnifying power is not necessarily the most important kind of power of a telescope. In principle, it would be possible for the student mentioned in the above paragraph to construct a telescope fully as powerful as the 200-inch telescope on Palomar Mountain. If the student bought or made, for example, a 3-inch diameter lens with 700-inch focal length, he or she could then place this lens at the end of a 700-inch tube (almost 60 feet long). By putting a 1-inch focal length eyepiece at the other end and pointing it at the moon, he or she would see an image of the moon magnified 700 times. However, there would be all kinds of difficulties that would make the experience a high unsatisfactory one. If the student could manage to get the tube to be stable enough, and if he or she could properly compensate for the motion of the earth on its axis, the biggest disappointment would be the brightness of the image that would be seen. With only a small telescope magnifying so large an amount, the brightness would be very feeble compared to what would be seen at the 200-inch telescope. This fact can be understood if we imagine an even more foolhardy person who constructs a telescope that has an objective no bigger than the human eye (approximately $1/3$ of an inch in diameter). If this person builds this telescope so that it has some very large magnification (say 700 times) and then looks into it, the eye would receive as much light from a tiny spot on the moon as it receives from that same tiny spot without the telescope. However, that light is now magnified to occupy a very large portion of the apparent vision of the viewer—an immense area 700 squared times the area without the telescope. He or she would be seeing an image of a section of the moon that would be nearly 500,000 times fainter than the image of the moon seen without a telescope. It is quite unlikely that the image would be seen at all.

Since astronomy deals with exceedingly distant and very faint objects, it is important that the light-gathering power of telescopes be as large as possible. Light-gathering power is determined simply by the amount of light received from the distant object, which depends on the area of the objective of the telescope.

In order to compare light-gathering powers of telescopes of different sizes, one can simply compare the squares of their diameters. For example, the light-gathering power of the 200-inch telescope compared to a 3-inch telescope is 200^2 divided by 3^2, a little more than 4,000 times. This means that the astronomer at a 200-inch telescope would see an image that would be 4,000 times brighter than that which a 3-inch telescope user would see. Most of the universe consists of objects that are so far away that they are invisible without a large telescope, and therefore astronomers of this century have been attempting to build as many large diameter telescopes as possible.

A third kind of power that is important in astronomy is the resolving power of a telescope. The resolution is limited by three primary effects. First is the quality of the lens or mirror. Defects in the shape or polishing of the optics will have the effect of distorting the image and of reducing the effective resolving power of the telescope. An imperfect telescope looking at a small feature, for example a crater on the moon, would see it blurred and perhaps indistinguishable compared to its image as seen through a perfectly made telescope. Astronomical telescopes used professionally have very nearly perfect optics, and therefore this problem is usually not a limitation for most research telescopes. Exceptions occur when a telescope is distorted in shape by changes in temperature, for example, on a cold night following a hot day. Astronomers are particularly concerned about such effects and take measures to prevent them by refrigerating astronomical domes, by building telescope mirrors out of special kinds of materials that do not change very much with temperature changes (such as fused quartz or Pyrex), and by opening the dome immediately after sunset so as to bring the temperature to an even value as soon as possible.

Resolving Power

The second limitation on the resolving power of a telescope is caused by atmospheric turbulence above the observatory. This problem, of course, does not exist for telescopes placed on space platforms, but it is usually the most important limitation for large telescopes at ground-based observatories. The atmosphere, because of differences in temperature and the turbulent motion of the air both close to the ground and at high altitudes, causes movement and blurring of an image when it is very highly magnified by a telescope. This limitation is familiar to anyone who has looked at the moon through a telescope with high magnification. The image of the moon often "dances" slightly in an irregular way and shows a slight blurring of the finest details. Different nights are often different in this regard, with some nights showing a steady and nearly perfect image, while others show particularly bad images. Astronomers call this effect *seeing*, and many important astronomical problems depend on the occurrence of optimal seeing.

Certain astronomical observatories have better records for seeing than others. Among the best are an observatory in Chile (built jointly by the governments of the United States and Chile at Cerro Tololo, 8,000 feet high in the foothills of the Andes) and the Mauna Kea Observatory (located by the

University of Hawaii on the top of the 13,750-foot high volcano, Mauna Kea). Average seeing at an observatory limits the size of detail that can be seen to approximately 2 seconds of arc. This is the equivalent of the angular size that a penny would subtend as seen without a telescope at a distance of 2 miles.

The third limitation on the resolving power of a telescope is imposed by the nature of light itself. Because light does have some of the properties of a wave, it can bend around edges the way waves in water do. This is called *diffraction*, and for a telescope, which has edges to its lenses, this diffraction limits the clearness of an image in a way that depends upon the diameter of the telescope and upon the wavelength of the light. The angular distance between two stars that can just barely be resolved (because of this limit) is smaller for telescopes of larger aperture and for light of shorter wavelengths. A 5-inch telescope theoretically can resolve two stars that are separated by only 1 second of arc, whereas a 200-inch telescope has a theoretical resolving power of 200/5 = 40 times smaller (or 0.025 second of arc). These figures are for visible wavelengths; but in the ultraviolet, even smaller separations can be resolved. If we examine these stars at longer wavelengths, such as in the red or infrared, however, they are less easily resolved for a given-size telescope. For the extreme case of radio waves, where the wavelengths are millions of times larger than the wavelengths of visible light, resolving powers are correspondingly millions of times worse for the same size telescope. At wavelengths of 50 centimeters, for instance, the 200-inch telescope would be able to resolve radio sources separated by 20,000 seconds, or $5^1/_2°$.

Radio Telescopes

Telescopes used in radio astronomy are similar in principle to optical telescopes, although their appearance is both different and varied. The appropriate design for a radio telescope depends very much upon the wavelengths of the radiation being observed. For the shorter wavelengths (millimeters or so), the telescopes are generally most similar to large optical telescopes in their appearance. They are often larger, however, for two reasons. First, because the wavelength is relatively large (e.g., 1 one-thousandth of a meter instead of 5 ten-millionths of a meter), it is not necessary to have as highly accurate and highly stable an optical surface as in the case of an optical telescope. Metal sheets can be used for millimeter telescope mirrors, and they need not be finely polished nor very heavy. The second reason that radio telescopes are large is because the resolving powers of radio telescopes are so very poor. A 200-inch optical telescope could resolve a penny at a distance of 10 miles or more (under ideal conditions). A 200-inch radio telescope, operating at radio wavelengths, could not resolve a football field at that distance. Astronomers attempt to make the largest radio telescopes possible, given mechanical possibilities and funds. For short-wavelength radio telescopes, the

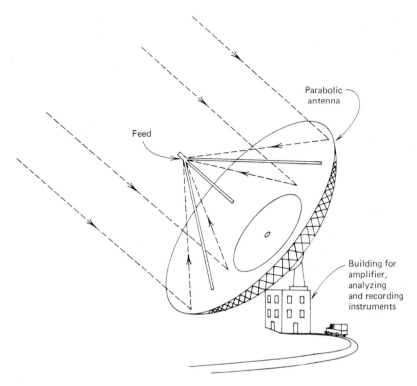

Feed

Parabolic antenna

Building for amplifier, analyzing and recording instruments

Fig. A 5-6. The design of a parabolic radio antenna.

diameter of the collecting mirror is usually at least 30 feet, and often bigger. This is larger than the largest optical telescopes in the world.

For radio telescopes that use mirrors like large reflecting optical instruments, the prime focus contains not an eyepiece or photometer, but a radio feed that collects the radio signal and transmits it by wires down from the telescope to various analyzing instruments in an adjacent building. Most modern radio telescopes have the capability of receiving radiation of several different wavelengths at the same time, and of analyzing these so that the telescopes are used with the greatest possible efficiency. Many of the instruments are mounted equatorially, but the largest (ranging in diameter from 200 to 300 feet) are often too large and cumbersome for equatorial mountings and have been given altazimuth mounts.

For long radio wavelengths, the resolution problem is exceedingly serious. Radio astronomers have developed arrays of many telescope mirrors, called interferometers, covering as large an area as possible. This arrangement can give better resolution than a single reflector if all the signals of the different dishes (mirrors) of the array are brought together and analyzed simultaneously. It is possible to realize a resolution equivalent to the size of the array, rather than to the size of an individual dish. The signal is not as strong as it would be for a telescope of comparable total diameter, but for many problems the receivers are nevertheless sensitive enough to study many thousands of radio sources in the sky.

The largest permanent arrays of radio telescopes are usually a few miles apart. Recently, however, astronomers have worked out a means of using two radio telescopes separated by thousands of miles to observe simultaneously certain objects and analyze the signals together, making it possible to achieve almost incredibly good resolution. This process is called *intercontinental interferometry*, and it has given astronomers radio-wavelength resolutions that are ten to a hundred times better than can be obtained optically. For example, simultaneous observations of certain very small and intense radio sources called *quasars* have been carried out interferometrically with radio telescopes located in California and in Australia.

Some radio telescopes have designs that are very different in appearance from any of the designs of optical instruments, though similar in general principle. For example, a simple antenna, such as a home television aerial, can be set up so as to receive radio radiation from space. Many telescopes specifically designed for long-wavelength radio telescopes consist of a maze of such aerials arranged in various advantageous patterns.

Radio telescopes normally do not provide a "picture" of the sky like an optical telescope. They are usually used to "look" at one point at a time in the sky and to measure the amount of signal at various wavelengths. By slowly moving a radio telescope with respect to an object, it is possible to build up a radio equivalent of a "photograph" of the object. This photograph can be analyzed by the radio astronomer in terms of the energy emitted in various places and at various wavelengths.

Space Telescopes

The earth's atmosphere provides two different kinds of difficulties for astronomical observations. The first is the fact that it absorbs radiation of certain wavelengths, including all those with wavelengths shorter than blue and ultraviolet, most of those with wavelengths longer than the near infrared (somewhat longer than the limit of red that the eye can detect), and shorter than radio wavelengths. These regimes of wavelength have always been blocked from the view of earth-based astronomers, and the information they might provide about astronomical objects has until recently remained highly speculative.

The second difficulty the atmosphere produces is a blurring of starlight (called *seeing*), which is the result of atmospheric turbulence and temperature variations. Theoretically, the largest telescopes should be able to see details some ten times smaller than if there were no atmosphere.

One of the most beneficial aspects of the space program for astronomy was the fact that it allowed scientists to place telescopes in orbit around the earth and above the atmosphere to avoid these two problems. The first space observatories were simple instruments placed on rockets, and enabled scientists to have only brief glimpses of the sky during the few minutes or seconds that the rockets flew above the atmosphere before falling back to earth. When satellites made it possible to place astronomical instruments in orbit for relatively extended periods of time, however, much greater possibilities were available. In the succeeding years, astronomers have proceeded to map the sky in the ultraviolet, to study the sun in wavelengths hitherto unavailable for study, to examine the ultraviolet and infrared properties of our Galaxy and others, to detect and study x-rays (shorter wavelengths than the far ultraviolet) and gamma rays (shorter wavelengths still) from astronomical sources, and to make many important and surprising discoveries about astronomical objects.

Many orbiting telescopes are of special design, depending upon the wavelength region for which they are intended. For example, the x-ray telescopes have had to be designed in an unusual way because of the extremely short wavelengths of the x-rays, which would pass through an ordinary mirror.

One of the more important projects of the post-Apollo phase of the space effort is a space telescope, the ST. This instrument is designed to be the space equivalent of the 200-inch Palomar telescope. Although it has a diameter of only 90 inches, it can exceed the Palomar instrument in performance. Its resolving power is designed to be at least 0.04 seconds of arc, meaning that it will be able to resolve objects as small as 7 miles across on the surface of Mars at closest approach. This is more than ten times smaller than the best we can do from the surface of the earth. Because of the high resolving power and the relative darkness of the sky from space, it can record stars many times fainter than the faintest that can be seen from the earth. It can also observe astronomical objects, both in the ultraviolet and in the infrared wavelength regions, which are invisible to earth-based telescopes.

Appendix F

PLANETARY DATA

ORBITAL PROPERTIES

Name	Semimajor Axis AU	Semimajor Axis 10⁶km	Sidereal period (years)	Eccentricity	Inclination to ecliptic
Mercury	0.3871	57.9	0.24084	0.2056	7°00'26"
Venus	0.7233	108.2	0.61515	0.0068	3°23'40"
Earth	1	149.6	1.00004	0.0167	0°00'14"
Mars	1.5237	227.9	1.8808	0.0934	1°51'09"
Jupiter	5.2028	778.3	11.867	0.0483	1°18'29"
Saturn	9.5388	1427.0	29.469	0.0560	2°29'17"
Uranus	19.1914	2871.0	84.074	0.0461	0°48'26"
Neptune	30.0611	4408.1	164.82	0.0100	1°46'27"
Pluto	39.5294	5913.5	248.53	0.2484	17°09'03"

PHYSICAL PROPERTIES

Name	Equatorial radius (km)	Mass (earth masses)	Mean density (gm/cm³)	Oblate-ness	Surface gravity (earth = 1)	Sidereal rotation period	Inclination of equator to orbit
Mercury	2,439	0.0553	5.44	0.0	0.378	59^d	<28°
Venus	6,052	0.8150	5.24	0.0	0.894	244.3^d R	177°
Earth	6,378.140	1	5.497	0.0034	1	$23^h56^m04.1^s$	23°27'
Mars	3,397.2	0.1074	3.9	0.009	0.379	$24^h37^m22.6^s$	23°59'
Jupiter	71,398	317.89	1.3	0.063	2.54	9^h50^m to $> 9^h55^m$	3°05'
Saturn	60,000	95.17	0.7	0.098	1.07	10^h14^m to $> 10^h38^m$	26°44'
Uranus	25,400	14.56	1.2	0.06	0.919	$\sim22^h$	97°55'
Neptune	24,300	17.24	1.7	0.021	1.19	$\sim23^h$	28°48'
Pluto	2,500	0.11	?	?	?	$6^d9^h17^m$	?

SATELLITES

Name	Semimajor axis of orbit (km)	Sidereal period (days)	Orbital eccen-tricity	Orbital incli-nation (degrees)	Radius (km)	Discoverer
Satellite of Earth						
The Moon	385,000	27.322	0.055	18–29	1738	—
Satellites of Mars						
Phobos	9,380	0.3189	0.018	1.0	14	Hall (1877)
Deimos	23,500	1.262	0.002	1.3	8	Hall (1877)
Satellites of Jupiter						
V Amalthea	181,000	0.4982	0.003	0.4	80	Barnard (1892)
I Io	422,000	1.769	0.000	0	1830	Galileo (1610)
II Europa	671,000	3.551	0.000	0	1550	Galileo (1610)
III Ganymede	1,070,000	7.155	0.001	0	2640	Galileo (1610)
IV Callisto	1,880,000	16.69	0.01	0	2500	Galileo (1610)
XIII Leda	11,110,000	239	0.147	26.7	8	Kowal (1974)

Name	Semimajor axis of orbit (km)	Sidereal period (days)	Orbital eccen- tricity	Orbital incli- nation (degrees)	Radius (km)	Discoverer
VI Himalia	11,500,000	250.6	0.158	27.6	60	Perrine (1904)
VII Elara	11,700,000	259.7	0.207	24.8	20	Perrine (1905)
X Lysithea	11,900,000	263.6	0.130	29.0	7	Nicholson (1938)
XII Ananke	21,200,000	631.1 R	0.169	147	6	Nicholson (1951)
XI Carme	22,600,000	692.5 R	0.207	164	7	Nicholson (1938)
VIII Pasiphae	23,500,000	738.9 R	0.378	145	6	Melotte (1908)
IX Sinope	23,700,000	758 R	0.275	153	7	Nicholson (1914)
XIV						Kowal (1975)
Satellites of Saturn						
Janus	169,500	0.749	0.0	0.0	100	Dollfus (1966)
Mimas	186,000	0.942	0.020	1.5	200	W. Herschel (1789)
Enceladus	238,000	1.370	0.004	0.0	300	W. Herschel (1789)
Tethys	295,000	1.888	0.000	1.1	500	Cassini (1684)
Dione	377,000	2.737	0.002	0.0	400	Cassini (1684)
Rhea	527,000	4.518	0.001	0.4	750	Cassini (1672)
Titan	1,220,000	15.95	0.029	0.3	2900	Huygens (1655)
Hyperion	1,480,000	21.28	0.104	0.4	200	Bond (1848)
Iapetus	3,560,000	79.33	0.028	14.7	750	Cassini (1671)
Phoebe	13,000,000	550.5 R	0.163	150	100	W. Pickering (1898)
Satellites of Uranus						
Miranda	130,000	1.414	0.017	0	200	Kuiper (1948)
Ariel	191,000	2.520 R	0.003	0	700	Lassell (1851)
Umbriel	260,000	4.144 R	0.004	0	500	Lassell (1851)
Titania	436,000	8.706 R	0.002	0	900	W. Herschel (1787)
Oberon	583,000	13.46 R	0.001	0	800	W. Herschel (1787)
Satellites of Neptune						
Triton	354,000		0.000	160.0	1900	Lassell (1846)
Nereid	5,570,000		0.76	27.4	300	Kuiper (1949)
Satellite of Pluto						
I	20,000	6.3867	0	105	?	Christie (1978)

(R indicates retrograde motion.)

Appendix G

SPACECRAFT DATA

MAJOR SPACECRAFT MISSIONS 1959–1981

Spacecraft	Launch date	Mission destination	Accomplishment
Sputnik 1	1957	Earth	Orbital, first satellite
Explorer 1	1959	Earth	Orbital, discovered radiation belts
Luna 3	1959	Moon	Flyby, first photos of far side
Mariner 2	1962	Venus	Flyby, surface temperature measured
Ranger 7	1964	Moon	Hard landing, close-up photos of surface
Mariner 4	1964	Mars	Flyby, first Mars pictures
Venera 3	1965	Venus	Landed
Surveyor I	1966	Moon	Soft landing, surface analysis
Orbiters	1967	Moon	Extensive close-up photo coverage
Venera 4	1967	Venus	Landed, relayed measures
Mariner 5	1967	Venus	Flyby, measured atmosphere
Venera 5	1969	Venus	Landed, properties of atmosphere
Venera 6	1969	Venus	Landed, properties of atmosphere
Mariner 6	1969	Venus	Flyby, measured atmosphere
Mariner 7	1969	Mars	Flyby, pictures, atmosphere
Apollo 11	1969	Moon	First manned landing, exploration
Apollo 12	1969	Moon	Manned landing
Luna 16	1970	Moon	Unmanned landing
Venera 7	1970	Venus	Landed, measured atmosphere
Apollo 14	1971	Moon	Manned landing
Apollo 15	1971	Moon	Manned landing
Mars 3	1971	Mars	Landed, no data
Mariner 9	1971	Mars	Orbital, many pictures, data
Pioneer 10	1972	Jupiter	Flyby, pictures, atmosphere
Venera 8	1972	Venus	Landed, first picture
Luna 20	1972	Moon	Unmanned, samples returned
Apollo 16	1972	Moon	Manned exploration
Apollo 17	1972	Moon	Manned exploration
Pioneer 11	1973	Jupiter Saturn	Flyby, pictures, atmosphere Flyby in 1979
Mars 4	1973	Mars	Orbital, pictures, data
Mars 5	1973	Mars	Orbital, pictures, data
Mars 6	1973	Mars	Landed, no data
Mariner 10	1973	Venus Mercury	Flyby, pictures, data
Viking 1	1975	Mars	Landed, soil analysis
Viking 2	1975	Mars	Landed, soil analysis
Luna 24	1976	Moon	Unmanned exploration
Voyager 1	1977	Jupiter Saturn	Flyby (J in 1979, S in 1980)
Voyager 2	1977	Jupiter Saturn Uranus	Flyby (J in 1979, S in 1981, U in 1986)
Pioneer 12	1978	Venus	To orbit
Galileo	1981	Mars Jupiter	Flyby Probe

Appendix H

ASTRONOMICAL TIME

The common kind of timekeeping on earth bases the length of the day on the motion of the sun and starts counting its hours either at the point when the sun is highest in the sky (noon) or lowest below the horizon (midnight). Strictly speaking, the position of the sun in the sky provides what is called *apparent solar time*, which is not perfectly uniform. Irregularities in the length of the apparent solar day amount to as much as 30 seconds, and are due primarily to the ellipticity of the earth's orbit and to the inclination of the ecliptic to the celestial equator. For our use on earth it is more convenient to use a perfectly uniform time system, so many years ago a new system was devised that uses an *average* length for the day. That system is called *mean solar time*.

Solar Time

At any place on the earth the time depends upon the position of that place with respect to the sun. When it is noon in England, it is early morning in the United States and the middle of the night in Japan. Any place on the earth will have its own local time, depending upon its longitude. In order not to have a situation where almost every city and almost every person has an individual time system, the world was long ago divided into twenty-four time zones, each approximately 15° wide in longitude, within which all people use the same time. This is called *standard time* on land and *zone time* at sea. The time in each zone usually differs from that in the next by exactly 1 hour.

Local Time

For many purposes it is useful to have a time system that is the same throughout the world. This global time is called (somewhat optimistically) *universal time* and is defined as the mean solar time at the 0° meridian of longitude, which passes through Greenwich, England. At a given instant, any place on the earth will have the same time on clocks set to *universal time*, but only in the time zone passing through Greenwich will the universal time be the same as the ordinary standard time.

Universal Time

Because the earth, in addition to rotating on its axis, is also revolving slowly around the sun, the length of a solar day is different from the length of a day that would be defined by the apparent motion across the sky of a star. In the interval from one solar day to the next the earth has turned in its orbit a certain ways. Any star that might have been at a particular position at midnight on the first night would be a little farther west at midnight on the second night. The length of the interval from one star passage to the next is called the *sidereal day*, and it

Sidereal Time

is 4 minutes shorter than the mean solar day. In one year the earth turns with respect to the stars one full revolution more than it does with respect to the sun, so that there are 366 sidereal days in a year and only 365 solar days. (Actually, there are not an exact even number of days in a year; the period of revolution of the earth is 365.256 solar days; we partially compensate for the fraction of a day by adding a day, February 29, every four years, and call that year a leap year).

Appendix I
NAMES, CONSTELLATIONS, AND COORDINATES

By convention, the sky as seen from the earth is divided up into large, irregular sections called *constellations*. These are analogous to the geographical division of the earth's land into countries and states. The constellations have no physical significance; they are merely arbitrary divisions of the apparent sky. They do not delineate real groups of stars, but contain stars spread out equally vastly in depth. They generally follow the ancient division of the bright stars into patterned groups. The ancient mythological names have been retained in Latin form with Latin endings. Thus, Leo is a constellation that includes the chance juxtaposition of bright stars likened by the ancients to the figure of a lion. The brightest star in Leo is called alpha Leonis: alpha, because the first letter of the Greek alphabet usually applies to the brightest star in a constellation, and Leonis, because that is the Latin genitive form for Leo.

It is sometimes announced in the popular press that so-and-so, a great astronomer, has discovered a new constellation. This is nonsense. It makes just as little sense to say that a geographer, driving across the country, discovered a new state. Constellations are not physical entities, but are convenient divisions in the sky that allow us to describe approximately where the object is located.

A much more specific designation than the constellation must be used to identify an object in the sky if it is small and faint. You can say "the Milky Way in Cygnus" and it will be found as easily as if you said "the Rocky Mountains in Colorado". But the constellation name would be as inadequate a description for identifying a certain faint star in Cygnus as if you gave the address of a house as "New York." Cities often use addresses built on a coordinate scheme such that a house might be identified by an address like "2515 20th Ave. East." Similarly, the sky is laid out in a coordinate grid that allows an accurate description of position. The brightest spiral galaxy, for instance, has the address "$0^h 40^m, + 41°6'$."

The most commonly used coordinate system in the sky is merely a projection of the longitude-latitude system of the earth. The celestial equator lies in the sky directly above the earth's equator, and the poles lie above the poles of the earth. The coordinates are plotted out on the *apparent* sphere of the sky, named the *celestial sphere*. The names of the coordinates are *right ascension* (instead of longitude) and *declination* (instead of latitude). Zero longitude on the earth is at Greenwich, England, and zero right ascension in the sky is pinpointed as the position, on the celestial equator, of the sun when it crosses the equator in the spring (on about March 21). This point in the sky is called the *vernal equinox*.

Using the system of celestial coordinates, any point in the sky can be

Fig. A 9-1. Celestial coordinates are defined much like coordinates on the surface of the earth, with the vernal equinox being the reference point from which all positions are measured.

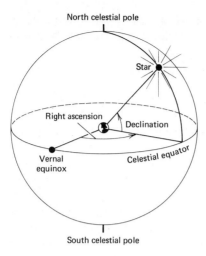

specified. For instance, the position given above for the brightest spiral galaxy is $0^h 40^m$ in right ascension. This means that it is 40 minutes east of the *vernal equinox* (considering the whole sky as 24 hours of right ascension). It is 41°6′ north of the equator (+ means north and - means south). The greek letters α and δ are used as abbreviations for right ascension and declination, respectively. The brightest star in the sky is alpha Canis Majoris, with $\alpha = 6^h 42^m$, $\delta = -16°39′$.

Individual stars, clusters, galaxies, and other objects in deep space are often given specific names. An object may have several different names, depending

Fig. A 9-2. The ecliptic is a projection of the earth's orbital plane onto the sky. The planets and the Sun are found along the ecliptic, because the planetary orbits are all almost in the same plane as the earth's orbit.

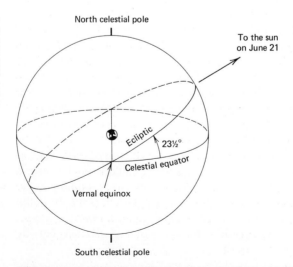

upon what catalogue or system is being used. For instance, the brightest star in the sky can be called "alpha Canis Majoris," "Sirius," "BD-16°1591," or "HD 48915."

The brightest stars visible in the sky were given common names by the ancients, and many of these names are still in use, at least for twenty or so very bright stars. For instance, *Sirius*, the name given by the ancients to the brightest is translated to mean *the scorcher*.

Another system for naming stars uses the Greek alphabet and names stars alphabetically—roughly in order of their brightness in a certain constellation. The second brightest star in Canis Major, for instance, is β (beta) Canis Majoris. Other systems were developed to extend this one to fainter stars, using arabic numbers, but these are not used frequently.

Usually, professional astronomers call faint stars by their catalogue names as given in certain important catalogues of stellar data. Most commonly used are the BD and HD catalogues. BD stands for the *Bonner Durchmusterung*, a catalogue (with accompanying maps) of all fairly bright stars visible from the northern hemisphere, assembled by astronomers in the nineteenth century in Bonn, Germany. The HD is the *Henry Draper* catalogue, assembled at the Harvard Observatory with funds given by a benefactor named Henry Draper. It contains almost all fairly bright stars and gives more information than the BD; specifically it lists the *spectral types*, which are indicative of the star's temperatures. A recent catalogue of stars was prepared by the Smithsonian Astrophysical Observatory. Stars listed in it have numbers preceded by SAO.

Stars with special characteristics are named by special systems. *Variable stars*, those that vary in brightness, are given letter or number names, followed by the constellation name with the genitive ending. The brighter variable stars have single or double letters, and the fainter have numbers preceded by the letter V. For example, the following are some famous variable stars: RR Lyrae, SS Cygni, RU Lupi, Z Andromedae, and V 603 Aquilae.

Many very bright nonstellar objects, such as star clusters, gas clouds, and galaxies, are called according to their number in a catalogue of 102 such objects published by the eighteenth-century French astronomer Messier. Thus the brightest spiral galaxy is called M 31. But more of these objects are called by the number given them in the *New General Catalogue*, compiled in Ireland by J. L. E. Dryer. Based on earlier catalogues published by Sir William Herschel and his son Sir John Herschel, the *New General Catalogue* listed 7,840 objects, and it has since been supplemented by two *Index Catalogues*. An object can be referred to by its number in these catalogues, preceded by the letters NGC or IC. Thus, the spiral galaxy M 31 may also be called NGC 224.

Radio sources are designated in a variety of ways. The brightest sources of radio waves in a constellation are given letter designations, such as "Cygnus A" or "Sagittarius B." Most other radio objects are called by their catalogue numbers; one frequently used catalogue is the Third Cambridge Catalogue of Radio Sources, abbreviated "3C." The brightest quasar is 3C 273.

Bright x-ray sources follow a similar naming pattern to that for bright radio sources. Scorpio X - 1 and Cygnus X - 1 are two famous examples.

In the solar system, all major planets and most of their natural satellites have names taken from classical mythology. Asteroids are numbered, though all the brighter asteroids have mythological names as well (e.g., Ceres, Juno, Vesta). Comets are usually named after their discoverers (e.g., Whipple's Comet and Enke's Comet); more officially they are designated according to their year of discovery, with a letter according to their order of discovery. Comet 1979a was the first comet discovered in 1979. Alternatively, if their orbits are known, comets can be called by the year of discovery and a Roman numeral designating their order of passage close to the sun (e.g., Comet 1979 II was the second new comet of the year to pass close to the sun, regardless of the order of discovery). Finally, meteorites that fall to the earth are named according to the nearest geographical name on the map (i.e., city, town, county, river, mountain, etc.).

There are so many objects in the universe that it is inevitable that most names in astronomy are numbers. But, just as citizens of the earth sometimes resent the use of too many numbers for too many things in the human experience, so do astronomers resist too strict an adherance to the number names of astronomy. The spiral galaxy M 31, more properly called NGC 224, will probably always be called just plain the *Andromeda galaxy.*

Appendix J
STAR CHARTS

Maps of the sky are appended at the end of this book on perforated pages, so that you may remove them for use outdoors at night. The first set of four maps shows the entire sky as it is seen by someone located at 35° N latitude, observing the sky during the four seasons at the hours specified. To account for differences in observing time or in month, remember that the stars advance westward a little each night, so that after 1 month's time, they will appear in a given position 2 hours earlier in the night. Thus the chart for fall constellations, which is drawn to correspond to the view at 9 P.M. on September 1, will be also applicable at 7 P.M. on October 1 (or at 11 P.M. on August 1). Notice that in the full sky you will see many of the same constellations for several months, so that if you can recognize their star patterns you can follow them for a substantial fraction of a year.

The second set of charts is for certain constellations and is intended for more detailed use, in connection with the suggested observations listed at the end of most chapters. Specific stars, star clusters, nebulae, and galaxies are identified so that you can find them. Most require the aid of binoculars or small telescopes.

Appendix K

THE TWENTY BRIGHTEST STARS

Star	Name	Apparent magnitude (V)	Spectral type		Absolute magnitude	Distance (ly)
αCMa A	Sirius	−1.47	A1	V	+1.45	9
αCar	Canopus	−0.72	FO	Ib−II	−3.1	98
αBoo	Arcturus	−0.06	K2	IIIp	−0.3	36
αCen A	Rigil Kentaurus	0.01	G2	V	+4.39	4
αLyr	Vega	0.04	AO	V	+0.5	26
αAur	Capella	0.05	G8	III+F	−0.6	45
βOri A	Rigel	0.14var.	B8	Ia	−7.1	900
αCMi A	Procyon	0.37	F5	IV−V	+2.7	11
αOri	Betelgeuse	0.41var.	M2	Iab	−5.6	520
αEri	Achernar	0.51	B3	Vp	−2.3	118
βCen AB	Hadar	0.63var.	B1	III	−5.2	490
αAql	Altair	0.77	A7	IV−V	+2.2	16
αTau A	Aldebaran	0.86var.	K5	III	−0.7	68
αVir	Spica	0.91var.	B1	V	−3.3	220
αSco A	Antares	0.92var.	M1	Ib+B	−5.1	520
αPsA	Fomalhaut	1.15	A3	V	+2.0	25
βGem	Pollux	1.16	K0	III	+1.0	35
αCyg	Deneb	1.26	A2	Ia	−7.1	1600
βCru	Beta Crucis	1.28var.	B0.5	III	−4.6	490
αLeo A	Regulus	1.36	B7	V	−0.7	84

Appendix L

MESSIER OBJECTS

M	NGC	$_h\alpha_m$ 1980	$°\delta'$	Mag.	Kind of object
1	1952	5 33.3	+22 01	11.3	Crab Nebula
2	7089	21 32.4	−00 54	6.3	Globular cluster
3	5272	13 41.3	+28 29	6.2	Globular cluster
4	6121	16 22.4	−26 27	6.1	Globular cluster
5	5904	15 17.5	+02 11	6	Globular cluster
6	6405	17 38.9	−32 11	6	Open cluster
7	6475	17 52.6	−34 48	5	Open cluster
8	6523	18 02.4	−24 23		Lagoon Nebula
9	6333	17 18.1	−18 30	7.6	Globular cluster
10	6254	16 56.0	−04 05	6.4	Globular cluster
11	6705	18 50.0	−06 18	7	Open cluster
12	6218	16 46.1	−01 55	6.7	Globular cluster
13	6205	16 41.0	+36 30	5.8	Globular cluster
14	6402	17 36.5	−03 14	7.8	Globular cluster
15	7078	21 29.1	+12 05	6.3	Globular cluster
16	6611	18 17.8	−13 48	7*	Open cluster
17	6618	18 19.7	−16 12	7	Omega Nebula
18	6613	18 18.8	−17 09	7	Open cluster
19	6273	17 01.3	−26 14	6.9	Globular cluster
20	6514	18 01.2	−23 02		Trifid Nebula
21	6531	18 03.4	−22 30	7	Open cluster
22	6656	18 35.2	−23 55	5.2	Globular cluster
23	6494	17 55.7	−19 00	6	Open cluster
24	6603	18 17.3	−18 27	6	Open cluster
25	IC4725	18 30.5	−19 16	6	Open cluster
26	6694	18 44.1	−09 25	9	Open cluster
27	6853	19 58.8	+22 40	8.2	Dumbbell Nebula; planetary nebula
28	6626	18 23.2	−24 52	7.1	Globular cluster
29	6913	20 23.3	+38 27	8	Open cluster
30	7099	21 39.2	−23 15	7.6	Globular cluster
31	224	0 41.6	+41 09	3.7	Andromeda Galaxy (Sb)
32	221	0 41.6	+40 45	8.5	Elliptical galaxy
33	598	1 32.8	+30 33	5.9	Spiral galaxy (Sc)
34	1039	2 40.7	+42 43	6	Open cluster
35	2168	6 07.6	+24 21	6	Open cluster
36	1960	5 35.0	+34 05	6	Open cluster
37	2090	5 51.5	+32 33	6	Open cluster
38	1912	5 27.3	+35 48	6	Open cluster
39	7092	21 31.5	+48 21	6	Open cluster
40	—	—	—		Double star
41	2287	6 46.2	−20 43	6	Open cluster
42	1976	5 34.4	−05 24		Orion Nebula (part)
43	1982	5 34.6	−05 18		Orion Nebula
44	2632	8 38.8	+20 04	4	Praesepe; open cluster
45	—	3 46.3	+24 03	2	The Pleiades; open cluster
46	2437	7 02.0	−14 46	7	Open cluster
47	2422	7 35.6	−14 27	5	Open cluster
48	2548	8 12.5	−05 43	6	Open cluster
49	4472	12 28.8	+08 07	8.9	Elliptical galaxy
50	2323	7 02.0	−08 19	7	Open cluster
51	5194	13 29.0	+47 18	8.4	Whirlpool Galaxy; spiral galaxy (Sc)
52	7654	23 23.3	+61 29	7	Open cluster
53	5024	13 12.0	+18 17	7.7	Globular cluster
54	6715	18 53.8	−30 30	7.7	Globular cluster
55	6809	19 38.7	−31 00	6.1	Globular cluster

M	NGC	$_h\alpha_m$ 1980	$°\delta'$	Mag.	Kind of object
56	6779	19 15.8	+30 08	8.3	Globular cluster
57	6720	18 52.9	+33 01	9.0	Ring Nebula; planetary nebula
58	4579	12 36.7	+11 56	9.9	Spiral galaxy (SBb)
59	4621	12 41.0	+11 47	10.3	Elliptical galaxy
60	4649	12 42.6	+11 41	9.3	Elliptical galaxy
61	4303	12 20.8	+04 36	9.7	Spiral galaxy (Sc)
62	6266	16 59.9	−30 05	7.2	Globular cluster
63	5055	13 14.8	+42 08	8.8	Spiral galaxy (Sb)
64	4826	12 55.7	+21 48	8.7	Spiral galaxy (Sb)
65	3623	11 17.8	+13 13	9.6	Spiral galaxy (Sa)
66	3627	11 19.1	+13 07	9.2	Spiral galaxy (Sb)
67	2682	8 50.0	+11 54	7	Open cluster
68	4590	12 38.3	−26 38	8	Globular cluster
69	6637	18 30.1	−32 23	7.7	Globular cluster
70	6681	18 42.0	−32 18	8.2	Globular cluster
71	6838	19 52.8	+18 44	6.9	Globular cluster
72	6981	20 52.3	−12 39	9.2	Globular cluster
73	6994	20 57.8	−12 44		Open cluster
74	628	1 35.6	+15 41	9.5	Spiral galaxy (Sc)
75	6864	20 04.9	−21 59	8.3	Globular cluster
76	650	1 40.9	+51 28	11.4	Planetary nebula
77	1068	2 41.6	−00 04	9.1	Spiral galaxy (Sb).
78	2068	5 45.8	+00 02		Small emission nebula
79	1904	5 23.3	−24 32	7.3	Globular cluster
80	6093	16 15.8	−22 56	7.2	Globular cluster
81	3031	9 54.2	+69 09	6.9	Spiral galaxy (Sb)
82	3034	9 54.4	+69 47	8.7	Irregular galaxy (Irr)
83	5236	13 35.9	−29 46	7.5	Spiral galaxy (Sc)
84	4374	12 24.1	+13 00	9.8	Elliptical galaxy
85	4382	12 24.3	+18 18	9.5	Elliptical galaxy (SO)
86	4406	12 25.1	+13 03	9.8	Elliptical galaxy
87	4486	12 29.7	+12 30	9.3	Elliptical galaxy (Ep)
88	4501	12 30.9	+14 32	9.7	Spiral galaxy (Sb)
89	4552	12 34.6	+12 40	10.3	Elliptical galaxy
90	4569	12 35.8	+13 16	9.7	Spiral galaxy (Sb)
91	—	—	—		Identity not known
92	6341	17 16.5	+43 10	6.3	Globular cluster
93	2447	7 43.6	−23 49	6	Open cluster
94	4736	12 50.1	+41 14	8.1	Spiral galaxy (Sb)
95	3351	10 42.8	+11 49	9.9	Barred spiral galaxy (SBb)
96	3368	10 45.6	+11 56	9.4	Spiral galaxy (Sa)
97	3587	11 13.7	+55 08	11.1	Owl Nebula; planetary nebula
98	4192	12 12.7	+15 01	10.4	Spiral galaxy (Sb)
99	4254	12 17.8	+14 32	9.9	Spiral galaxy (Sc)
100	4321	12 21.9	+15 56	9.6	Spiral galaxy (Sc)
101	5457	14 02.5	+54 27	8.1	Spiral galaxy (Sc)
102	—	—	—		Identity not known
103	581	1 31.9	+60 35	7	Open cluster
104	4594	12 39.0	−11 35	8	Sombrero Nebula; spiral galaxy (Sa)
105	3379	10 46.8	+12 51	9.5	Elliptical galaxy
106	4258	12 18.0	+47 25	9	Spiral galaxy (Sb)
107	6171	16 31.8	−13 01	9	Globular cluster
108	3556	11 10.5	+55 47	10.5	Spiral galaxy (Sb)
109	3992	11 56.6	+53 29	10.6	Barred spiral galaxy (SBc)

Appendix M

ASTRONOMY VS. ASTROLOGY

There is sometimes a certain amount of confusion in people's minds about astronomy and astrology—which is which, and which are you supposed to believe? The difference in the two is easily enough described, but it is not so easy to answer the second of these questions.

Astronomy is the science that attempts to study and understand the universe of stars, planets, and galaxies, by using careful scientific methods and the basic laws of physics. Astrology is the practice of using the positions of the heavenly bodies as perceived on earth to attempt to understand human personalities and human destiny. Astrologers believe that somehow a person's characteristics are set by the planets and stars that happened to be above the horizon at just the moment he or she was born, and that this heaven-made pattern sticks throughout an individual's life. To be able to establish just what heavenly influences might be involved in a given personality, the astrologer must go through a certain amount of calculations, using various technical-sounding names. The resulting mumbo-jumbo may be difficult for a layperson to distinguish from some of the technical mumbo-jumbo of astronomy. After all, some of the astronomer's fantastic stories of black holes and possible other universes are just as farfetched-sounding as the idea that a certain planet or star can somehow reach down and form our future.

One way of distinguishing between two such very different enterprises as astronomy and astrology is to consider which one of them subjects itself continually to tests of its correctness, tests that lead to new and different concepts and methods. These are characteristics of a science; astronomy, which is continually being revised and amplified, is the one which does these things. Then ask which of these is based on precepts and principles that were laid down by the ancients as unarguable truths; which uses concepts and methods that remain unchanged over the centuries. Those are the characteristics of a religion, where truth is a matter of faith; astrology fits that description well. Astrology is a type of ancient religion, one that seems to instill the kind of devotion and mystical faith that is found in many ancient religions. It seems to be a help to many modern people, particularly people who do not want to feel responsible for the way they are and who do not want to make difficult daily or long-term decisions without some external guidance, no matter how arbitrary. As a religion, astrology seems to serve its purpose. As a science, astronomy serves its purpose, and the two really have no common meeting ground or overlapping area. People who argue that astrology has some mysterious scientific basis are undermining its basic religious precepts. People who argue that astronomy proves astrology wrong are failing to realize that proof of rightness or wrongness is quite irrelevant to an ancient religious system of belief and human behavior.

Appendix N

OBSERVATORIES IN THE UNITED STATES AND CANADA

1. Observatories in the northeastern United States

Name	Location
Agassiz Station, Harvard University	Harvard, Mass.
Flower and Cook Observatory	Philadelphia, Pa.
Harvard College Observatory	Cambridge, Mass.
Maria Mitchell Observatory	Nantucket, Mass.
Princeton University Observatory	Princeton, N.J.
Sproul Observatory	Swarthmore, Pa.
Van Vleck Observatory	Middletown, Conn.
Yale Observatory	Bethany, Conn.

2. Observatories in the southeastern United States

Name	Location
Arecibo National Radio Observatory	Arecibo, Puerto Rico
Dyer Observatory	Nashville, Tenn.
Leander McCormick Observatory	Charlottesville, Va.
National Radio Astronomy Observatory	Greenbank, W. Va.
University of Louisiana Observatory	Baton Rouge, La.
U.S. Naval Observatory	Washington, D.C.

3. Observatories in the midwestern United States

Name	Location
Dearborn Observatory	Evanston, Ill.
Goethe Link Observatory	Brooklyn, Ind.
McMath-Hurlbut Observatory	Lake Angeles, Mich.
Perkins Observatory	Delaware, Ohio
University of Michigan Observatory	Ann Arbor, Mich.
Warner and Swasey Observatory	Cleveland, Ohio
Washburn Observatory	Madison, Wisc.
Yerkes Observatory	Williams Bay, Wisc.

4. Observatories in the western United States

Name	Location
Chabot Observatory	Oakland, Calif.
Griffith Observatory	Los Angeles, Calif.
Haleakala Observatory	Haleakala, Hawaii
Hat Creek Radio Observatory	Hat Creek, Calif.
Kitt Peak National Observatory	Tucson, Ariz.
Leuschner Observatory	Lafayette, Calif.
Lick Observatory	Mt. Hamilton, Calif.
Lowell Observatory	Flagstaff, Ariz.
Mauna Kea Observatory	Mauna Kea, Hawaii

Manastash Ridge Observatory	Ellensberg, Wash.
McDonald Observatory	Fort Davis, Tex.
Mt. Hopkins Observatory	Amado, Ariz.
Mt. Wilson Observatory	Mt. Wilson, Calif.
Owens Valley Radio Observatory	Bishop, Calif.
Palomar Observatory	Palomar Mt., Calif.
Steward Observatory	Tucson, Ariz.
U.S. Naval Observatory (station)	Flagstaff, Ariz.

5. Observatories in Canada

Name	Location
David Dunlap Observatory	Richmond Hill, Ontario
Dominion Astrophysical Observatory	Victoria, B.C.
Dominion Radio Astrophysical Observatory	Penticton, B.C.
Dominion Observatory	Ottawa, Ontario
Algonquin Radio Observatory	Lake Traverse, Ontario
Quebec Observatory	Quebec, Quebec
University of Edmonton Observatory	Edmonton, Alberta
University of New Brunswick Observatory	Fredricton, N.B.

Appendix O

BIBLIOGRAPHY

BOOKS   There are hundreds of good books that deal with various aspects of astronomy at the popular level. Most book stores and libraries will have a good selection of them. Beware of any that are more than about 10 years old, however, as recent progress in astronomy has been so rapid that certain parts of older books will be badly out of date.

The following are only a few of the many good books on astronomy that will give a nonspecialist reader information on a variety of astronomical subjects.

GENERAL
Calder, N. *The Violent Universe* (Viking, 1969)
Hodge, P. *The Revolution in Astronomy* (Holiday House, 1970)
Jastrow, R. *Red Giants and White Dwarfs* (Signet, 1971)
Ronan, C. *Invisible Astronomy* (Lippincott, 1972)
Sagan, C. *The Cosmic Connection* (Dell, 1973)
Shipman, H. *Black Holes, Quasars and the Universe* (Houghton-Mifflin, 1976)
Weinberg, S. *The First Three Minutes* (Basic Books, 1977)

SOLAR SYSTEM
French, B. *The Moon Book* (Penguin, 1977)
Henderson, A., and J. Grey (eds.). *Exploration of the Solar System* (NASA, 1974)
Kopal, Z. *The Solar System* (Oxford, 1973)
Lewis, R. *The Voyages of Apollo* (Quadrangle, 1975)
Mason, B., and W. Melson *The Lunar Rocks* (Wiley, 1970)
Scientific American. *The Solar System* (Freeman, 1975)

STARS
Aller, L. *Atoms, Stars and Nebulae* (Harvard, 1971)
Meadows, A. *Stellar Evolution* (Pergamon, 1967)
Menzel, D. *Our Sun* (Harvard, 1972)
Vershuur, G. *The Invisible Universe* (Springer, 1974)

GALAXIES AND THE UNIVERSE
Bok, B., and P. Bok. *The Milky Way* (Harvard, 1974)
Hodge, P. *Galaxies and Cosmology* (McGraw-Hill, 1966)
John, L. (ed.). *Cosmology Now* (Taplinger, 1973)
Schatzman, E. *The Structure of the Universe* (McGraw-Hill, 1968)
Shapley, H. *Galaxies* (Harvard, 1972)
Whitney, C. *The Discovery of Our Galaxy* (Knopf, 1971)

There are several magazines that have articles on astronomy, as well as a few that are totally devoted to the subject. In the United States, the principal journals containing elementary-level articles are the following:

Astronomy
Mercury
Natural History
Scientific American
Sky and Telescope
Smithsonian

Appendix P
AUDIO-VISUAL MATERIALS

SLIDES   A collection of slides covering all of astronomy and accompanied by a book that describes each slide is published by the McGraw-Hill Book Co. under the title SLIDES FOR ASTRONOMY, by P. Hodge.

Other slides are available from a few observatories and several commercial outlets:

Hale Observatories, Pasadena, Calif., 91109
Lick Observatory, Santa Cruz, Calif., 95060
Kitt Peak National Observatory, Tucson, Ariz., 94720
Yerkes Observatory, Williams Bay, Wisc., 53191
Norton Scientific, Tucson, Ariz., 85717
MMI Corp., Baltimore, Md., 21211
Astrophoto Laboratory, Neustadt, W. Germany
Educational Materials, Orange, Calif., 92667
Hansen Planetarium, Salt Lake City, Utah, 84111

MOVIES   There are now reasonably large numbers of good films about astronomical topics. A comprehensive catalog of science films as of 1975 has been published by the American Association for the Advancement of Science; it gives data on 6000 films available in the U.S.

A catalog of nearly 200 NASA films available for educational use is published each year. Films are sent out from regional film libraries. The current catalog can be obtained from the National Aeronautics and Space Administration, EPSO, Washington, D.C., 20546.

AUDIO TAPES   Several astronomically oriented audio tapes are available and can sometimes be used effectively, especially for historical topics. Some sources are:

Center for Cassette Studies, North Hollywood, Calif., 91605
Spring Green Multimedia, Washington, D.C., 20003
Astronomical Society of the Pacific, San Francisco, Calif., 94122.

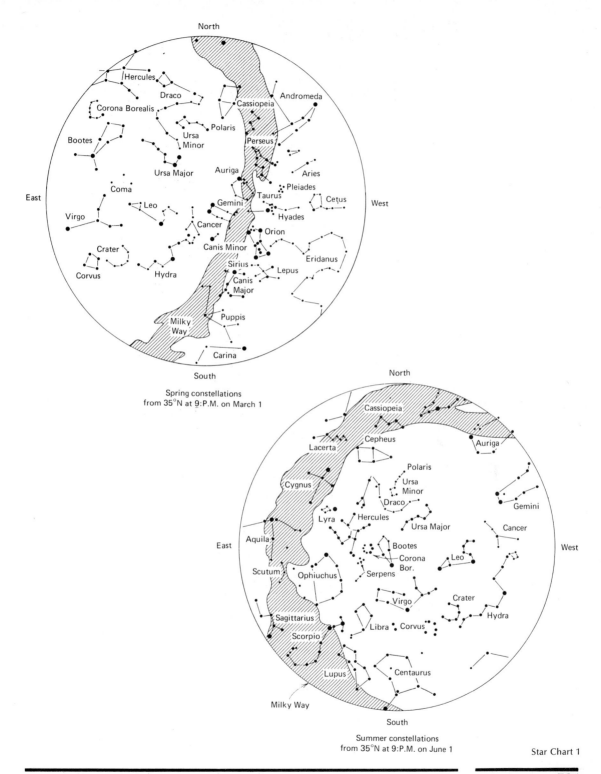

Spring constellations
from 35°N at 9:P.M. on March 1

Summer constellations
from 35°N at 9:P.M. on June 1

Star Chart 1

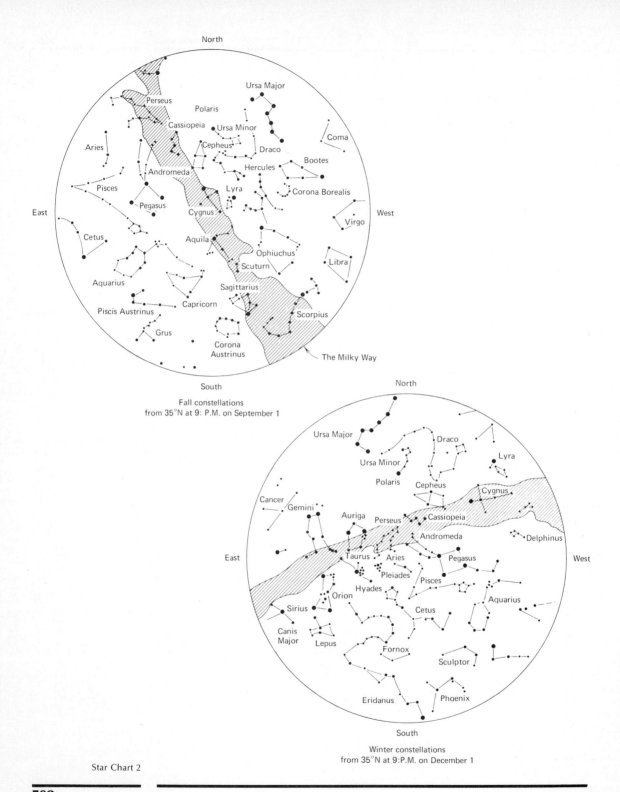

North

Ursa Major

Polaris

Perseus

Cassiopeia

Ursa Minor

Cepheus

Draco

Coma

Aries

Andromeda

Hercules

Bootes

Pisces

Lyra

Corona Borealis

Pegasus

Cygnus

East

West

Cetus

Aquila

Ophiuchus

Virgo

Aquarius

Scuturn

Libra

Piscis Austrinus

Capricorn

Sagittarius

Scorpius

Grus

Corona
Austrinus

The Milky Way

South

Fall constellations
from 35°N at 9: P.M. on September 1

North

Ursa Major

Draco

Lyra

Ursa Minor

Polaris

Cepheus

Cygnus

Cancer

Cassiopeia

Gemini

Auriga

Perseus

Andromeda

Delphinus

East

Taurus

Aries

Pegasus

West

Pleiades

Pisces

Hyades

Orion

Cetus

Aquarius

Sirius

Canis
Major

Lepus

Fornox

Sculptor

Eridanus

Phoenix

South

Winter constellations
from 35°N at 9:P.M. on December 1

Star Chart 2

502

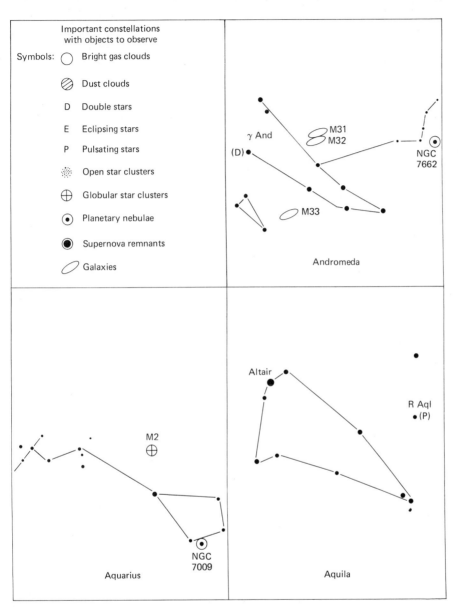

Symbols:

◯   Bright gas clouds

◑   Dust clouds

D   Double stars

E   Eclipsing stars

P   Pulsating stars

⁘   Open star clusters

⊕   Globular star clusters

⊙   Planetary nebulae

◉   Supernova remnants

⬭   Galaxies

Important constellations with objects to observe

γ And
(D)
M31
M32
NGC 7662
M33
Andromeda

M2 ⊕
NGC 7009
Aquarius

Altair
R Aql (P)
Aquila

Star Chart 3

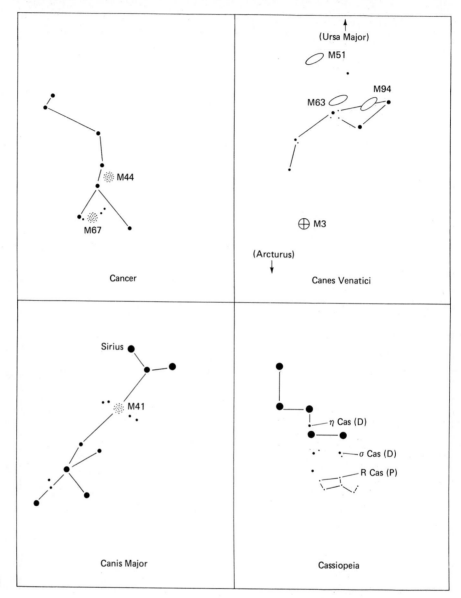

Star Chart 4

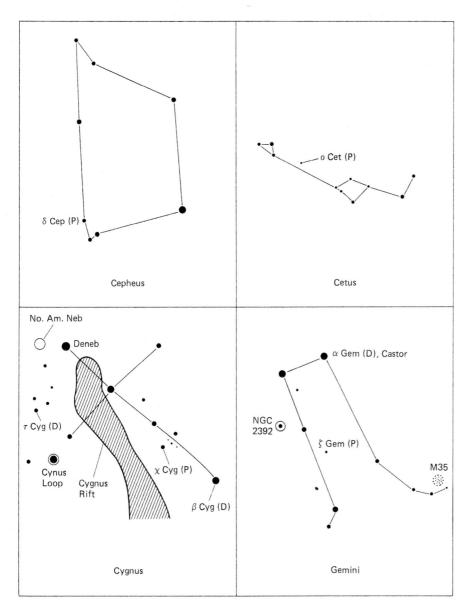

Cepheus

Cetus

Cygnus

Gemini

Star Chart 5

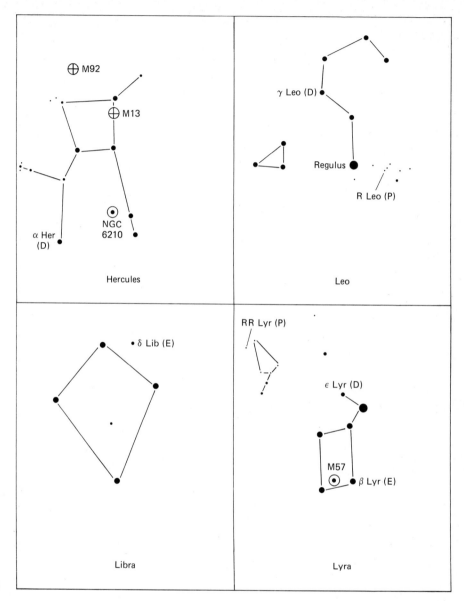

Hercules

Leo

Libra

Lyra

Star Chart 6

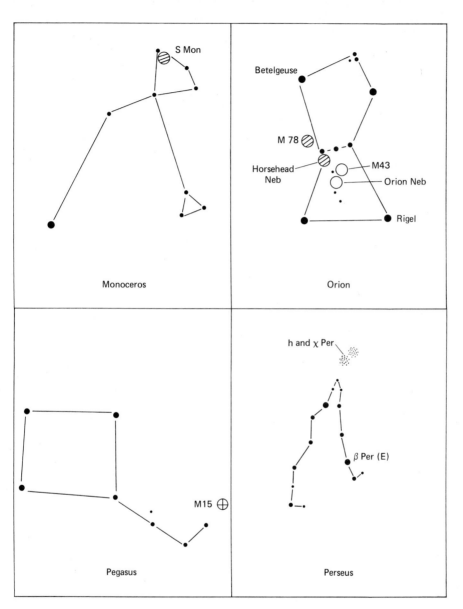

Monoceros

Orion

Pegasus

Perseus

Star Chart 7

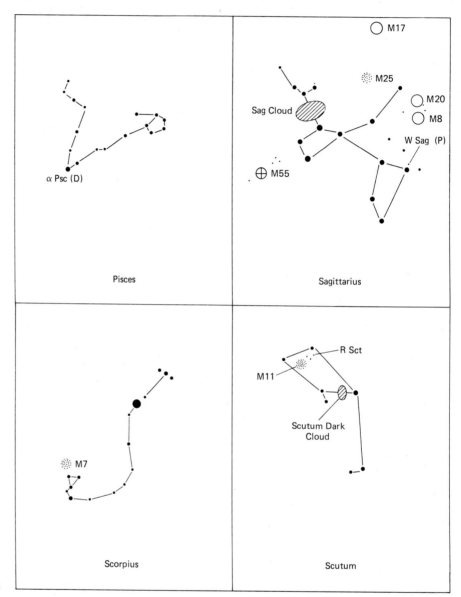

α Psc (D)

Pisces

M17

M25

Sag Cloud

M20

M8

W Sag (P)

M55

Sagittarius

M7

Scorpius

R Sct

M11

Scutum Dark
Cloud

Scutum

Star Chart 8

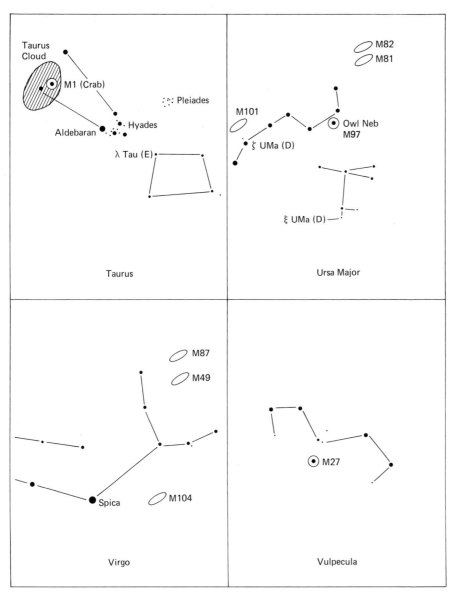

Taurus Cloud

M1 (Crab)

Pleiades

Hyades

Aldebaran

λ Tau (E)

Taurus

M82

M81

M101

ζ UMa (D)

Owl Neb
M97

ξ UMa (D)

Ursa Major

M87

M49

Spica

M104

Virgo

M27

Vulpecula

Star Chart 9

INDEX

Date Due

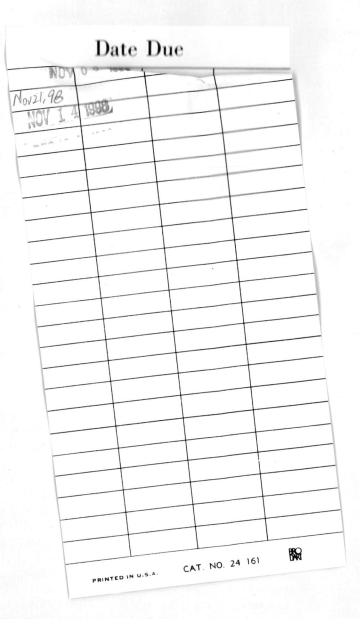

NOV 0 8

Nov 21, 98

NOV 1 4 1998